BIRDS
of
Southwestern Australia

BIRDS

of Southwestern Australia

An atlas of changes in the distribution
and abundance of the wheatbelt avifauna

Denis A. Saunders and John A. Ingram

Surrey Beatty & Sons

in association with

Western Australian Laboratory
CSIRO Division of Wildlife and Ecology

The National Library of Australia
Cataloguing-in-Publication entry;
Saunders, D. A. (Denis Allan), 1947–.
Birds of Southwestern Australia; an
atlas of changes in the distribution and
abundance of wheatbelt avifauna.

Bibliography.
Includes index.
ISBN 0 949324 57 4.

1. Birds — Western Australia. I. Ingram, John A. (John Alexander), 1946–.
II. CSIRO. Division of Wildlife and Ecology. Western Australian Laboratory.
III. Title.

598.29941

Published April 1995

Printed in Australia by

SURREY BEATTY & SONS PTY LIMITED
43 Rickard Road, Chipping Norton, NSW 2170

This book is dedicated to the many people who provided data or otherwise contributed to this project, and especially to V Saunders for her support during the conduct of the project and the preparation of the book.

Contents

Acknowledgements

A venture such as this to plot the distribution and abundance of birds at this scale is dependent on the goodwill, support and hard work of many people who collect data and make those data freely available for analyses. We would like to thank; Edward Abe, Heather Adamson, Bob Allen, Jeanette Allen, Murray Arnold, Dawn Atwell, Robert Badger, Graham Barrett-Leonard, Peter Beaton, Darryl and Susan Bell, S. L. Bennier, Bill Bialas, Mark Blewett, Fred Boase, Peter Boase, Rosemary Bond, Philippa Bostock, Stephen Bowes, Katrina Bowron, Elaine and Rodney Boyce, Phillip Brace, Mary Bristow, R. B. Brockman, Michael Brooker, Vernon Brown, Justin Bruce, M. A. Burges, Brad Burns, Graeme and Sharon Butler, Peter Cale, Mrs N. Carew-Hopkins, Sandy Carnduff, Ann Carr, Alan Carter, Dan Carter, Kevin Chick, Bev and Robert Clare, Fran Coad, Janet Colbourne, Brian Colegate, Barry Cornish, Peter Coyle, A. H. B. Dawes, Gloria De Campi, Peter Denton, Perry de Rebeira, John Dixon, Alison and John Doley, Marilyn Dunne, Glynne Dyer, George Evans, Jack Ewert, Fiona Falconer, Kelly Fare, Murray Field, Mrs E. Fleay, Bill Flint, Chris Florides, Belinda Fraser, Darren Fraser, Elizabeth Frusher, R. Galloway, Gwen Gath, Susie Gearon, Heather Gibson, Moira Girando, Peter Godfrey, Mal Graham, Shirley Green, Judith Gundill, Richard Hamilton, R. Hammond, L. A. Harris, Lesley Harrison, Judith Harvey, Dorothy Harwood, Arpita Hayes, M. Higham, Elizabeth Hill, John Hill, Sheila Hill, Les and Margo Hodge, Ian and Lynda Hudson, Neville Humphries, Charles and Gwen Hyde, Y. Jenkins, Grant Jenks, Monica Jenks, Narelle Jenks, Stan and Wendy Johnson, Kevin Joss, A. J. Keefe, Judith Kleinig, Simon Kniveton, Robert Lambeck, D. H. and I. S. Lane, Cherie Lasky, Pam Last, Prue Leake, Christina Lee, W. S. Llewellyn, Rebecca Lloyd, Jim Lynch, Anne and Ted Lyneham, John Lynn, R. Malone, R. J. Marold, Sally Marsh, A. E. Martin, John Martin, Jim Masters, John Masters, Tony McAleer, Joan McDowall, Denise McGillivray, Natalie McGlinn, Reg McInnes, Heather McKenzie, Gordon McNeill, Simon Merewether, Janice Metcalf, P. S. Mills, Tony Mitchell, Penny Moir, Ray Morgan, Greg Moulton, Brenda and Ken Newby, Robin Newman, J. P. Nilson, Peter Nilson, Mrs L. North, Phil Pauley, Ray Paynter, Kerryn Piper, Marion Pitcher, Lawry Pitman, Jean Poole, David Price, B. Pritchett, Mrs W. Prnich, Barbara Read, Margaret Reed, Jim Riches, Mrs A. Ridgeway, Elva Rolinson, David Rose, Laurence Rose, Ted Rose, V Saunders, Carol Scheer, Gwen Schildmann, Don Shepherd, Pauline Short, A. J. Sibley, Basil and Mary Smith, Graeme Smith, Mrs K. Snell, Mary Squire, Evelyn Stasinowsky, Betty Stokes, Tammy Stone, Dawn Swinbank, Helen Taylor, Mrs L. S. Taylor, Barbara Tierney, Maxine Tierney, Judy Towers, Dale Tyler, Katherine and Ron Van Delft, Brother W. M. Van Veen, Thomas Vigilante, Angela Waters, George and Rita Watkins, John Watkins, Richard Watkins, Rosemary and Roy West, Dennis Whisson, Joyce White, Kath White, Mary White, Noreen White, Mrs G. L. Williams, Bill Willis, Ian and Sylvia Willox, Doug Wilson, John Wilson, Maxine Woodhouse, Kate York, Kevin Young and Mrs D. D. Zis for collecting data, and Clare Taplin for maintaining address lists and organizing newsletters and postage. We are also grateful to Michael Brooker, Peter Cale, Jim Masters and Harry Recher who patiently went through the manuscript and freely gave us the benefit of their considerable knowledge of the birds of southwestern Australia, Eric Saunders and V Saunders who provided advice on how to make the manuscript easier to read, and Ivor E. Beatty and staff at Surrey Beatty and Sons for support in publishing this book.

chapter one

Introduction

Outside my door when I was young
There flowed a shining river
Gleaming in the summer sun
It used to shine like silver
The banks were lined with willow trees
And tall green waving rushes
Songbirds sang in the summer breeze
And nested in the bushes.

Don't you think it's time we got together
To save our shining river?
It will soon be gone forever
Don't you think it's time?

The willow trees have long since gone
The birds are getting fewer
And where my river used to run
There's just an open sewer.

Eric Bogle, "Shining River"
(Copyright© *Larrikin Music Publishing Pty Ltd,*
Sydney. Reproduced with permission).

ECOLOGICALLY Sustainable Development became a catchphrase which was widely used in Australia and many other parts of the world in the late 1980s and early 1990s. The concept is simple and involves using natural resources in such a way that current users do not deplete those resources. The result should be that future generations will enjoy the same options to use those resources that we currently enjoy. The problem that we face at present is that much of our development is at variance with the principles of ecologically sustainable development and has resulted in widespread changes in the distribution and abundance of the native flora and fauna, major changes in ecosystem processes and function, and degradation of the resources on which that development is based. Individually, and as a Nation, we need to examine our current landuses and modify them so that they do conform to the principles of ecologically sustainable development.

The southwest of Western Australia is a classic example of our failure to follow the principles of ecologically sustainable development. The region is rich biologically, and over the past 165 years has been subjected to development for intensive agriculture which has imposed extensive changes upon the landscape. Widespread, rapid and selective clearing of native vegetation for the development of agriculture, together with the introduction of exotic species have seen the disappearance of many species of native plants and animals. Changes in ecosystem processes have also resulted, as for example, rising water tables with associated increases in the area affected by soil salinity. These detrimental changes, together with other forms of land degradation affect agricultural productivity and have resulted in strong community support for the development of land management practices that reverse the degrading processes and do not further compromise our remaining native biota (plants and animals).

In order to develop sustainable land management practices it is necessary to understand ecosystem processes operating in the region and the way those processes have been altered by development. From a nature conservation point of view, one way to look at these alterations is to examine the changes in the biota. In 1984 the Western Australian Laboratory of CSIRO Division of Wildlife and Ecology began a study into the dynamics of remnant vegetation and associated fauna. The aim of this research was to establish the conservation potential of remnants of native vegetation in the extensively cleared areas of the central wheatbelt, to examine ways nature conservation could be integrated into land management within agricultural landscapes, and to provide guidelines for the management of remnants of native vegetation.

Birds were one of the groups chosen for study and this book presents the results of a community-based observer scheme to map the distribution of birds in the southwestern agricultural area (called the wheatbelt), to gauge their abundance in the period 1987–1990, and to examine changes in their distribution and abundance this century. The book also presents the results of an analysis of changes in status of wheatbelt birds and discusses the major factors responsible for the changes. It concludes with suggestions of ways of incorporating nature conservation into management of the wheatbelt landscape, so that our declining birds may be offered some protection.

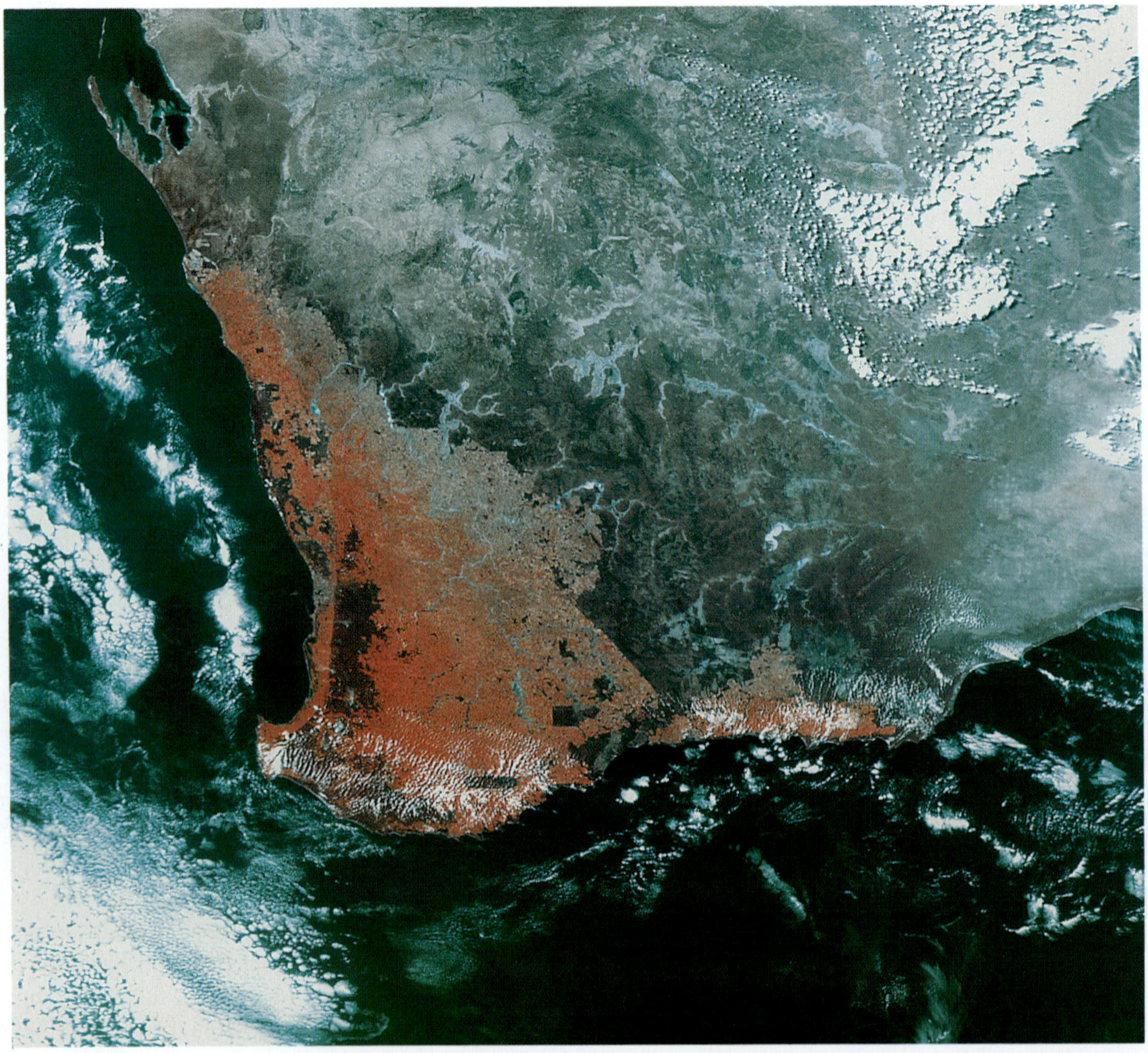

Fig. 1. NOAA Satellite image of southwestern Australia. Cleared land is indicated by shades of red, uncleared land by dark colour, and drainage systems by pale blue. Image supplied by Department of Land Administration, Remote Sensing Applications Centre, and used with their permission.

chapter two

Study region, development history and environmental changes

THIS survey of the distribution and abundance of birds focused on the region of southwestern Australia known as the agricultural area or wheat-sheep zone, which throughout this book is called the wheatbelt. This region lies within the area receiving between 300 and 600 mm of annual average rainfall. It has a mediterannean climate with hot, dry summers and cool, wet winters. About 90 per cent of the rainfall occurs between April and September.

The region is roughly bounded inland by a line from Kalbarri in the north, to Esperance in the southeast and, closer to the southwest corner, by a line between Perth and Albany (Figs 1, 2 and 3). It does not include the southwest forests which occur to the southwest of the line between Perth and Albany (Figs 1 and 2). The extent of the wheatbelt is shown clearly by the pale red areas in Figure 1 which indicate cleared land.

Southwestern Australia has been free of glaciation for about 200 000 000 years. It has undergone a long period of weathering and erosion. This has resulted in little relief in the landscape, and convoluted drainage patterns. The thin soils are nutrient-poor and heavily leached (McArthur 1993). The area has evolved an extremely rich biota. The flora in particular is renowned for its species richness and for the large number of endemic species (i.e., they occur only in southwestern Australia). There are about 8 000 species of higher plant in the region, with about three quarters of these found nowhere else (Hopper 1992). The region's importance as regards nature conservation is illustrated by the fact that it contains over 40 per cent of Australia's higher plants. However, they are distributed in an area that occupies less than 5 per cent of the continent. In terms of plant species richness the region ranks as one of the world's biological "hotspots". The vegetation varies from species-rich heathlands and shrublands on the deep sands and laterites of the uplands, through mallee and woodland associations on the valley slopes, down to fresh water and salt-lake systems in the valley floors.

After Europeans established themselves in the Swan River Colony (now called Perth), they moved into the area now known as the wheatbelt and started grazing sheep on native vegetation. This pastoral phase of development took place in the latter half of the nineteenth century and resulted in the introduction of fencing and disturbances to native vegetation. These disturbances were more acute around water holes and the dwellings of settlers (Main 1993), where for example, domestic livestock and poultry grazed on native vegetation and disturbed litter, and dogs, cats and humans preyed on native animals.

Towards the end of last century, the government encouraged closer settlement in the pastoral areas of southwestern Australia and divided what became the wheatbelt into small lots (64.8 ha) for intensive agriculture. To service the settlements, the government developed a system of roads and railways. This phase of development resulted in rapid and extensive changes throughout the region as native vegetation was cleared and non-native species including domestic plants and animals, weeds, and feral animals were brought into the area and released.

In the first half of the twentieth century clearing was slow. However, after the Second World War with the availability of heavy machinery, clearing of native vegetation accelerated (Main 1993). Fifty-four per cent of all land cleared in southwestern Australia was cleared after 1945. The release and clearing of land reached its peak in the 1960s when the government made 405 000 ha (1 million acres) of land available annually for wheat and sheep farms under what was known as "conditional purchase"; the conditions being that the land had to be fenced and cleared. No consideration was given to nature or soil conservation. As pointed out by Don Couper (1993), a farmer in the central wheatbelt,

> "Land was released with conditions for clearing set down, which in most cases led to total removal of the native vegetation. Those concerned about erosion and those who sought to retain areas of native vegetation were largely ignored. Little heed was paid to anyone with the courage to speak out against this wholesale clearing or warning of impending environmental problems. Their voices were drowned in the clamour and euphoria emanating from politicians and the people responsible (or irresponsible) for overseeing development. Development at this scale and pace was justified because Western Australia was 'a state on the move'."

By the late 1970s most of the large scale development was complete and the wheatbelt occupied over 16 million ha. In total, around 90 per cent of the

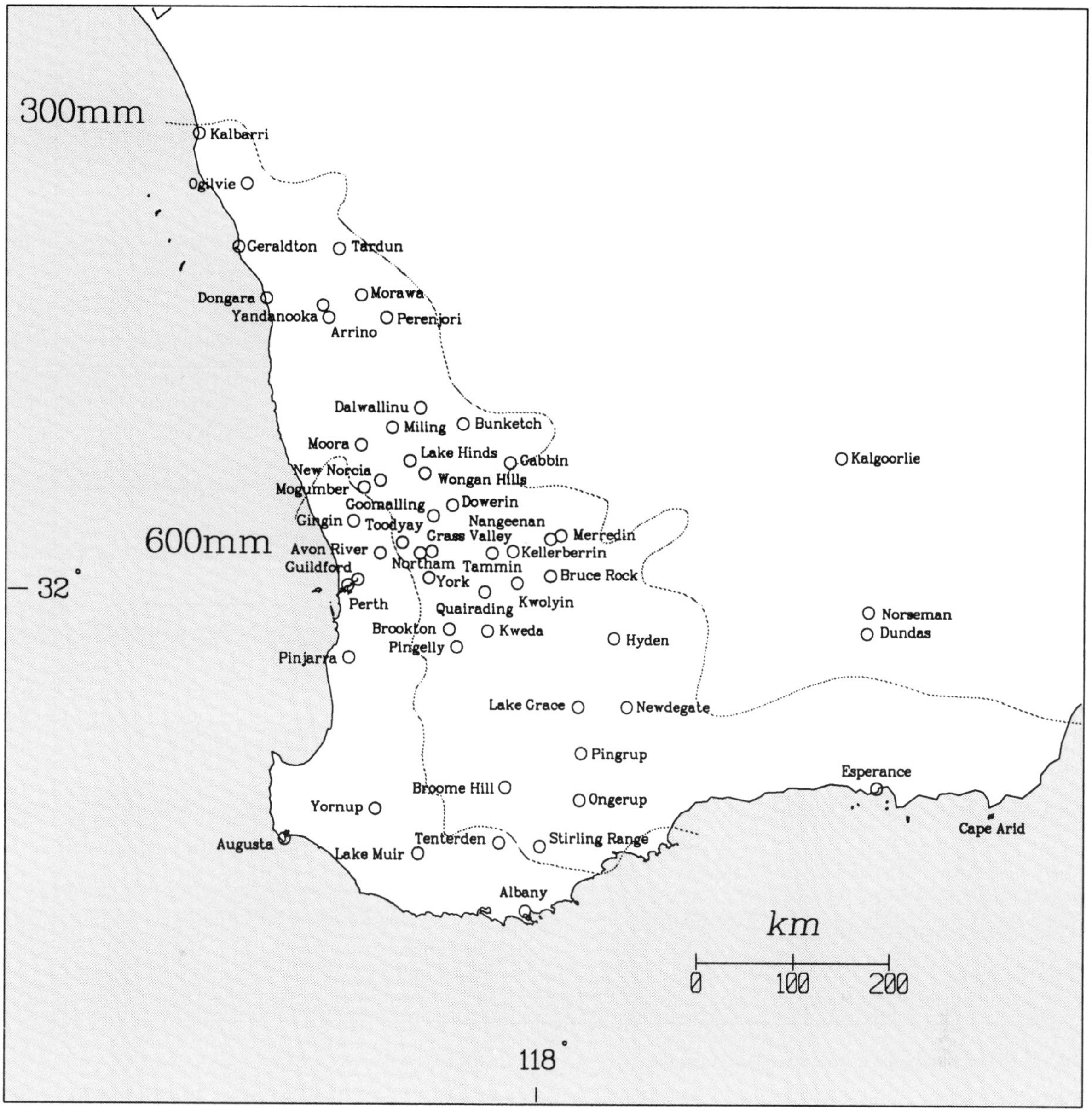

Fig. 2. Southwestern Australia showing localities mentioned in text and the 300 and 600 mm isohyets. The wheatbelt is the area lying between the two isohyets (see Fig. 1).

area of original vegetation had been removed. The remainder was distributed among literally thousands of patches of varying sizes, shapes, positions in the landscape, biotic composition, ownership, management histories, and degrees of degradation. A detailed account of the settlement, development and ecological changes in the central wheatbelt is given in Hobbs and Saunders (1993).

Patterns of clearing vary throughout the wheatbelt, depending on the period during which development took place. Those areas along main roads and railway lines which were developed during the pioneering phase of land clearing have been more extensively cleared than areas developed more recently (Coates 1987). For example, clearing in the Shire of Cunderdin in the central wheatbelt began in the late nineteenth century, was completed by the 1960s and about 3 per cent of the area of original vegetation remains. Similarly, in the neighbouring Shire of Tammin about 7 per cent of its area is retained under native vegetation (Coates 1987) (Fig. 1 and Plate 1a). In these and similar areas, road verges are narrow, often completely cleared, and where there is some native vegetation, it is usually degraded. The majority of remnant patches of native vegetation have been degraded by grazing (Plate 1b) with little or no understorey and changed microclimatic conditions (Scougall *et al.* 1993).

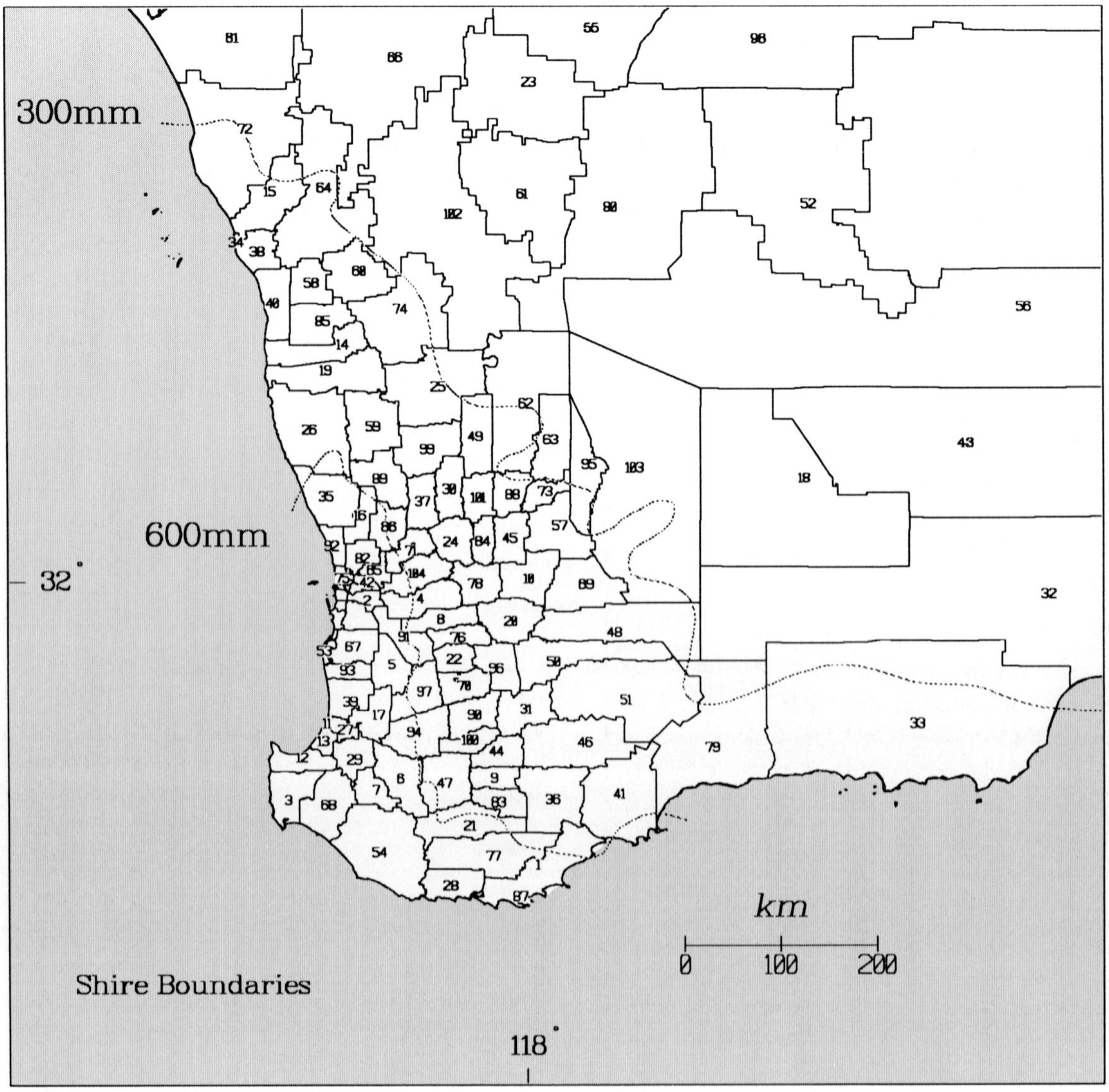

Fig. 3. Shire boundaries in southwestern Australia.

1 Albany
2 Armadale
3 Augusta-Margaret River
4 Beverley
5 Boddington
6 Boyup Brook
7 Bridgetown-Greenbushes
8 Brookton
9 Broomehill
10 Bruce Rock
11 Bunbury
12 Busselton
13 Capel
14 Carnamah
15 Chapman Valley
16 Chittering
17 Collie
18 Coolgardie
19 Coorow
20 Corrigin
21 Cranbrook
22 Cuballing
23 Cue
24 Cunderdin
25 Dalwallinu
26 Dandaragan
27 Dardanup
28 Denmark
29 Donnybrook-Balingup
30 Dowerin
31 Dumbleyung
32 Dundas
33 Esperance
34 Geraldton
35 Gingin
36 Gnowangerup
37 Goomalling
38 Greenough
39 Harvey
40 Irwin
41 Jerramungup
42 Kalamunda
43 Kalgoorlie-Boulder
44 Katanning
45 Kellerberrin
46 Kent
47 Kojonup
48 Kondinin
49 Koorda
50 Kulin
51 Lake Grace
52 Leonora
53 Mandurah
54 Manjimup
55 Meekatharra
56 Menzies
57 Merredin
58 Mingenew
59 Moora
60 Morawa
61 Mount Magnet
62 Mount Marshall
63 Mukinbudin
64 Mullewa
65 Mundaring
66 Murchison
67 Murray
68 Nannup
69 Narembeen
70 Narrogin
71 Northam
72 Northampton
73 Nungarin
74 Perenjori
75 Perth
76 Pingelly
77 Plantagenet
78 Quairading
79 Ravensthorpe
80 Sandstone
81 Shark Bay
82 Swan
83 Tambellup
84 Tammin
85 Three Springs
86 Toodyay
87 Town of Albany
88 Trayning
89 Victoria Plains
90 Wagin
91 Wandering
92 Wanneroo
93 Waroona
94 West Arthur
95 Westonia
96 Wickepin
97 Williams
98 Wiluna
99 Wongan-Ballidu
100 Woodanilling
101 Wyalkatchem
102 Yalgoo
103 Yilgarn
104 York

In areas which have been more recently developed more native vegetation has been retained. For example, the eastern half of the Shire of Lake Grace was only recently opened up for agriculture, with the last release of land in 1984. About 31 per cent of the shire has been retained under native vegetation and road verges are wider and less degraded than those in the central wheatbelt (Coates 1987) (Plates 1c and d). Those areas in the higher rainfall zones of the wheatbelt have also been less cleared. For example, the Shire of Collie in the transition zone between the wheatbelt and the southwest forests has 89 per cent of its area under native vegetation. These different patterns of clearing in the southwest show up on the satellite image in Figure 1 where uncleared vegetation is indicated by dark areas and cleared land by pale red.

With the loss of such a high percentage of the original vegetation, it is inevitable that there have been major impacts on the region's biota. Three hundred and forty-eight species of plant are listed as rare and endangered in the wheatbelt (Hopper *et al.* 1990). This is one of the highest number for any region in Australia. In addition, at least 24 species of plant are believed to be extinct (Leigh *et al.* 1984). While many species may have existed in small or isolated populations before clearing for agriculture, now they are surrounded by an agricultural matrix and the remnants on which they survive are degrading (Plate 1b). The long term viability of most of these isolated populations is extremely poor.

The remaining native vegetation in the wheatbelt is distributed in small patches, most of which are less than 20 ha. These remnants do not represent the original vegetation. Woodlands for example, grew on the heavy soils and were regarded by the early land surveyors as indicative of good agricultural land. As a result, they were selectively taken up and cleared. Less than 3 per cent of the original area of York gum *Eucalyptus loxophleba*/wandoo *E. wandoo*/salmon gum *E. salmonophloia* woodlands remain (Beard and Sprenger 1984), and are poorly represented on all but a few of the wheatbelt conservation reserves.

Less than 15 per cent of the area of remnant vegetation in the wheatbelt is in conservation reserves. The other 85 per cent is on private property and is still being degraded because of weed invasion, grazing by rabbits and domestic livestock, particularly sheep, and by changed ecological conditions in the surrounding agricultural land. One consequence of agricultural development has been the changing hydrologic balance and rising water tables are threatening both remnant vegetation and agricultural land. A recent assessment of the conservation value of remnant vegetation in the central wheatbelt using satellite imagery showed that only 3 per cent of the area of original vegetation could be considered to be in good condition (Lambeck and Wallace 1993).

With such an enormous loss of habitat, it is inevitable that the region's animals must have been dramatically affected. Intuitively, we suspect that with over 90 per cent of available habitat being removed, somewhere approaching that percentage of the total number of animals dependent upon that habitat must have been lost. For example, the highest rates of decline and extinction of native mammals has occurred in the wheatbelt (Burbidge and McKenzie 1989). Of 43 species of native mammal (excluding bats) believed to have occurred there at European settlement, 13 species had disappeared from the region and only 12 were abundant or moderately common by the 1970s (Kitchener *et al.* 1980). Mammals are still being lost from remnants of native vegetation and more species will disappear given present trends. Apart from mammals and birds, there are limited data available on other animal groups, particularly invertebrate taxa. We have little idea of which species were present before development for agriculture, and we have little idea of what occurs in the region now. Thus we have inadequate knowledge of what we have lost and the ecological consequences of those losses.

The loss of our native biota indicates that we do not have sustainable nature conservation. Ecological changes imposed on the wheatbelt have also adversely affected agricultural production, indicating a failure to adhere to the principles of ecologically sustainable development. Agriculture is an important industry in Western Australia, with wheat being a significant contributor to the State's earnings. The extensive clearing of native vegetation and its replacement with agricultural crops has altered the hydrologic balance of the wheatbelt. Water tables are rising, in some cases over 20 m since the 1920s (George *et al.* 1995), bringing salt to the surface. Salt was always a feature of the landscape, and extensive salt-lake systems existed before clearing of native vegetation began. Ocean-derived salt is continuously transported inland and deposited with rainfall. In the central wheatbelt, the amount of salt stored in the soil profile varies from an average of 247 tonnes ha^{-1} upslope to around 13 500 tonnes ha^{-1} in the valley floors (McFarlane *et al.* 1993). At present, about 3 per cent of all cleared land is affected by soil salinity to the extent that it is useless for cereal cropping. It is estimated that within the next 30 years, 15 per cent of all cleared land could be affected (Nulsen 1993). The problems of salinity extend to the drainage systems as saline discharge results in intrusion of salt water into formerly good quality water supplies. As an example, George *et al.* 1995 point out that Lake Toolibin, an important freshwater conservation reserve will become saline, and its associated vegetation will die unless it is protected from rising groundwaters and surface runoff.

In addition to changes in the hydrologic balance, it is possible that changes imposed upon the agricultural landscape have resulted in climatic changes. For example, between 1913 and 1986 parts of the wheatbelt had 4–5 per cent decreases of winter rainfall (Pittock 1988), while the uncleared Goldfields woodland to the east had a 6 per cent increase. Examination of satellite imagery reveals that annual surface albedo (or the fraction of incident radiation reflected by the surface) has increased by 14–25 per cent over cleared areas, and surface temperatures in summer over cleared agricultural land have increased by 1–5°C (Smith *et al.* 1992). The lower albedo of native vegetation may be due to the greater absorption of solar energy compared with the agricultural system which replaced it. The differences in temperature are due to year-round evapotranspiration of native vegetation and the senescence of agricultural systems. Smith *et al.* (1992) thought that these differences may result "in greater convective activity over native vegetation and the evolution of a higher planetary boundary layer during the day". Satellite imagery of the area has shown that convective cumulus clouds were more common over large areas of native vegetation than adjacent agricultural land. Possibly, the more extensive mixing of thermal layers over native vegetation is sufficient to reach the condensation level, but insufficient mixing takes place over cleared land. It may be that rainfall, which before clearing would have been induced to fall on the wheatbelt, may now be passing further inland and falling on the Goldfields.

chapter three

Community-based observer atlas of birds of the wheatbelt

As POINTED out earlier, in Western Australia the distribution and abundance of the native biota is not well known and information on the basic biology of all but a few species is lacking. While birds have been better studied than most groups, there were insufficient data available to examine the effects the changes imposed upon the landscape have had on their distribution and abundance. To rectify this, a community-based observer survey was established to map the distribution and abundance of birds in the wheatbelt. In the species accounts given in Chapter 4, the results of this survey are compared with historic records to examine changes in distribution and abundance of species in the wheatbelt and to establish the effects changing landuse has had on them.

Historic records

Comparisons between past and present distributions are easier when made between two periods spaced as widely apart as possible. As previously stated, well over half of the wheatbelt was cleared after the Second World War, so changes in the distribution and abundance of birds since then may well be related to changes in landuse and attendant changes to the landscape. Accordingly, records of the historic distribution and abundance of birds of the wheatbelt were taken from detailed bird lists based on records collected or published before 1937. The pastoral phase before clearing began may have had an impact on the avifauna of the wheatbelt, but there is no way of establishing this because there are no published records for the region before 1900.

The first published bird list for the wheatbelt was from Stirling Range (Fig. 2) and included an annotated list of birds seen or collected during a three-week trip in late September and early October 1902, when Milligan's party travelled from Tenterden to Stirling Range and back (Milligan 1903). In early October 1903, Milligan (1904) spent two weeks in the Wongan Hills which his party reached by way of Mogumber and New Norcia. He produced a detailed account of this expedition, including an annotated list of the birds encountered and collected. In October 1903 Lawson (1905) spent several weeks around Mogumber along the Moore River and published an annotated list of the birds. In 1909 Crossman (1909) published the first list of birds and their status based on an extended period of residence in one locality. He lived at Cumminin Station, an area south of Bruce Rock, over a three year period. Orton and Sandland were residents of the Moora district, and in 1913 they published an annotated list, including status, of the birds found within a 48 km radius of Moora. In 1923 Carter produced a detailed annotated list of birds of the Broome Hill district based on records made between 1905 and 1913. This was followed in 1931 by Jenkins who produced an annotated bird list based on four years spent in the Northam district. In 1933 Carnaby published an annotated list of the birds of the Lake Grace district based on his residency. In 1937 Sedgwick produced a list, which was mostly not annotated, based on four years he spent in the Nangeenan area during the early 1930s. In 1962 Leake published a book privately which gave an account of the natural history of an area just north of Kellerberrin. This book included a chapter on the birds of the area during the early part of this century, together with comments on the changes in the avifauna and some of the reasons for those changes.

Data obtained from these bird lists have been used to calculate the percentage of observers recording each species, thus giving an indication of how widespread it was before 1937. Where the author(s) made some observation about how common or otherwise a particular species was, these comments have been interpreted to provide an indication of the status of that species.

Current distribution of birds of the wheatbelt

The current distribution of wheatbelt birds was established by enrolling the support of a number of observers scattered throughout the region. Early in 1987 a 16-page booklet was produced which listed all species of bird known to have occurred in the central wheatbelt. Each species was listed by common name and beside each species were 12 monthly columns, each of which was divided into four weekly columns (Fig. 4). This represented 48 recording weeks in each year. In May 1987 a radio and newspaper campaign was conducted to enrol observers in the scheme. Once enrolled, each observer was asked to note in the booklet all species of birds seen each week during the year in their nominated collection area. At the end of the year a new booklet was provided for each observer and a stamped addressed envelope included to allow them to return the completed booklet.

Common Name	Status	Habitat	Jan	Feb	Mar	Apr	May	Jun	Jul	Aug	Sep	Oct	Nov	Dec	Notes
Grey Fantail	Breeding Migrant	W,M,S													
Willie Wagtail	Resident	W,M,S,H,F													
Chestnut Quail-thrush	?	W,M,S,H													
White-browed Babbler	Resident	W,M,S,H													
Rufous Song Lark	Vagrant	F													
Brown Song Lark	Breeding Migrant	F													
Blue-breasted Wren	Resident	W,M,S,H													
White-winged Wren	?	M,H													
Southern Emu Wren	?	H													
White-browed Scrub Wren	?	M,S,H													
Shy Hylacola	Resident	S,H													
Redthroat	Resident	W,M													
Field Wren	Resident	S,H													
Southern Whiteface	Vagrant	W													
Weebill	Resident	W,M													

Common Name	Status	Habitat	Jan	Feb	Mar	Apr	May	Jun	Jul	Aug	Sep	Oct	Nov	Dec	Notes
Western Warbler	Non-breeding Migrant?	W,M													
Broad-tailed Thornbill	Resident	W,M,S,H													
Chestnut-rumped Thornbill	Resident	W,M,S													
Yellow-rumped Thornbill	Resident	W,M,S													
Varied Sittella	Resident	W													
Rufous Treecreeper	?	W													
Red Wattlebird	Vagrant	W,M,S,H													
Little Wattlebird	Vagrant	W,S,H													
Spiny-cheeked Honeyeater	Vagrant?	W,M,S,H													
Yellow-throated Miner	Resident	W,M,S,F													
Singing Honeyeater	Resident	W,M,S,H													
White-eared Honeyeater	Resident	W,M,S,H													
Purple-gaped Honeyeater	Vagrant	W,M,H													
Yellow-plumed Honeyeater	Vagrant	W,M													
Grey-fronted Honeyeater	?	W,M													

Fig. 4. Example page from data collection booklet.

Each observer was required to describe the location in which they were collecting data, including the extent of cleared land, the type of vegetation, with sufficient information to allow the locality to be plotted on a map and the co-ordinates determined. The data from each completed booklet were then entered on to a computer data base and verified. The original completed booklets were photocopied and then returned to the observers. Photocopies of the completed data booklets are stored at the Western Australian laboratory of CSIRO Division of Wildlife and Ecology, as is the computer data base.

Data collection commenced in May 1987 and continued through to the end of 1990. Data were supplied by 127 observers in 1987, 142 in 1988, 117 in 1989 and 104 in 1990. One hundred and eighty-seven people submitted data for the survey and 79 of these submitted data for each year of the scheme. Although there were 48 recording weeks per year, observers were not necessarily able to supply data for every week because of periods of absence. Data were collected during a total of 2 910 recording weeks in 1987, 5 212 in 1988, 4 671 in 1989 and 4 283 in 1990. Observers were located throughout the wheatbelt and there were also observers collecting data from the higher rainfall area to the southwest of the wheatbelt (Fig. 5).

Presentation of data

By comparing data from the community-based observer survey conducted from 1987 to 1990 with data from the region during the early part of this century, it is possible to assess changes in the distribution and abundance of birds and suggest the reasons for these changes. The data are presented in Chapter 4 as individual species accounts which are given under the common (Readers' Digest 1976) and scientific names. Some species have several names in common usage. These names are listed in the species account and cross-referenced in the index.

Each species account has a table which summarizes data from the historic records 1900–37 and from the community-based observer survey each year from 1987 to 1990. The line in the table titled *% of observers recording species* indicates how widespread a species was during the periods 1900–37 and 1987–90. These data were affected by how conspicuous each species was, its abundance, what types of habitats the observer was collecting in, the area over which data were collected, the observer's ability to recognize species, and how much time the observer spent collecting data. Comparisons between data collected in the period pre-1937 with those collected after 1987 can be made by looking for any change in the percentage of observers recording a species. With only 10 observers in the earlier period, and the data being collected in a different way, it is necessary to be conservative about assessing changes between the two periods. However, examination of other published material about birds of southwestern Australia helps interpret the differences.

For a few locations there are published bird lists from two widely spaced periods. For example, Milligan's (1904) account of the birds of the Wongan Hills can be compared with a comprehensive survey of the same area by the Western Australian Naturalists' Club over 21 visits between August 1973 and July 1976 (de Rebeira and de Rebeira 1977). There are also data available from the Northam area for the period 1930 to 1973 based on records of two residents (Masters and Milhinch 1974). This material, as well as other published data, is presented as appropriate in the individual species' narratives and each source acknowledged.

Most of the 10 lists of birds published before 1937 were annotated by the author(s), who often commented on the status of each species. Where this information has been provided it has been interpreted, summarized and presented in the data table for each species. These data are given in the line titled *% of recording period species seen*. These data are also presented on the historic map for each species. Where there was a range of estimates of abundance, this has been expressed by noting the minimum and maximum estimates. For example, the emu, which was seen by 80 per cent of observers pre-1937, appeared as uncommon in some locations, but common in others (U-C in the historic map in Figure 8 and in the table presented in the species account).

An estimate of the status of a species for each year of the period 1987–90 can be gauged by calculating the percentage of the total observation period (i.e., total number of recording weeks) of each year that the species was seen (shown on

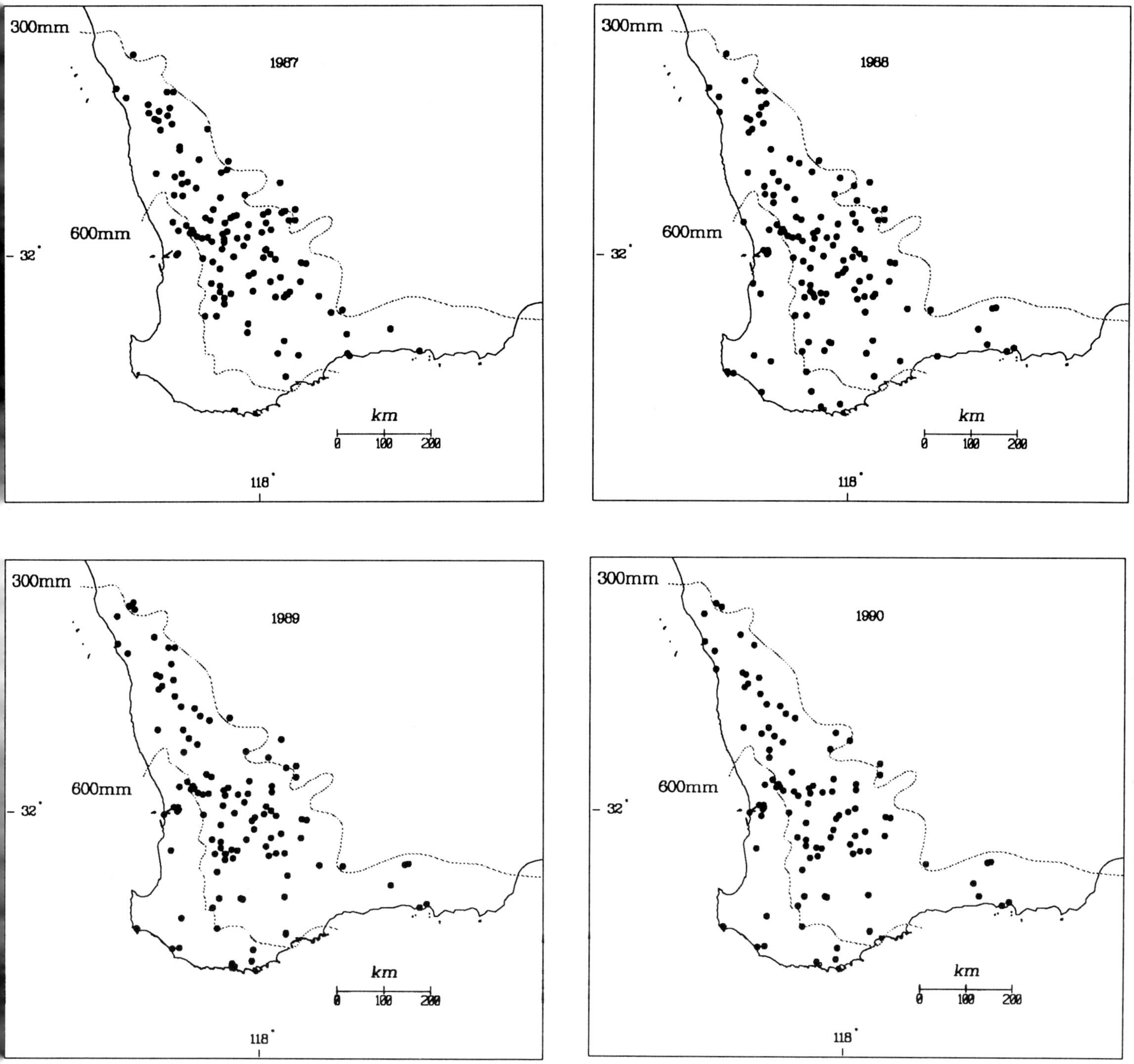

Fig. 5. Location of observers 1987–1990.

each species table as % *of recording period species seen*). A species like the Port Lincoln parrot which was recorded by 98 per cent of observers during 1987 and noted during 91 per cent of the total 1987 recording time (i.e., 91% of 2 910 recording weeks) was widespread and abundant. In contrast, the ground cuckoo-shrike which was recorded by only 2.4 per cent of observers in 1987 and noted during only 1.1 per cent of the recording period, was rare and restricted in distribution.

In Chapter 4, a narrative is provided for each species which summarizes its abundance, status, distribution, the habitats it is known to frequent, its diet (from Barker and Vestjens 1989, 1990), nest site and number of eggs laid (from Readers' Digest 1976; Beruldsen 1980). The historic and 1987–90 distributions are presented in the figures which accompany each narrative. The historic data for each of the 10 locations for which there are data pre-1937 are shown together with the status of the species at each location, if known (see Fig. 6). The distribution during the period 1987–90 is shown on one map if there were no marked changes in distribution during the observation period, or presented on several different maps if there were annual or seasonal differences in distribution. The size of the circle marking each location expresses the recording rate; the larger the circle, the more often the species was recorded at that location (Fig. 6).

The abundance of each species is summarized as either common, uncommon or rare, and the status as resident, migrant, nomad or vagrant. A species was regarded as:
common if it was recorded more than 30 per cent of the total recording period;
uncommon if recorded between 1–30 per cent of the recording period; and
rare if less than 1 per cent.

A species was classified as:
resident if it remained in the area all year;
migrant if it moved between locations in a regular annual cycle, usually breeding in one area and spending the non-breeding season in another area;
nomad if it wandered from one area to another in search of optimal conditions; and
vagrant if its normal range did not include the wheatbelt.

The wheatbelt does not constitute a well-defined natural region and despite consisting of a highly modified landscape that has a similar appearance throughout, it is in fact a transitional zone between the mesic southwestern forest and the arid zone. Accordingly, the wheatbelt avifauna has elements which are characteristic of both zones. For example, Kellerberrin in the central wheatbelt shares 88 species with the Murchison River catchment in the arid zone (Saunders and Curry 1990) and probably a similar number with the forest zone. Because it is a transition zone, the abundance and status of species may vary over the region. For example, the western warbler is essentially a species of the mesic zone, a common resident in the higher rainfall districts of the wheatbelt, but an uncommon migrant in the drier districts and does not occur in the southeastern wheatbelt. By way of contrast, the budgerigar is a bird of the arid zone which is a seasonal nomad in the drier districts of the wheatbelt and does not occur in the higher rainfall districts of the wheatbelt. These differences in status and abundance are summarized in the species narratives and in the accompanying figures.

There are seasonal variations in the abundance of some species in the wheatbelt. For those species, this variation is shown in a figure which presents the percentage monthly recording rates for that

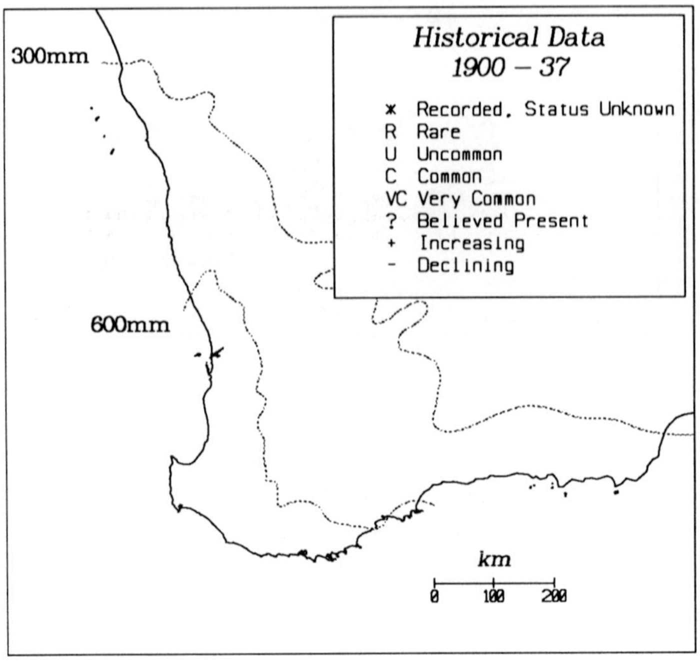

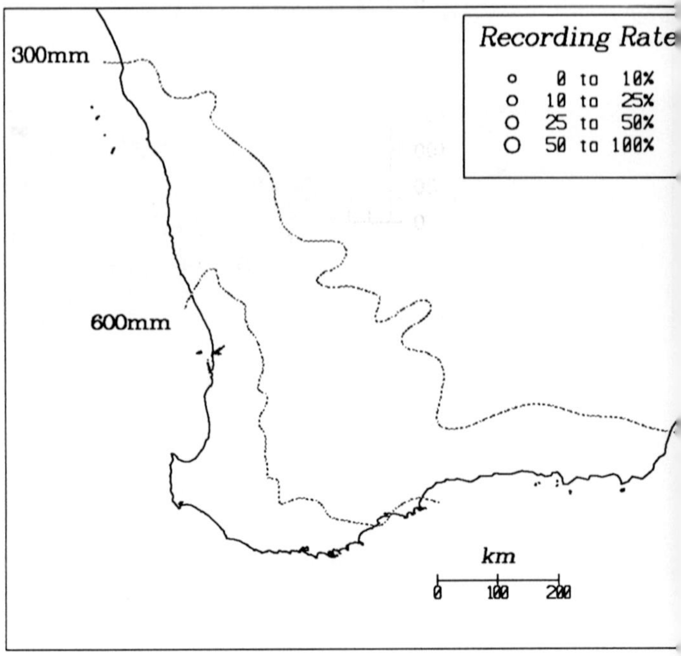

Fig. 6. Illustration of how data are displayed on distribution maps in individual species accounts in Chapter 4. The locations of the 10 observers for the period 1900–1937 are shown on the distribution map (1900–1937) in Figure 68 on page 90. Recording rates for the period 1987–1990 are indicated by the size of the circle; the larger the circle, the more often the species was recorded at that location.

particular species for 1988, 1989 and 1990, the three complete years of the survey. The distribution of some species, particularly water birds, was influenced by rainfall. Monthly rainfall figures for the period 1987–1990 for four locations throughout the wheatbelt or its periphery (Arrino in the north, Guildford to the west of the central wheatbelt, Bruce Rock in the east of the central wheatbelt, and Pingrup in the south) are shown in Figure 7.

Two transparent overlays have been provided with this book. One shows the locations of all observers who collected data for this survey (1987–1990) and the other the locations of the 10 historic records. By placing these over distribution maps for individual species they may help the reader to interpret the distibution maps by aiding orientation or showing clearly where the gaps in distribution are in relation to observers.

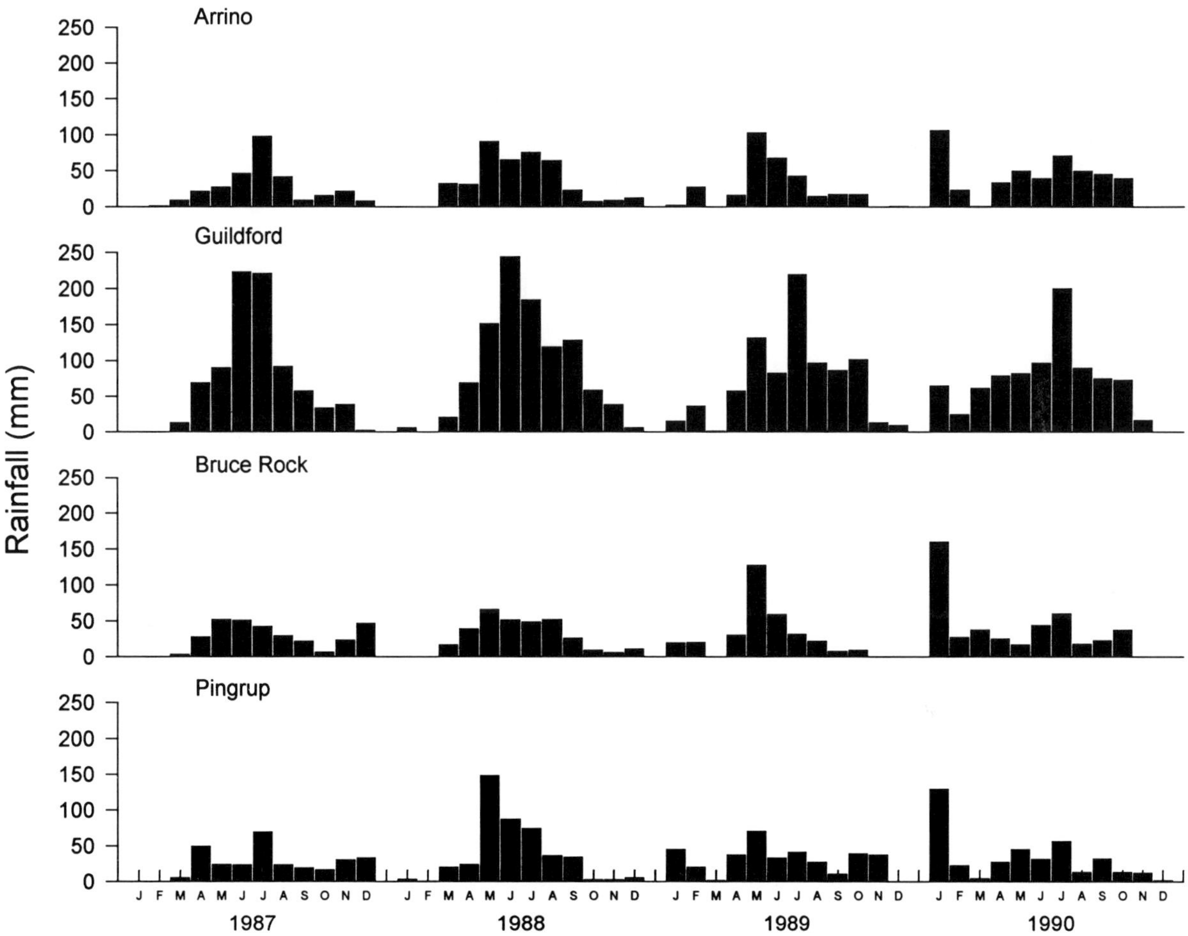

Fig. 7. Monthly rainfall for four locations in southwestern Australia 1987–1990.

Plate 1a

Plate 1b

Plate 1c

Plate 1d

PLATE 1a. Aerial oblique photograph of Tammin Shire in the central wheatbelt showing degraded narrow road verges and extensive nature of the clearing (photo D. A. Saunders).

PLATE 1b. Remnant patch of native vegetation in Kellerberrin Shire in the central wheatbelt illustrating degradation from grazing. Note the lack of understorey (photo D. E. Steven).

PLATES 1c and d. Aerial obliques of Lake Grace Shire in the southeastern wheatbelt showing road verges with wider strips of native vegetation than occurs in the central wheatbelt and clearing patterns more typical of recently cleared areas (photos D. A. Saunders).

chapter four

Species Accounts

NON-PASSERINES

Emu

Dromaius novaehollandiae

The emu is an uncommon resident in woodland, mallee, shrubland, heath and farmland over much of the wheatbelt (Fig. 8). It feeds on arthropods including grasshoppers, crickets, ladybirds, caterpillars and ants, seed, plant matter and fruit. It breeds in winter, nests on the ground and lays 8–10 eggs.

The emu has decreased in range and abundance in the wheatbelt over the last 80 years. Milligan (1904) saw fresh emu tracks in many places around the Wongan Hills, but de Rebeira and de Rebeira (1977) did not record it in four years of surveys in the same area. Orton and Sandland (1913) noted that the emu was once plentiful in the Moora district, but was becoming rarer every year. Carter (1923) noted that as the Broome Hill district became more settled the birds disappeared. It is rare or no longer occurs in the older more extensively cleared areas of the wheatbelt, particularly the central wheatbelt.

Fig. 8. Distribution 1900–1937

Emu

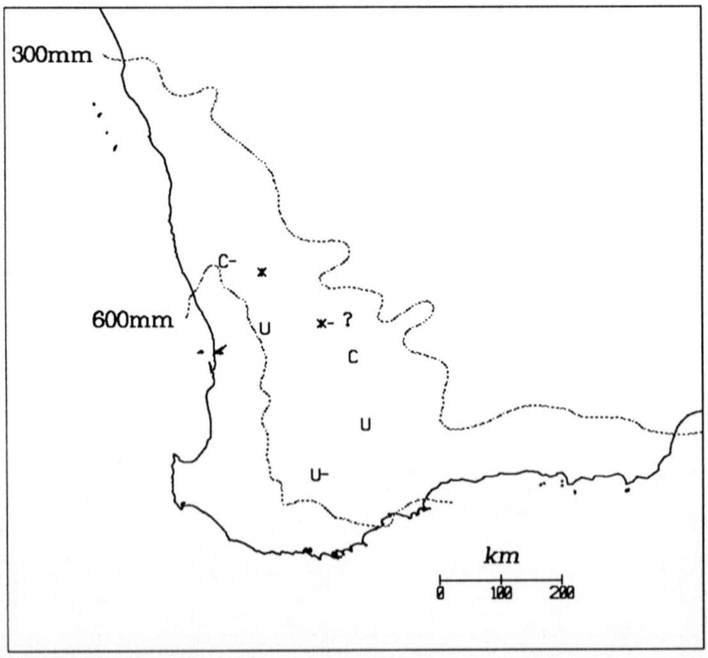

Fig. 8. Distribution 1987–1990

Emu

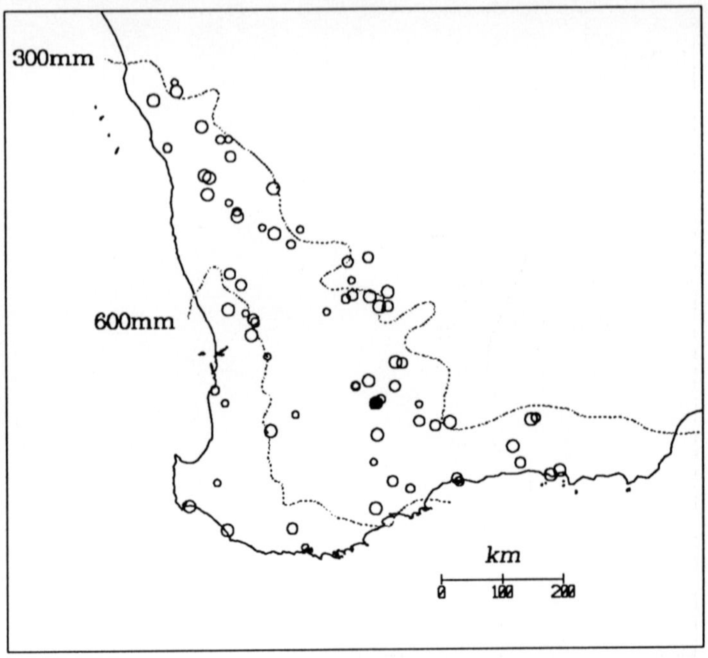

	1900/37	1987	1988	1989	1990
% of observers recording species	80	32	36	26	30
% of recording period species seen	U-C	13	16	12	13

Fig. 9. Distribution 1900–1937

Hoary-headed grebe

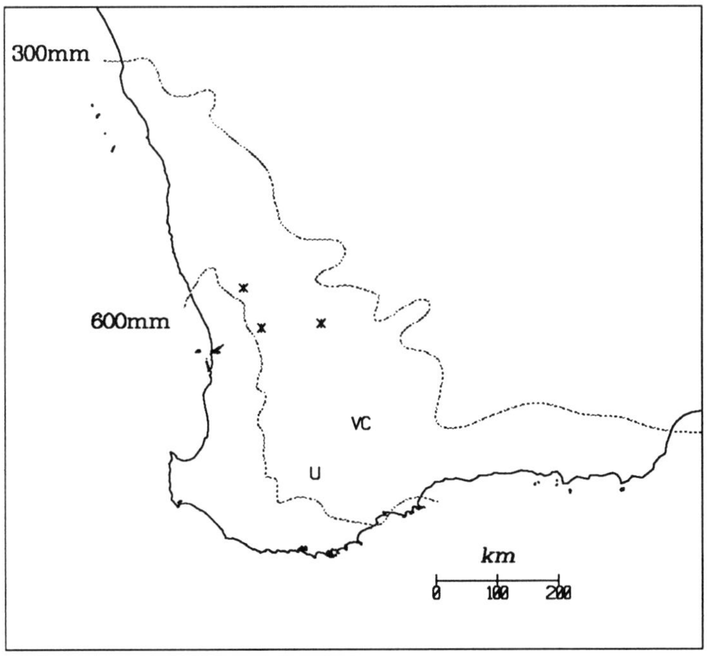

Fig. 9. Distribution 1987–1990

Hoary-headed grebe

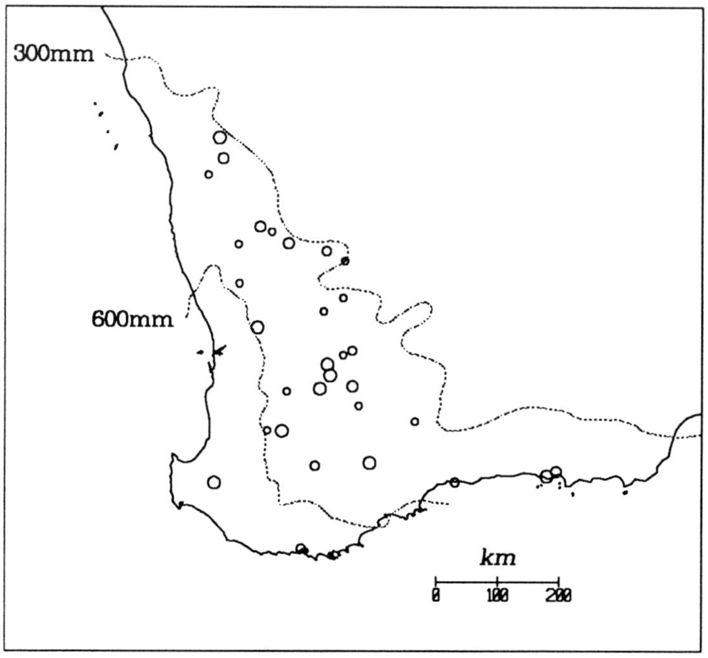

	1900/37	1987	1988	1989	1990
% of observers recording species	50	5.5	15	9.4	14
% of recording period species seen	U-VC	3.2	4.5	3.5	5.3

Hoary-headed grebe

Poliocephalus poliocephalus

The hoary-headed grebe is an uncommon nomad on aquatic habitats including farm dams throughout the wheatbelt (Fig. 9). It may be resident along the Avon River and other favourable habitats (J. Masters, in litt.). It feeds on freshwater snails, grasshoppers, back-swimmers, water boatmen, beetles, weevils, caddis fly larvae, meat ants, brine shrimps, ostracods, freshwater shrimps, yabbies, spiders, mosquito fish, and seed. The hoary-headed grebe breeds in summer or after rains. It lays 4–5 eggs on a raft on water.

Formerly regarded as uncommon to very common, the hoary-headed grebe is nowhere common and has probably declined in abundance in the wheatbelt.

Australasian little grebe

Tachybaptus novaehollandiae

OTHER NAMES:

Australasian grebe, black-throated grebe

The Australasian little grebe occurs on aquatic habitats including farm dams throughout the wheatbelt (Fig. 10). It is resident, and may be common, on suitable habitats in the higher rainfall zones (J. Masters, in litt.), but a nomad in the drier areas. It feeds on freshwater mussels and snails, damselflies, dragonfly larvae, cockroaches, bugs, back-swimmers, water boatmen, beetles, weevils, flies, ants, freshwater shrimps, yabbies, mosquito fish, and plant matter including grass seed. It breeds in spring and early summer on a raft on water, and lays 4–5 eggs.

There is no evidence to suggest any change in abundance in the wheatbelt over the last 50 years.

Fig. 10. Distribution 1900–1937

Australasian little grebe

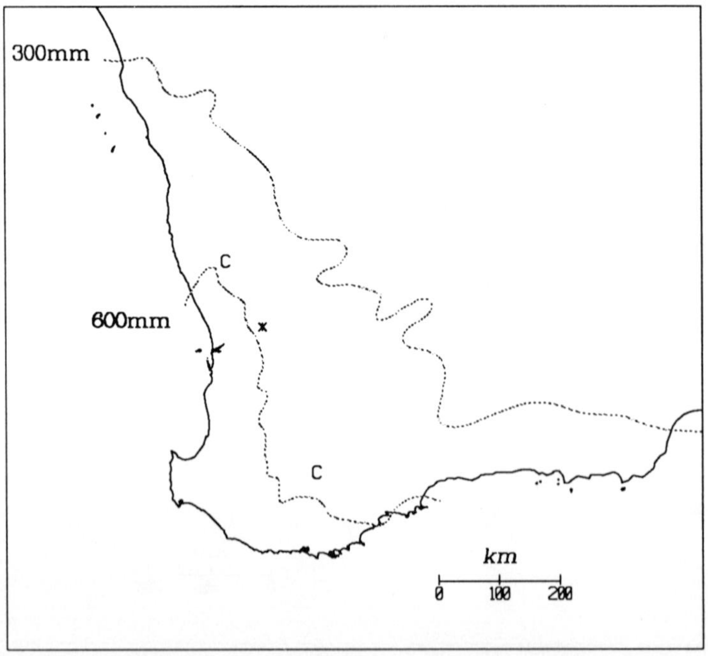

Fig. 10. Distribution 1987–1990

Australasian little grebe

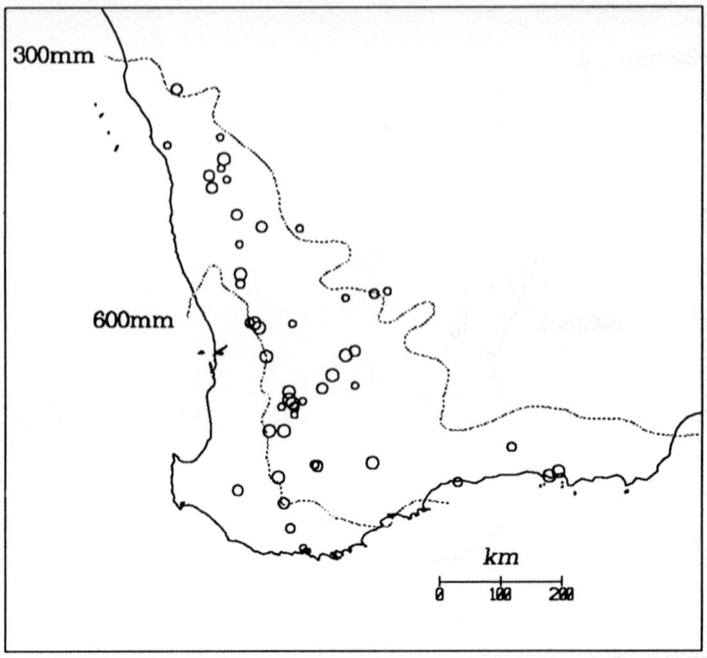

	1900/37	1987	1988	1989	1990
% of observers recording species	30	14	23	23	25
% of recording period species seen	C	6.6	6.3	9.6	10

Great crested grebe

Podiceps cristatus

The great crested grebe is a rare vagrant on aquatic habitats in the higher rainfall districts of the wheatbelt (Fig. 11). It eats fish, grasshoppers, beetles, shrimps and plant matter, including seed. It breeds in late spring and early summer on a raft on water. It lays 3–7 eggs.

Fig. 11. Distribution 1987–1990

Great crested grebe

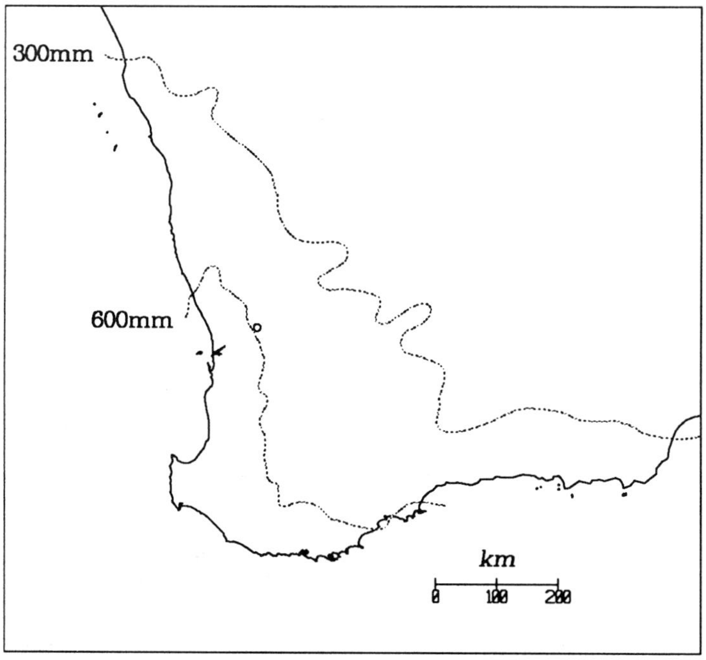

	1900/37	1987	1988	1989	1990
% of observers recording species	0	0	0.7	0.9	0
% of recording period species seen	NR	0	0.06	0.07	0

Australian pelican

Pelecanus conspicillatus

The Australian pelican is a nomad on aquatic habitats in the wheatbelt (Fig. 12). It eats fish, freshwater shrimp and crayfish, yabbies, and young birds, including ducks and gulls. The pelican may breed at any time of year, if conditions are suitable. It nests on the ground and lays 2 eggs.

There is no evidence to suggest any change in abundance over the last 50 years.

Fig. 12. Distribution 1987–1990

Australian pelican

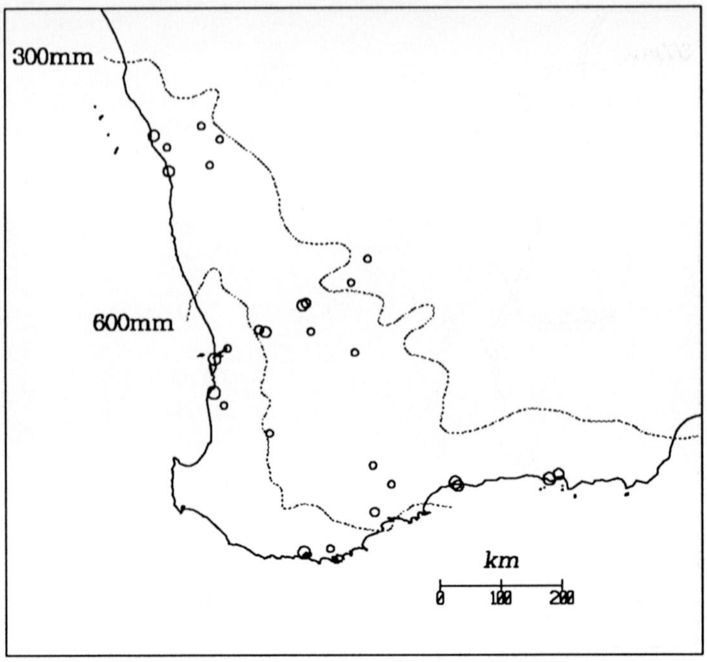

	1900/37	1987	1988	1989	1990
% of observers recording species	0	8.7	11	10	12
% of recording period species seen	NR	2.9	2.9	3.3	2.8

Fig. 13. Distribution 1900–1937

Darter

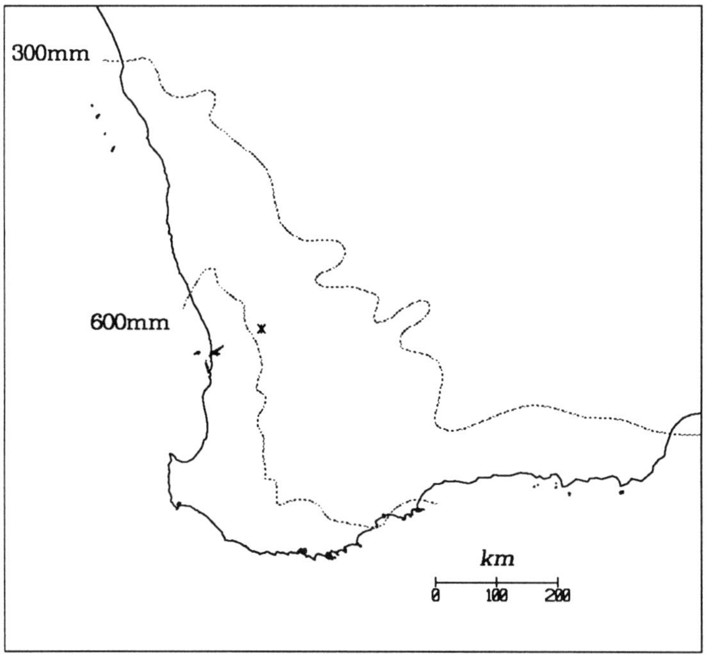

Fig. 13. Distribution 1987–1990

Darter

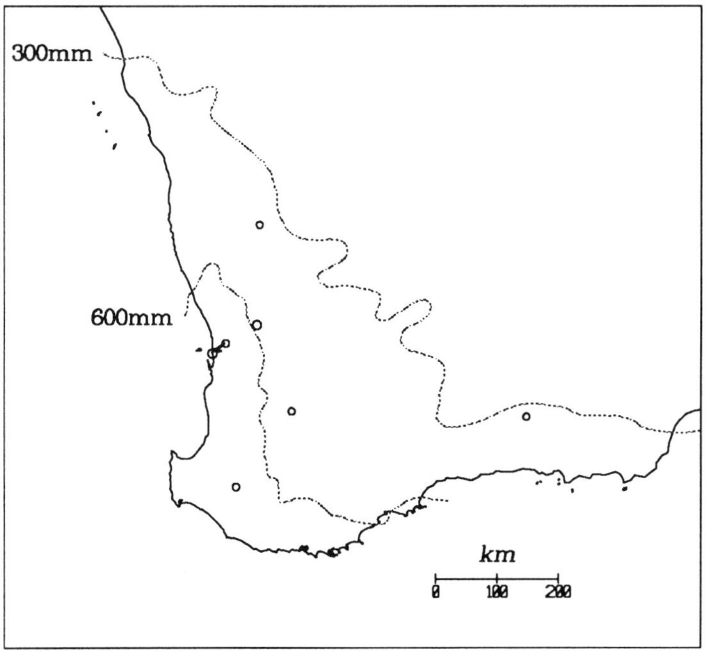

	1900/37	1987	1988	1989	1990
% of observers recording species	10	0.8	1.4	5.1	2.9
% of recording period species seen	—	0.1	0.2	0.4	0.3

Darter

Anhinga melanogaster

The darter is a nomad on aquatic habitats in the higher rainfall districts of the wheatbelt (Fig. 13) and a vagrant in the drier areas. It eats fish, small tortoises, molluscs, worms, backswimmers, water boatmen, beetles, flies, moths, freshwater shrimp, and plant material. The darter breeds in spring and summer on a platform in a tree and lays 3–4 eggs.

There is no evidence to suggest any change in abundance over the last 50 years.

Little black cormorant

Phalacrocorax sulcirostris

The little black cormorant is an uncommon nomad on aquatic habitats including farm dams throughout the wheatbelt (Fig. 14). It eats fish, molluscs, dragonfly larvae, beetles, shrimp, yabbies, marron, crayfish, tadpoles and plant material. It breeds in winter and spring on a stick nest in a tree or shrub, and lays 4–6 eggs.

There is no evidence to suggest any change in abundance over the last 50 years.

Fig. 14. Distribution 1900–1937

Little black cormorant

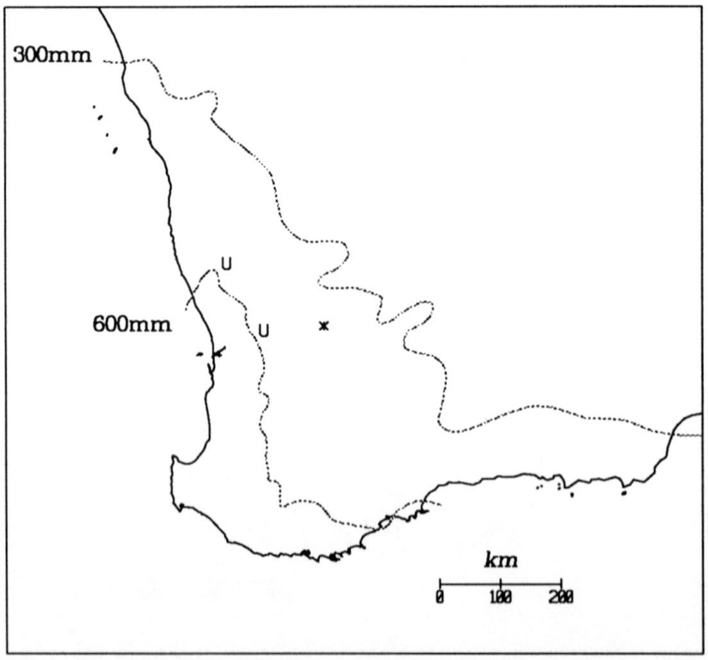

Fig. 14. Distribution 1987–1990

Little black cormorant

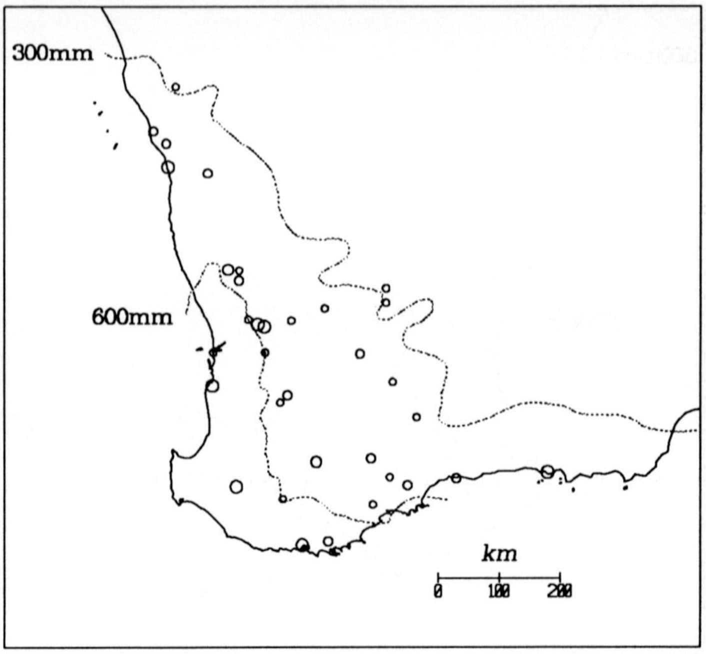

	1900/37	1987	1988	1989	1990
% of observers recording species	30	9.4	13	11	12
% of recording period species seen	U	3.6	3.7	3.6	4.0

Fig. 15. Distribution 1900–1937

Black cormorant

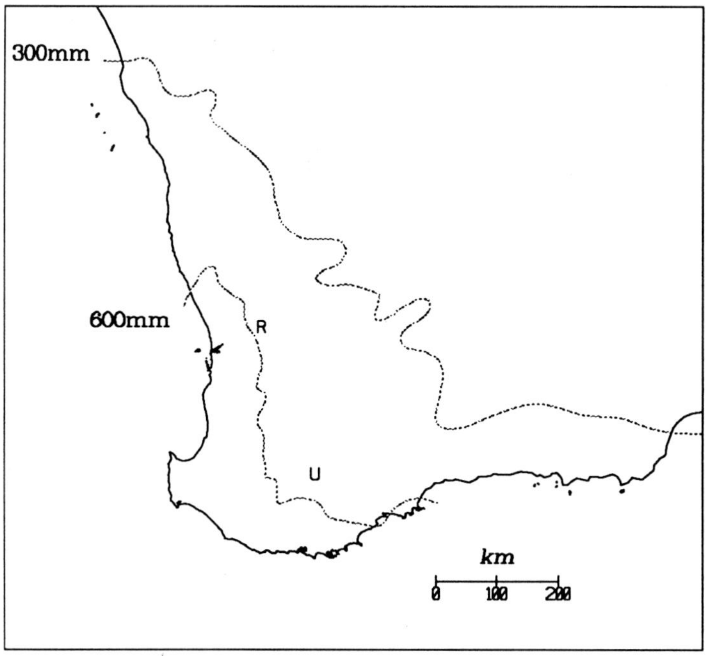

Fig. 15. Distribution 1987–90

Black cormorant

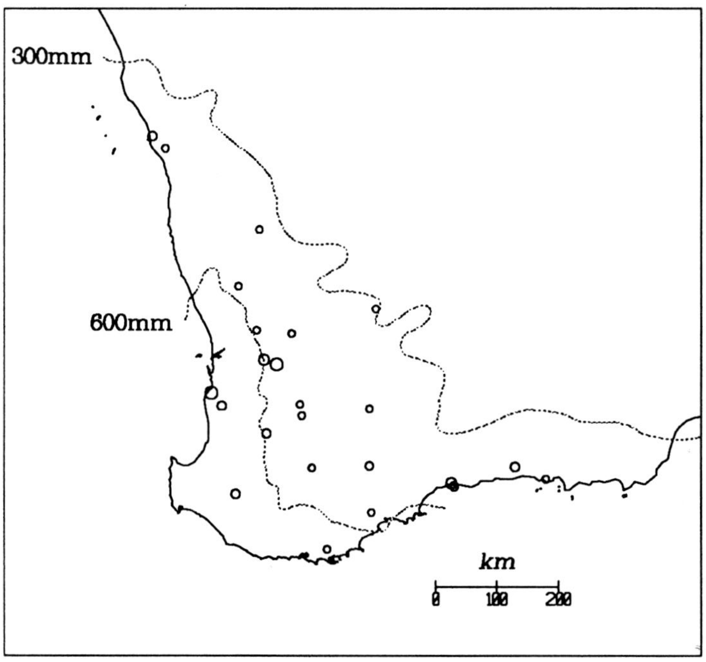

	1900/37	1987	1988	1989	1990
% of observers recording species	20	7.1	9.2	8.5	7.7
% of recording period species seen	R-U	1.3	1.1	1.1	1.1

Black cormorant

Phalacrocorax carbo

OTHER NAME:

Great cormorant

The black cormorant is an uncommon nomad on aquatic habitats including farm dams in the wheatbelt (Fig. 15). It eats fish, water bugs, shrimp, yabbies, crabs, spiders, frogs, and juvenile ducks. It breeds from autumn to spring on a stick nest in a tree, shrub, or on the ground. It lays 3–4 eggs.

There is no evidence to suggest any change in abundance over the last 50 years.

Little pied cormorant

Phalacrocorax melanoleucos

The little pied cormorant occurs on aquatic habitats including farm dams throughout the wheatbelt (Fig. 16). It is resident in the higher rainfall districts or along the coast, and a nomad in drier areas. It eats fish, freshwater snails, worms, dragonfly larvae, crickets, grasshoppers, bugs, beetles, weevils, caddis flies, ants, shrimp, yabbies, crayfish, marron, tadpoles and frogs. It breeds from winter to early summer on a stick nest in a tree and lays 4–5 eggs.

Masters and Milhinch (1974) recorded that the little pied cormorant had decreased along the Avon River between Northam and Toodyay over the previous twenty years.

Fig. 16. Distribution 1900–1937

Little pied cormorant

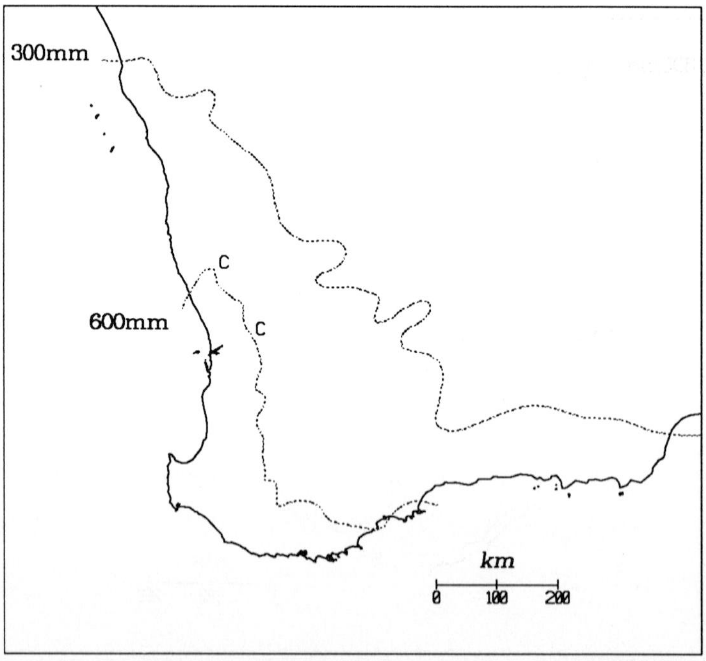

Fig. 16. Distribution 1987–1990

Little pied cormorant

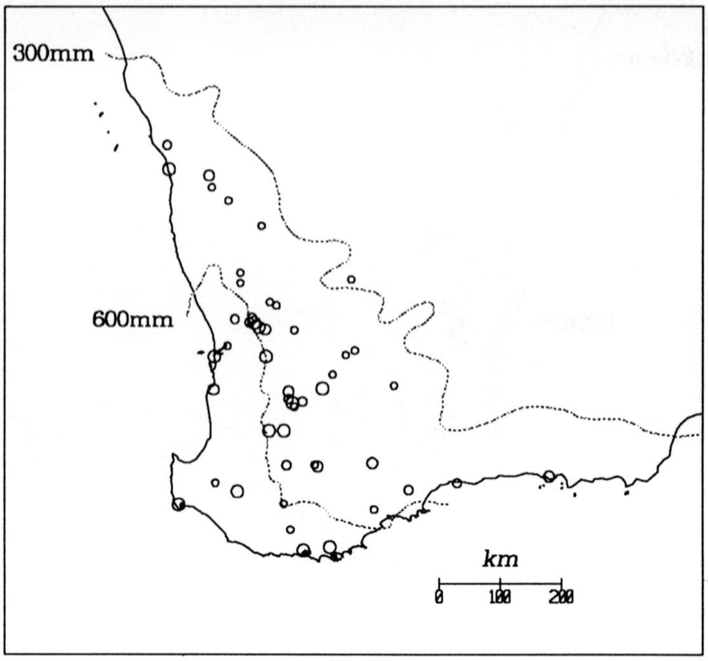

	1900/37	1987	1988	1989	1990
% of observers recording species	20	15	19	22	26
% of recording period species seen	C	5.3	5.0	7.3	9.0

Fig. 17. Distribution 1900–1937

White-necked heron

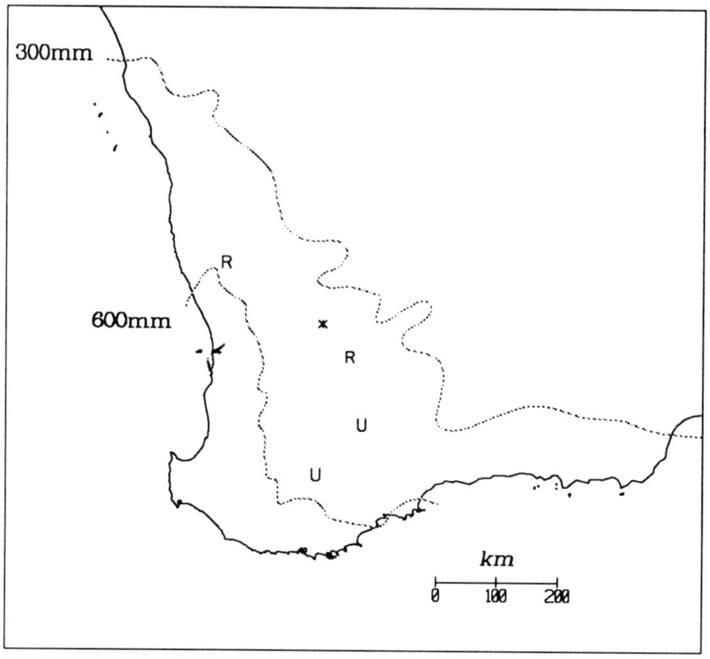

Fig. 17. Distribution 1987–1990

White-necked heron

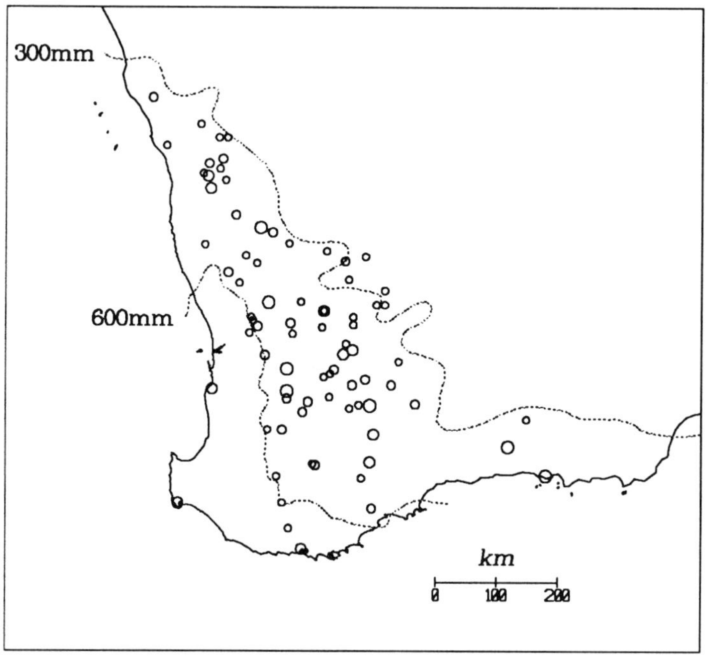

	1900/37	1987	1988	1989	1990
% of observers recording species	50	23	28	26	38
% of recording period species seen	R-U	4.8	4.7	4.3	6.1

White-necked heron

Ardea pacifica

OTHER NAME:

Pacific heron

The white-necked heron is a nomad on aquatic habitats and farmland throughout the wheatbelt (Fig. 17). It feeds on damselflies, dragonfly larvae, mantids, grasshoppers, bugs, back-swimmers, water boatmen, beetles, flies, caterpillars, freshwater shrimp and crayfish, spiders, and small mammals like the house mouse. It breeds in late winter and spring in a tree near water. It lays 3–5 eggs.

During the period 1988 to 1990 the white-necked heron was more commonly recorded in the wetter months of May to September (see Fig. below).

There is no evidence to suggest any change in abundance over the last 50 years.

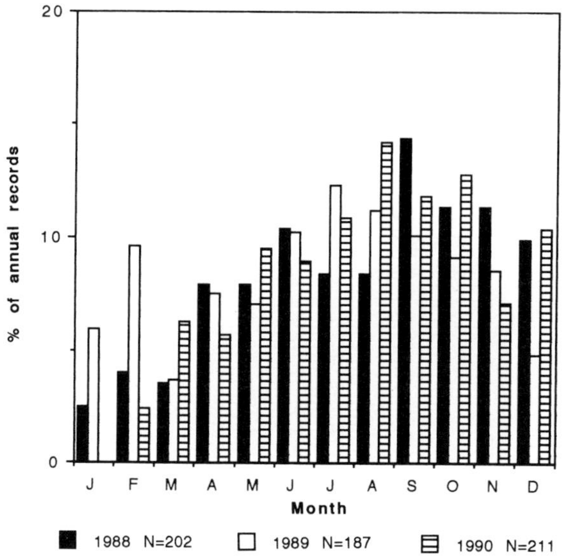

White-faced heron

Ardea novaehollandiae

The white-faced heron is an uncommon resident, usually seen singly, on aquatic habitats including farm dams, and farmland throughout the wheatbelt (Fig. 18). During the period of the survey 1987-90, it was recorded more often in the second half of the year. It was more widespread and more often recorded in 1990 than the earlier years of the survey (see Table). It feeds on fish, gastropods, damselflies, dragonfly larvae, earwigs, crickets, grasshoppers, bugs, backswimmers, water boatmen, beetles, fly larvae, caterpillars, moths, sawfly larvae, ants, slaters, shrimp, yabbies, freshwater crayfish, spiders, frogs, reptiles, and small mammals such as the house mouse. The white-faced heron may breed at any time of year on a stick nest in a tree. It lays 4 eggs.

Orton and Sandland (1913) regarded the white-faced heron as very common in the Moora district. It is nowhere common in the wheatbelt now. Given the salination of much of the drainage system, it is probable that this heron has decreased in abundance over the last 80 years.

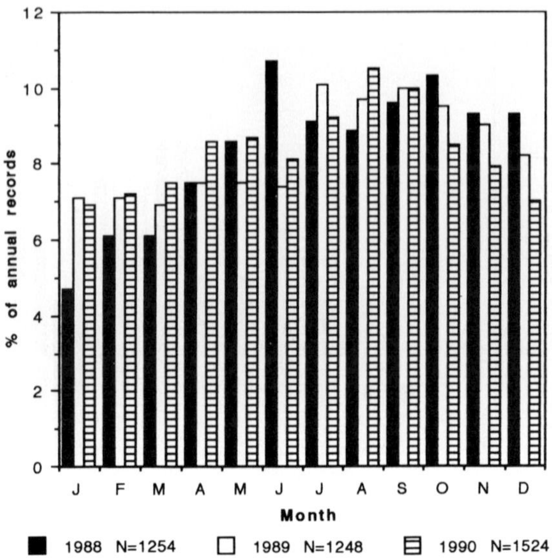

Fig. 18. Distribution 1900–1937

White-faced heron

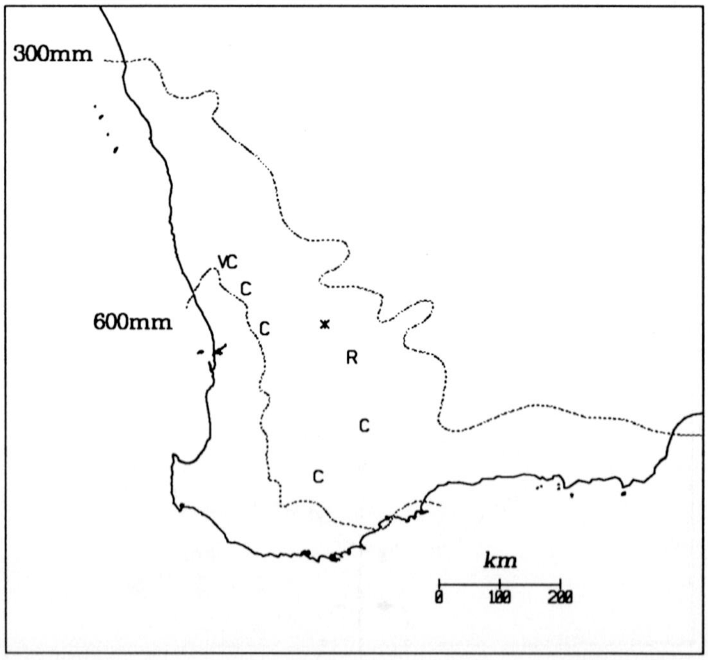

Fig. 18. Distribution 1987–1990

White-faced heron

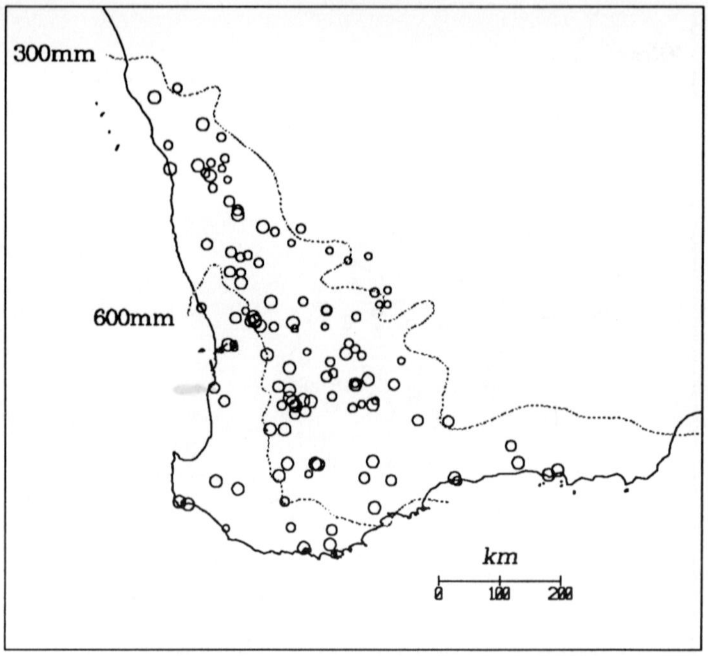

	1900/37	1987	1988	1989	1990
% of observers recording species	70	41	56	56	69
% of recording period species seen	R-VC	17	25	27	36

Fig. 19. Distribution 1900–1937

Nankeen night heron

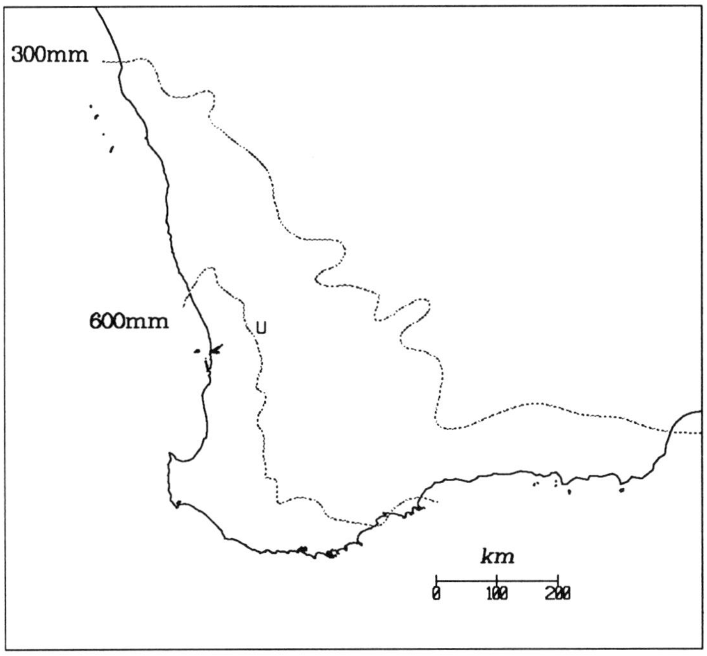

Fig. 19. Distribution 1987–1990

Nankeen night heron

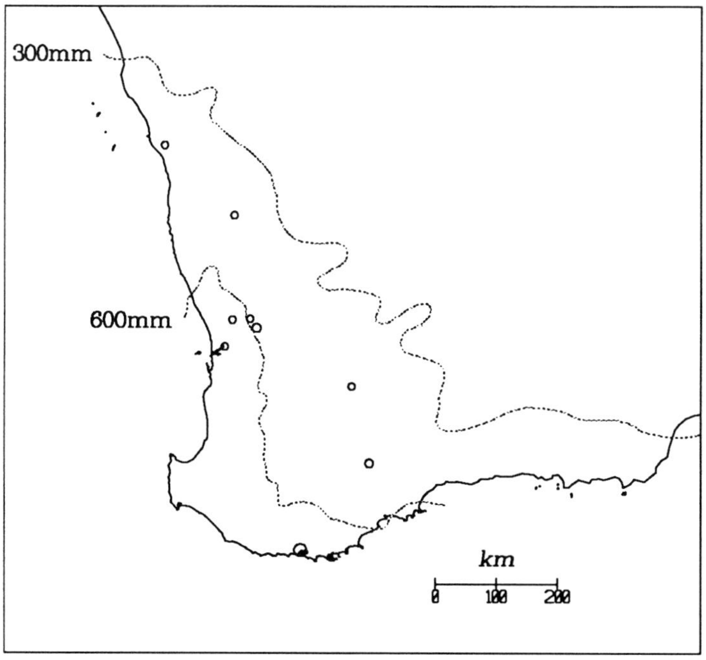

	1900/37	1987	1988	1989	1990
% of observers recording species	10	2.4	1.4	3.4	5.8
% of recording period species seen	U	0.4	0.3	0.8	0.6

Nankeen night heron

Nycticorax caledonicus

OTHER NAME:

Rufous night heron

The nankeen night heron is a rare nomad on aquatic habitats in the wheatbelt (Fig. 19). It feeds on seed and other plant material, freshwater mussels, centipedes, damselflies, dragonfly larvae, cockroaches, termites, earwigs, crickets, grasshoppers, bugs, beetles, weevils, fly larvae, caterpillars, wasps, ants, slaters, decapods, shrimp, yabbies, freshwater crayfish, spiders, fish, frogs, birds' eggs, small birds, and small mammals such as the house mouse. It breeds in spring on a stick nest in a tree, shrub, or on the ground. It lays 2–3 eggs.

There is no evidence to suggest any change in abundance over the last 50 years.

Cattle egret

Ardeola ibis

The cattle egret is a rare nomad on farmland in the wheatbelt (Fig. 20). It feeds on flies and also ecto-parasites of grazing cattle.. It occurs often in association with livestock. It breeds in spring and summer on a stick nest in a tree and lays 3–4 eggs.

Serventy and Whittell (1976) noted that the cattle egret was first recorded in the wheatbelt in 1952 at Three Springs. It has been expanding its range since then.

Fig. 20. Distribution 1987–1990

Cattle egret

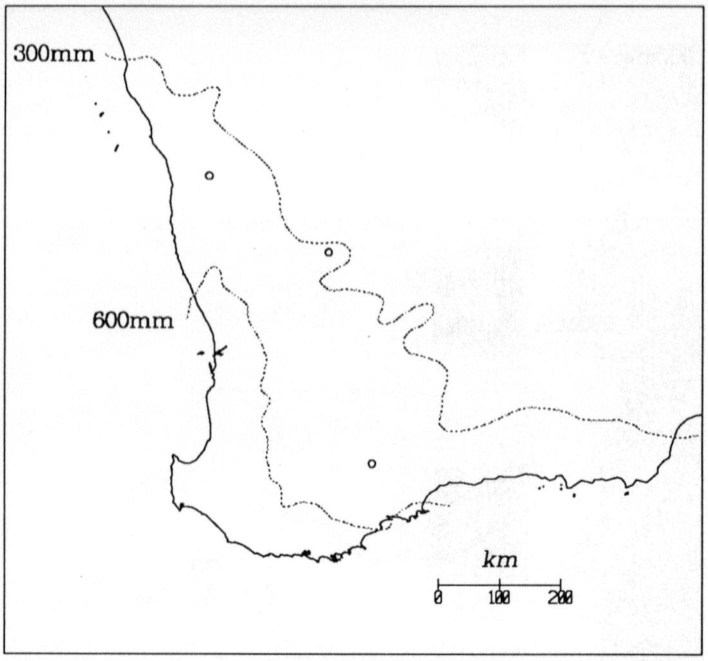

	1900/37	1987	1988	1989	1990
% of observers recording species	0	0	0.7	0	1.9
% of recording period species seen	NR	0	0.02	0	0.1

Woodland habitat dominated by wandoo *(above)* and gimlet *(below)*. Woodland was indicative of good agricultural soils and so was selectively taken up for agriculture and cleared (photos by D. A. Saunders).

Large egret

Egretta alba

OTHER NAME:

Great egret

The large egret was not recorded on any of the historic lists for the wheatbelt. It is a rare nomad on aquatic habitats including farm dams (Fig. 21). Along the Avon River between Northam and Toodyay, on the western edge of the wheatbelt, the large egret was first recorded in 1936 (Masters and Milhinch 1974). It was considered rare until 1955, and the increase had ceased by 1965. The large egret eats fish, frogs, freshwater snails, damselflies, dragonflies, crickets, water boatmen, beetles, flies, shrimp, and birds. It breeds in spring on a stick nest in a tree and lays 3–4 eggs.

There was an apparent increase in sightings of the large egret between 1987 and the end of 1990. Four times as many observers recorded them in 1990 as in 1988. The recording rate was four times higher in 1990 than in 1988. The data do not establish if this was as a result of the species becoming more common in the wheatbelt, or a result of small sample sizes. However, apart from the white ibis, no other species recorded during the survey demonstrated this pattern, so it is plausable that it represents a trend of increasing range and abundance.

Fig. 21. Distribution 1988

Large egret

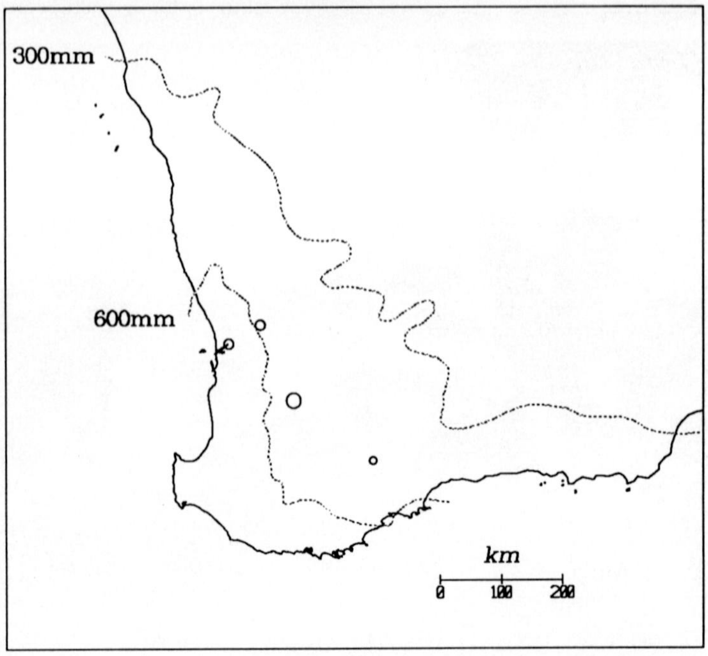

	1900/37	1987	1988	1989	1990
% of observers recording species	0	1.6	2.8	6.0	13
% of recording period species seen	NR	0.8	0.4	1.4	1.9

Fig. 21. Distribution 1989

Large egret

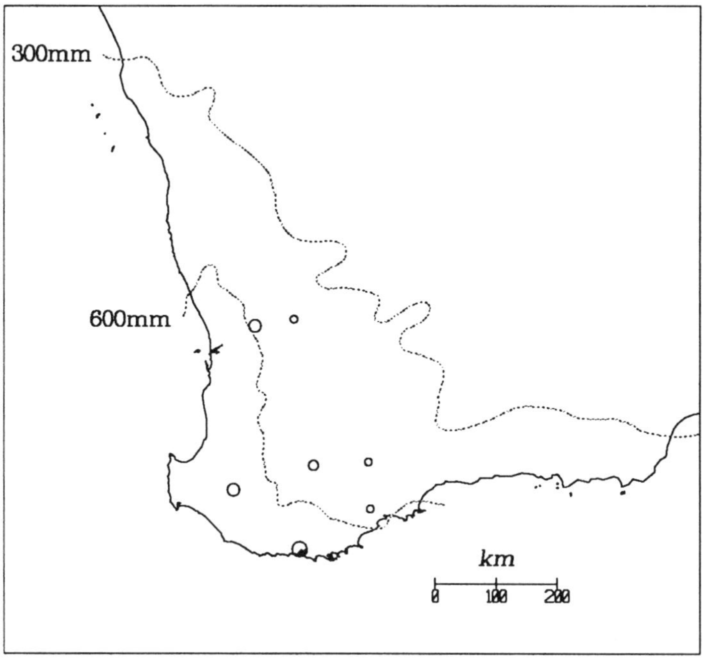

Fig. 21. Distribution 1990

Large egret

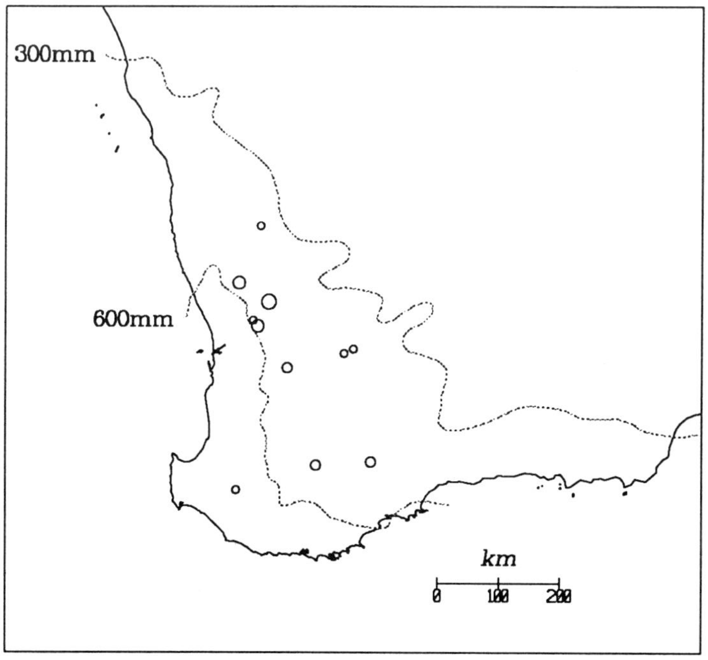

Glossy ibis

Plegadis falcinellus

The glossy ibis is a rare non-breeding vagrant on aquatic habitats in the northern wheatbelt (Fig. 22). It feeds on freshwater snails and mussels, grasshoppers, crickets, water bugs and boatmen, beetles, weevils, fly larvae, yabbies and frogs.

There is no evidence to suggest any change in abundance over the last 50 years.

Fig. 22. Distribution 1987–1990

Glossy ibis

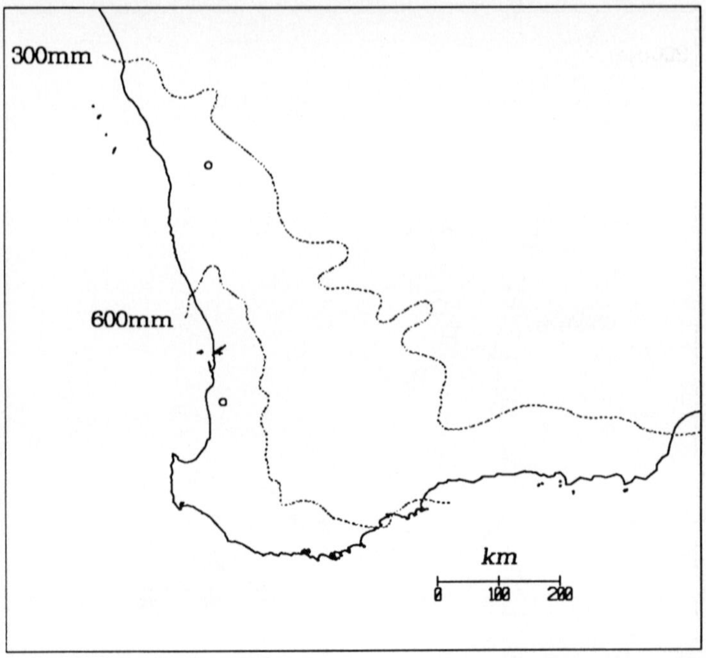

	1900/37	1987	1988	1989	1990
% of observers recording species	0	0.8	0.7	0	0
% of recording period species seen	NR	0.03	0.02	0	0

White ibis

Threskiornis molucca

OTHER NAME:

Sacred ibis

Formerly, the white ibis was a rare vagrant in southwestern Australia. A large irruption in 1952 saw it established in the southwest in small numbers (Serventy and Whittell 1976). At present it is an uncommon nomad on aquatic habitats in the higher rainfall regions of the wheatbelt (Fig. 23). It feeds on freshwater snails and mussels, earthworms, leeches, centipedes, dragonflies, cockroaches, earwigs, crickets, grasshoppers, bugs, water boatmen, beetles, fly larvae, caterpillars, moths, bees, ants, slaters, yabbies, freshwater crayfish, fish, frogs, reptiles, and small mammals like rats and mice. It breeds in spring on a stick nest in a tree, shrub or on the ground. It lays 2–3 eggs.

There was an apparent increase in sightings of the white ibis between 1987 and the end of 1990, with twice as many observers recording them in 1990 as in 1988. The recording rate was nearly four times higher in 1990 than in 1988. The data do not establish if this was as a result of the species becoming more common in the wheatbelt, or an artifact of small sample sizes. However, apart from the large egret, no other species recorded during the survey demonstrated this pattern, so it is plausable that it represents a trend of increasing range and abundance.

Fig. 23. Distribution 1987–1990

White ibis

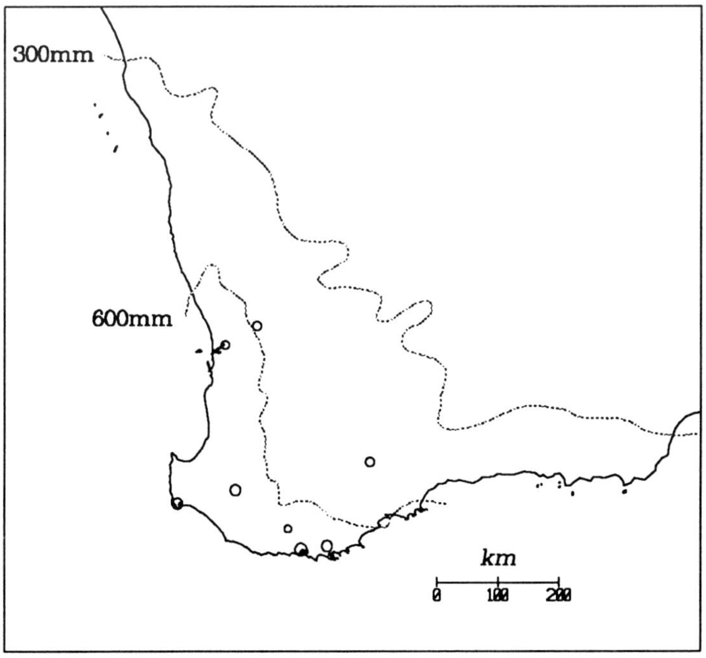

	1900/37	1987	1988	1989	1990
% of observers recording species	0	0.8	3.5	4.3	6.7
% of recording period species seen	NR	0.4	0.5	1.4	1.9

Straw-necked ibis

Threskiornis spinicollis

The straw-necked ibis is an uncommon nomad on aquatic habitats and farmland throughout the wheatbelt (Fig. 24). It feeds on seed, freshwater snails and mussels, earthworms, leeches, millipedes, centipedes, damselflies, dragonflies, cockroaches, earwigs, crickets, grasshoppers, bugs, water boatmen, beetles, flies and larvae, caterpillars, moths, bees, ants, freshwater shrimp, slaters, yabbies, freshwater crayfish, spiders, fish, frogs, reptiles, and small mammals such as rats and mice. The straw-necked ibis breeds in spring on a stick nest in a tree or shrub. It lays 2–3 eggs.

Numbers of the straw-necked ibis fluctuate over time (Serventy and Whittell 1976). Accordingly, it is hard to assess any change in status over the past 50 years. This fluctuation in numbers was demonstrated during the survey as the straw-necked ibis was recorded by nearly twice as many observers in 1990 as during the previous 3 years, and at a slightly higher recording rate. This change is revealed on the distribution maps for 1988, 1989 and 1990 (Fig. 24). The change in distribution in 1990 may have been associated with widespread cyclonic rains in January and a summer that was wetter than usual (Fig. 7).

Fig. 24. Distribution 1900–1937

Straw-necked ibis

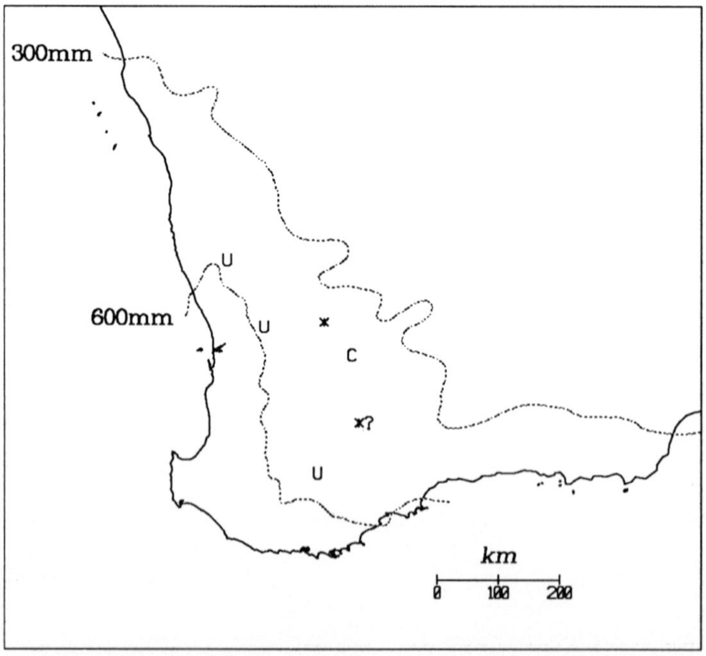

Fig. 24. Distribution 1988

Straw-necked ibis

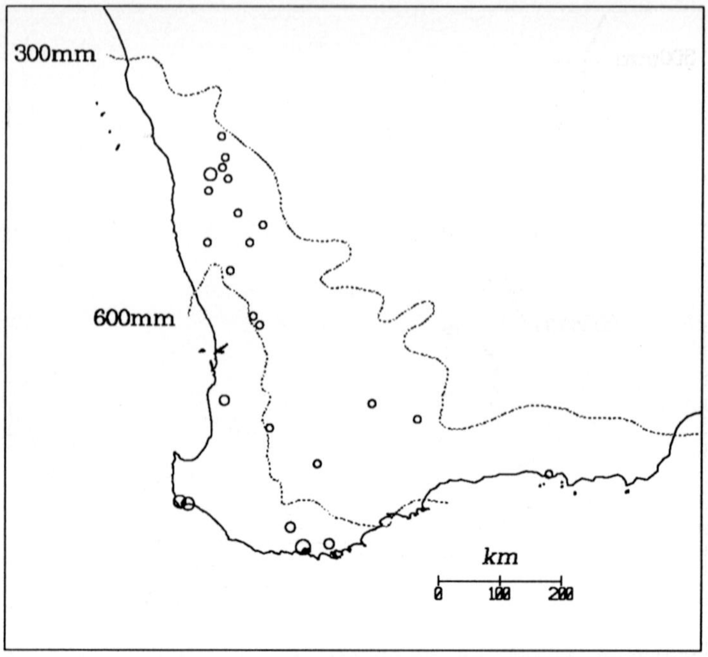

	1900/37	1987	1988	1989	1990
% of observers recording species	60	11	17	13	28
% of recording period species seen	U-C	1.2	2.5	3.8	4.6

Fig. 24. Distribution 1989

Straw-necked ibis

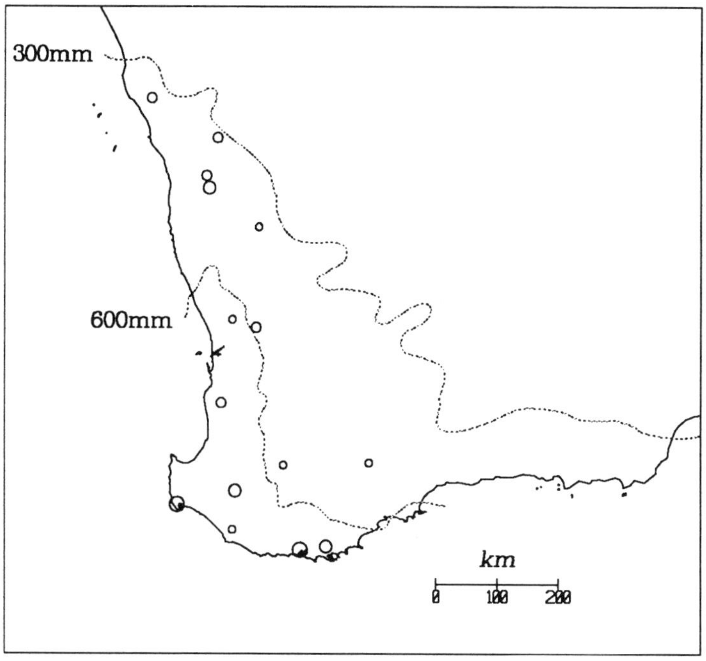

Fig. 24. Distribution 1990

Straw-necked ibis

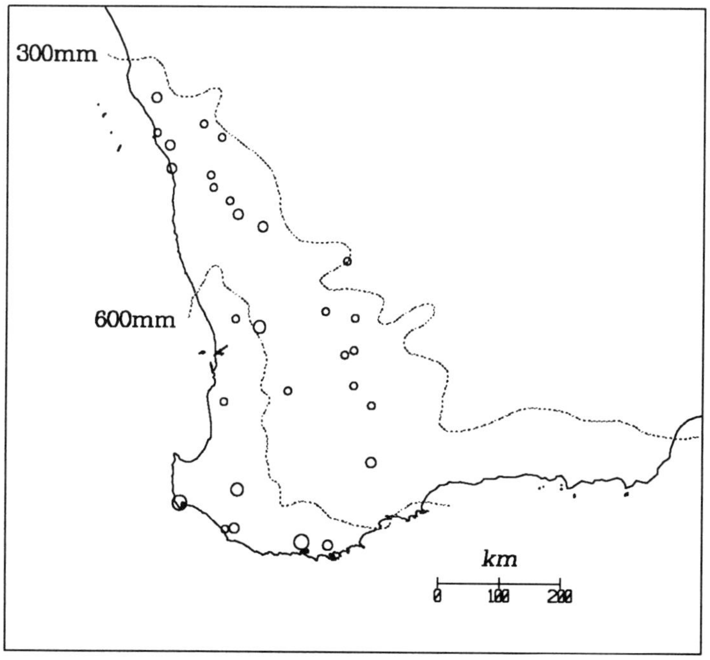

Australian brown bittern

Botaurus poiciloptilus

OTHER NAMES:

Australasian bittern, brown bittern

The Australian brown bittern is a rare nomad on aquatic habitats in the wheatbelt (Fig. 25). It feeds on plant material, insects and frogs. It breeds in spring on a platform of reeds and lays 4–5 eggs.

The Australian brown bittern has probably suffered from the loss of suitable habitat as a result of drainage, grazing and salinization (Garnett 1992). It is regarded as a threatened species. In the southwest it formerly occurred from Moora to Cape Arid, but is now much more restricted in distribution. Garnett quoted Marchant and Higgins (1990) as saying there are probably fewer than 100 pairs in southwestern Australia. It has disappeared from the Avon River (J. Masters, in litt.).

Fig. 25. Distribution 1900–1937

Australian brown bittern

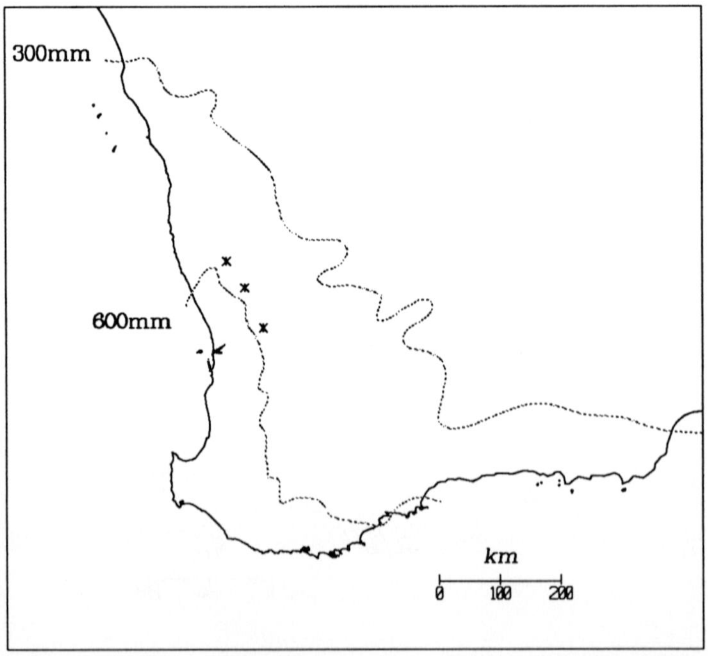

Fig. 25. Distribution 1987–1990

Australian brown bittern

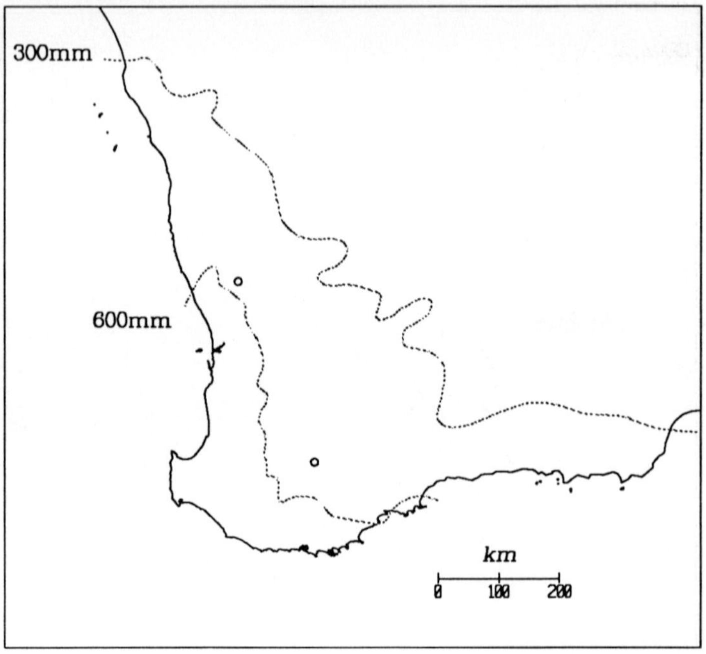

	1900/37	1987	1988	1989	1990
% of observers recording species	30	0	0	0.9	1.0
% of recording period species seen	—	0	0	0.02	0.1

Royal spoonbill

Platalea regia

The royal spoonbill is a rare non-breeding vagrant on aquatic habitats in the wheatbelt (Fig. 26). It feeds on seed and other plant material, molluscs, dragonfly nymphs, backswimmers, water boatmen, beetles, flies and larvae, caterpillars, moths and pupae, wasps, bees, ants, freshwater shrimp, yabbies, freshwater crayfish, and fish.

Fig. 26. Distribution 1987–1990

Royal spoonbill

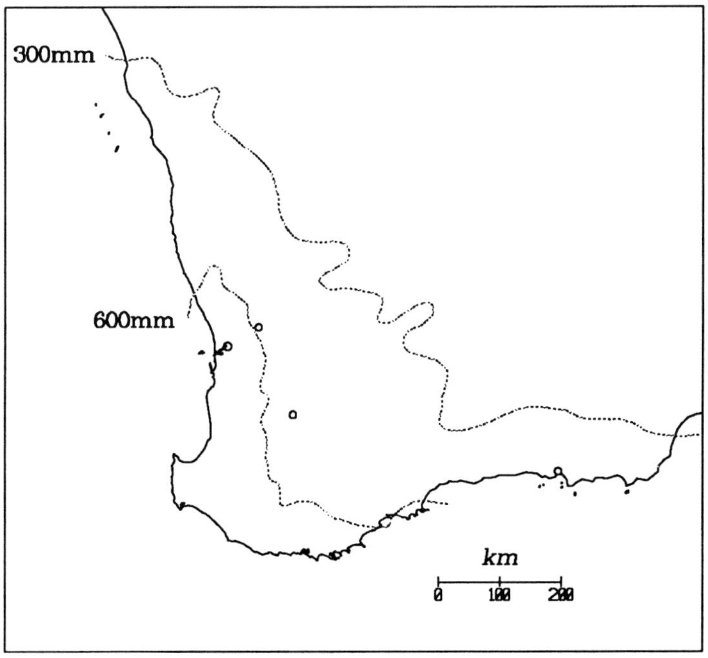

	1900/37	1987	1988	1989	1990
% of observers recording species	0	0	0.7	1.7	1.0
% of recording period species seen	NR	0	0.02	0.07	0.2

Yellow-billed spoonbill

Platalea flavipes

The yellow-billed spoonbill is an uncommon nomad on aquatic habitats in the wheatbelt (Fig. 27). It feeds on plant material, molluscs, damselfly nymphs, crickets, bugs including water boatmen, beetles, freshwater shrimp, yabbies, freshwater crayfish, spiders, and fish. It breeds in spring on a stick nest in a tree, shrub or on rushes. It lays 2–3 eggs.

J. Masters (in litt.) noted that the yellow-billed spoonbill has increased markedly along the Avon River from Toodyay to York during the last 10 years. This evidence, together with the wide distribution recorded during this survey indicates that this species has increased in range over the past 50 years.

Fig. 27. Distribution 1987–1990

Yellow-billed spoonbill

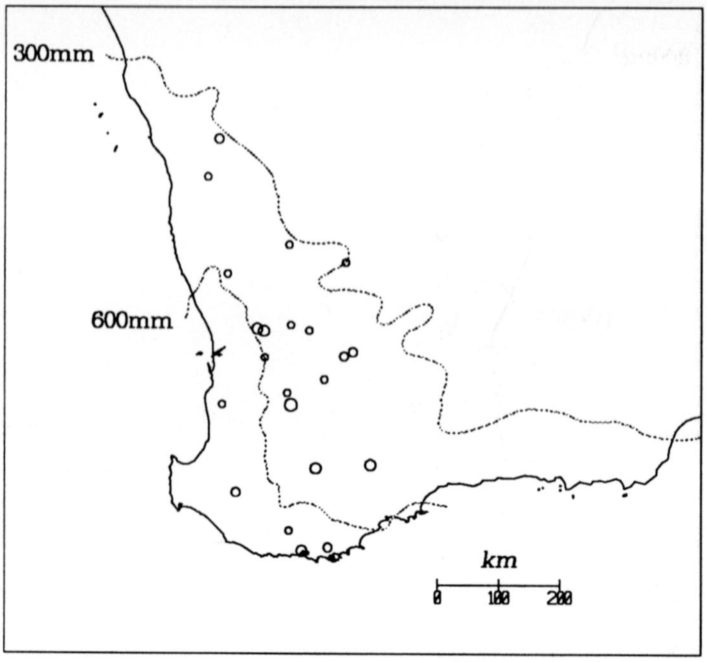

	1900/37	1987	1988	1989	1990
% of observers recording species	0	2.4	9.2	10	12
% of recording period species seen	NR	1.0	1.2	1.7	1.8

Fig. 28. Distribution 1900–1937

Black swan

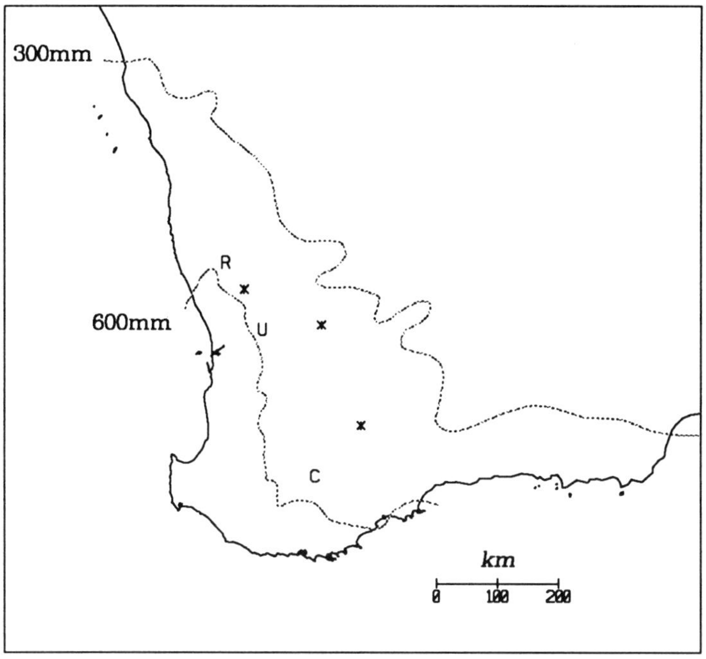

Fig. 28. Distribution 1987–1990

Black swan

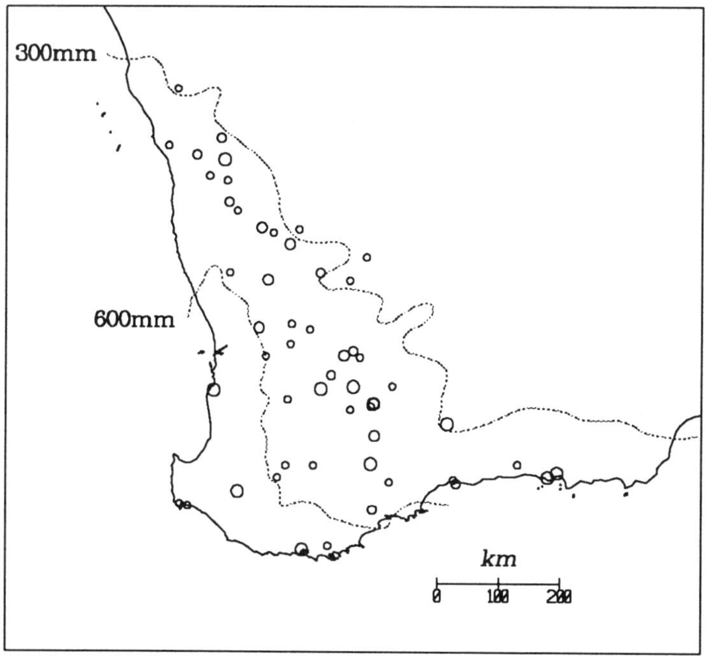

	1900/37	1987	1988	1989	1990
% of observers recording species	60	13	23	21	19
% of recording period species seen	R-C	2.0	5.8	6.9	6.7

Black swan

Cygnus atratus

The black swan is an uncommon nomad on aquatic habitats in the wheatbelt (Fig. 28). It feeds on seed and other plant material, beetles and shrimp. The black swan breeds from autumn to early summer and nests on the ground. It lays 4–6 eggs.

During 1988 and 1989, the black swan was more commonly recorded in the wetter months during winter. In 1990 it was more commonly recorded in summer and was probably more widely distributed in this period than during the previous two summers. It is possible that this was the result of widespread cyclonic rains in January and further rains in February (Fig. 7) providing more suitable habitat than is available in more typical years.

It is possible that this species has decreased in abundance in the wheatbelt over the last 50 years because of the loss of suitable wetlands.

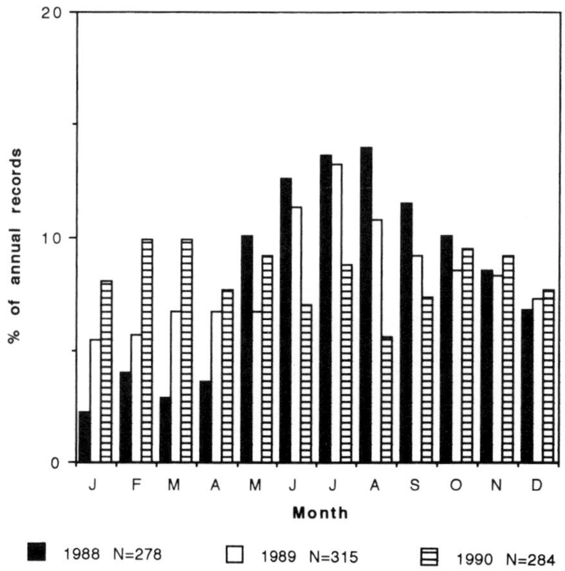

White swan

Cygnus olor

OTHER NAME:

Mute swan

The white swan was originally introduced to Western Australia by humans. During the period 1987–1990 the only records for this species were from Northam where it was a breeding resident on the Avon River. It feeds on plant material, fish, frogs and invertebrates. It breeds in spring and nests on the ground. It lays 3–5 eggs.

Freckled duck

Stictonetta naevosa

The freckled duck is a rare nomad on aquatic habitats in the wheatbelt (Fig. 29). It feeds on seed and other plant material, freshwater sponge, molluscs, small shells, bugs, beetles, copepods, ostracods, and small fish. It breeds in spring and builds a stick nest in a shrub or on the ground. It lays 5–7 eggs.

Regarded as one of the rarer waterfowl (Serventy and Whittell 1976; Garnett 1992), there is no evidence to suggest any change in abundance in the wheatbelt over the last 50 years, although drainage of wetlands on the coastal plain may have affected them adversely. Garnett cites Jaensch *et al.* (1988) who recorded 151 freckled ducks in 872 wetlands in south-western Australia in March 1986. They thought only about 10 pairs breed there regularly.

Fig. 29. Distribution 1900–1937

Freckled duck

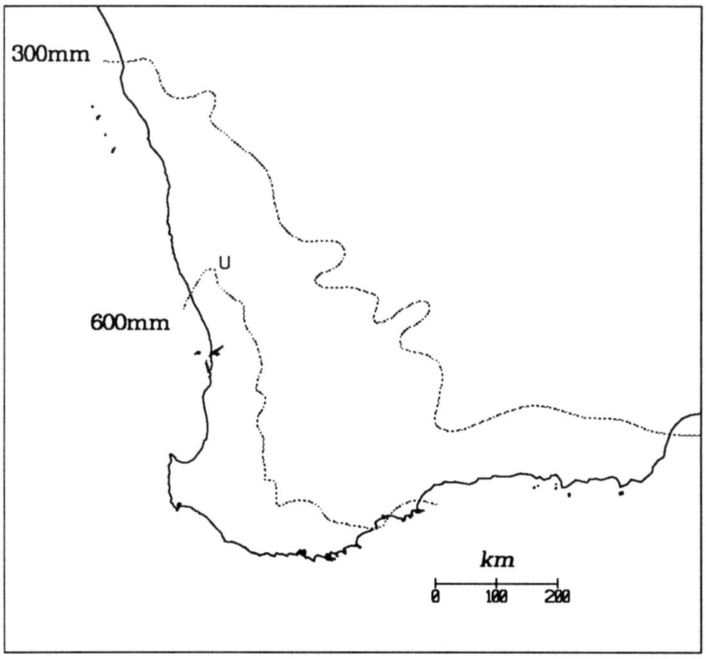

Fig. 29. Distribution 1987–1990

Freckled duck

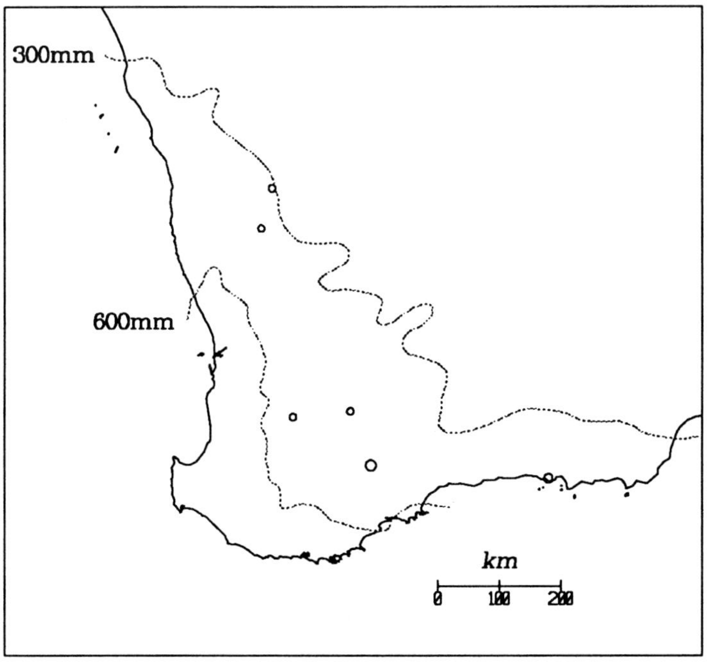

	1900/37	1987	1988	1989	1990
% of observers recording species	10	2.4	2.1	2.6	1.0
% of recording period species seen	U	0.2	0.3	0.2	0.1

Mountain duck

Tadorna tadornoides

OTHER NAMES:

Australian shelduck, chestnut-breasted shelduck

The mountain duck is resident in the higher rainfall districts of the wheatbelt, and on some suitable habitat through the remainder. Throughout the rest of the wheatbelt it is a breeding migrant between March and October (Fig. 30). It feeds on seed and other plant material, water boatmen, beetles, flies, ants, brine shrimp, water fleas, and ostracods. It breeds in winter and spring in a hollow in a eucalypt and lays 8–14 eggs.

Early this century the mountain duck was regarded as rare (Crossman 1909; Orton and Sandland 1913), although it was widely distributed. Subsequently it has become common on aquatic habitats including farm dams, and on farmland throughout the wheatbelt (Fig. 30). For example, Milligan (1904) saw one pair in the lake country around the Wongan Hills, but de Rebeira and de Rebeira (1977) stated it was the most common duck in the area. The mountain duck was not observed around Stirling Range by Milligan (1903), however, Whitlock (1911) found it breeding on Lake Balicup on the northwest corner of the Range. Bradshaw (1928) noted that it was "plentiful; really too numerous", Sedgwick (1964) noted that it had been seen on the edge of the Stirling Range National Park and Burbidge and Rose (1993) noted that it bred in small numbers in the Park. Leake (1962) noted that the mountain duck first appeared in the Kellerberrin district in the late 1920s. In the Northam district, Masters and Milhinch (1974) noted that it was a scarce visitor in the 1930s, but that it increased markedly in the mid-1960s and early 1970s when it was classified as common. The development of intensive agriculture has favoured this species as there are still many tree hollows for nest sites, and farm dams for brood territories. Sedgwick (1974) noted that the mountain duck arrived on a newly-cleared farm in the eastern wheatbelt within four years of the first clearing of native vegetation.

Fig. 30. Distribution 1900–1937

Mountain duck

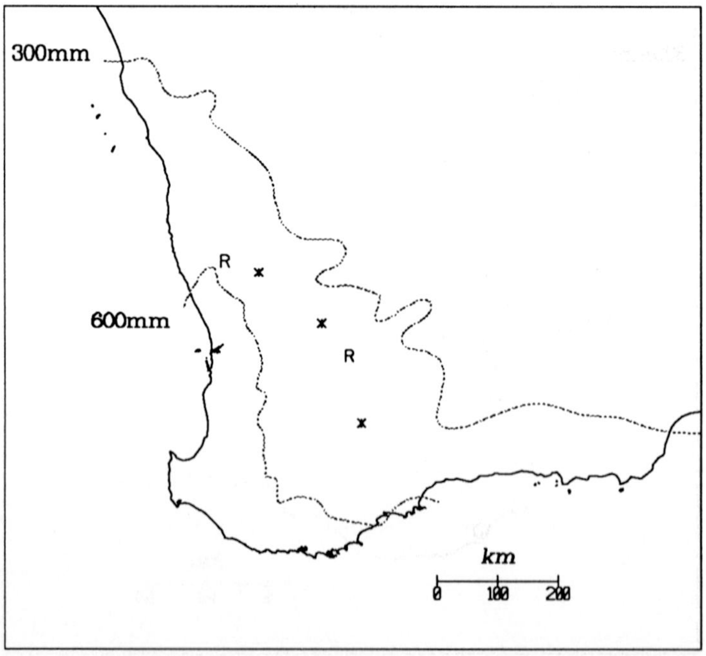

Fig. 30. Distribution 1987–1990

Mountain duck

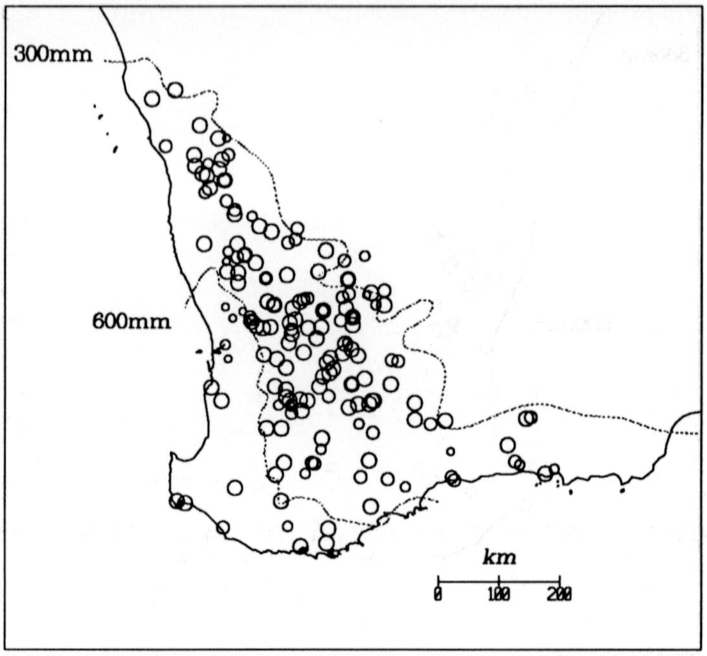

	1900/37	1987	1988	1989	1990
% of observers recording species	50	85	85	85	88
% of recording period species seen	R	41	47	47	51

Fig. 30. Distribution 1987–1990 — April–September

Mountain duck

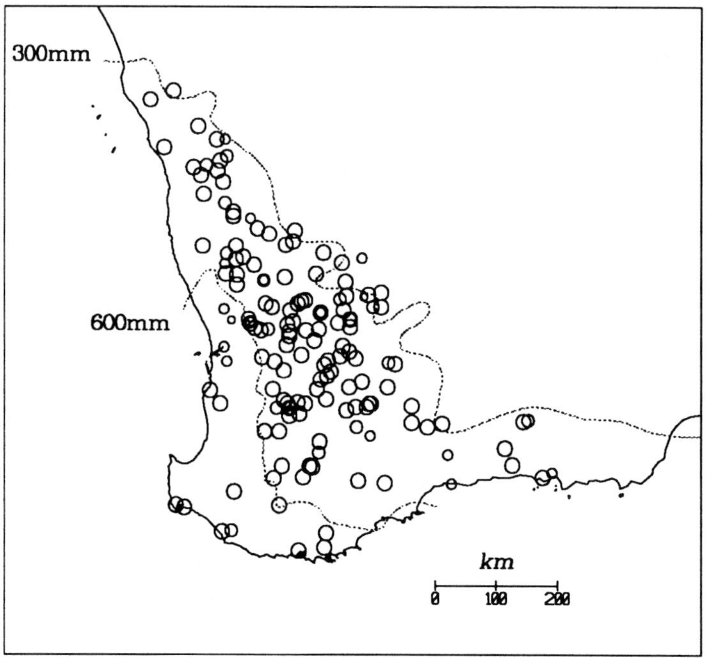

Fig. 30. Distribution 1987–1990 — November–February

Mountain duck

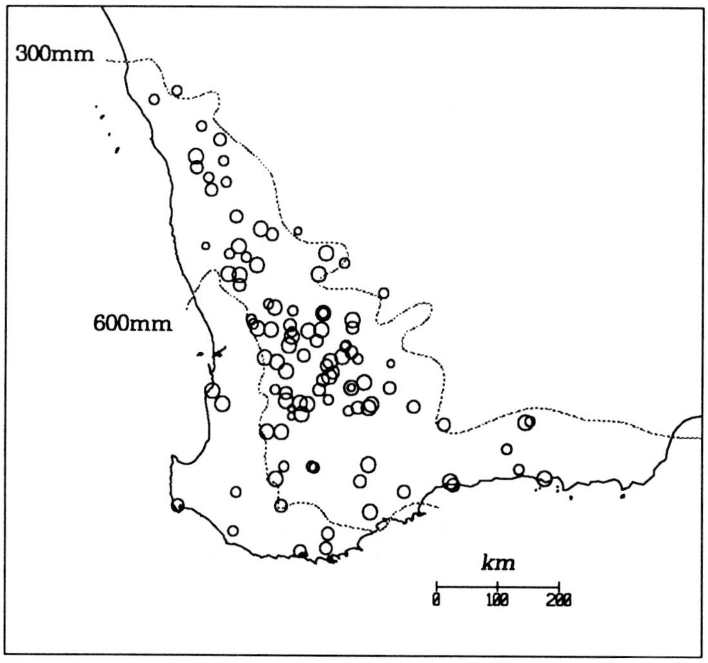

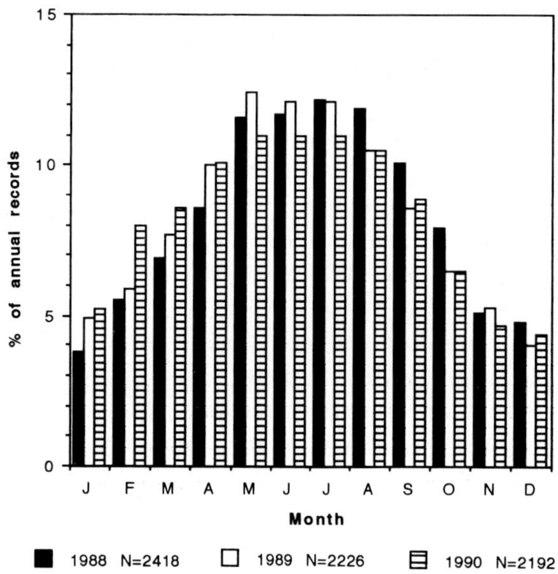

Black duck

Anas superciliosa

OTHER NAME:
Pacific black duck

The black duck is a common resident on aquatic habitats in the higher rainfall districts of the wheatbelt (Fig. 31). It is an uncommon nomad during the wetter periods (June to September) in other areas. It feeds on seed and other plant material, freshwater mussels, millipedes, mayflies, nymphs of damselflies and dragonflies, bugs, beetles, fly and mosquito larvae, water fleas, ostracods, copepods, shrimp, yabbies, and fish. The black duck breeds in winter and spring. It nests on the ground or in a tree hollow and lays 8–10 eggs.

The seasonality of occurrence of the black duck is shown by the percentage of total recordings for each month of the three complete years of this survey (see Fig. on facing page). In 1988 and 1989 it was more commonly recorded in the period June to September, however, in 1990 recording rates were higher from February to September. This is probably related to the fact that the summer of 1990 was wetter than usual with cyclonic rains in January and further rains in February (Fig. 7). The changes in distribution with season are also shown in the distribution maps which show that the black duck is more widely distributed in the period June to September than earlier in the year.

Although the black duck is still widespread and relatively common, it has decreased in abundance over the last 50 years. This is the result of loss of suitable habitat (e.g., draining wetlands and the salination of much of the remainder) and interaction with other species of duck which increased in range and abundance as a result of changes to their environment. Masters and Milhinch (1974) commented that the black duck had decreased in the 1960s and early 1970s. Leake (1962) wrote that the mountain duck was also partly responsible for displacing the black duck. Leake noted that initially, after clearing and provision of water supplies, the black duck took advantage of the changes (see Sedgwick 1986). However with the arrival of the mountain duck, these larger, more aggressive birds "completely ousted the black (duck) from dams."

Fig. 31. Distribution 1900–1937

Black duck

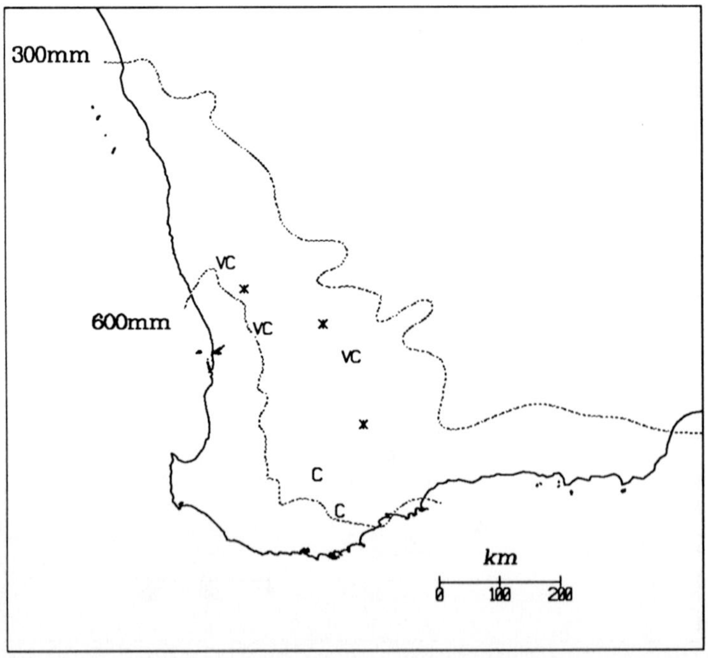

Fig. 31. Distribution 1987–1990

Black duck

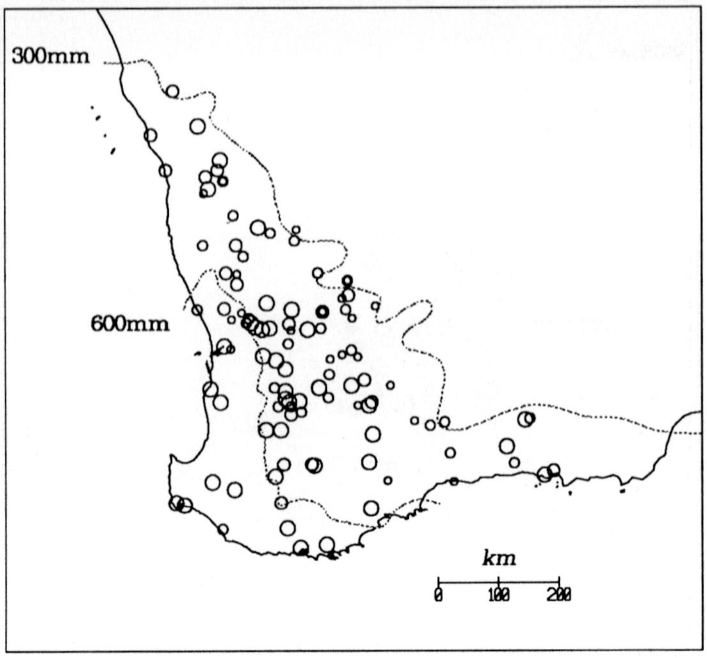

	1900/37	1987	1988	1989	1990
% of observers recording species	80	38	53	46	52
% of recording period species seen	C-VC	15	20	23	25

Fig. 31. Distribution 1987–1990 — January–April

Black duck

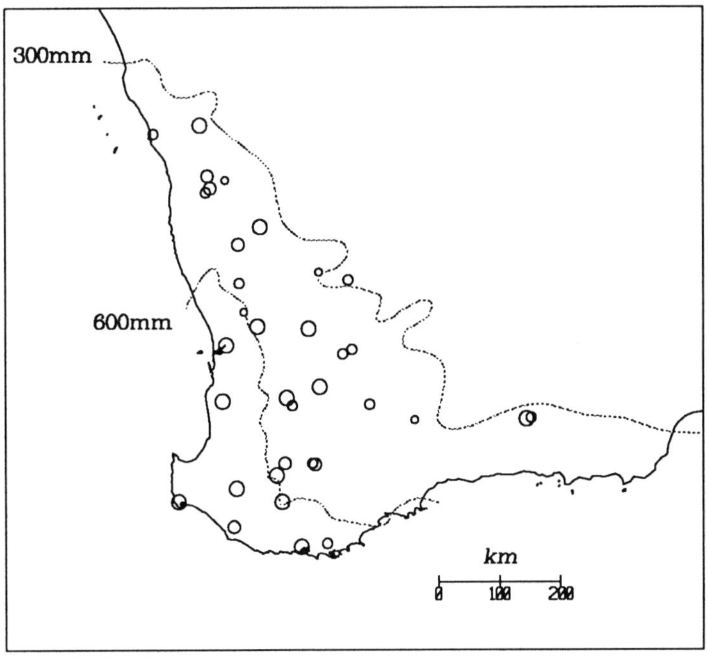

Fig. 31. Distribution 1987–1990 — June–September

Black duck

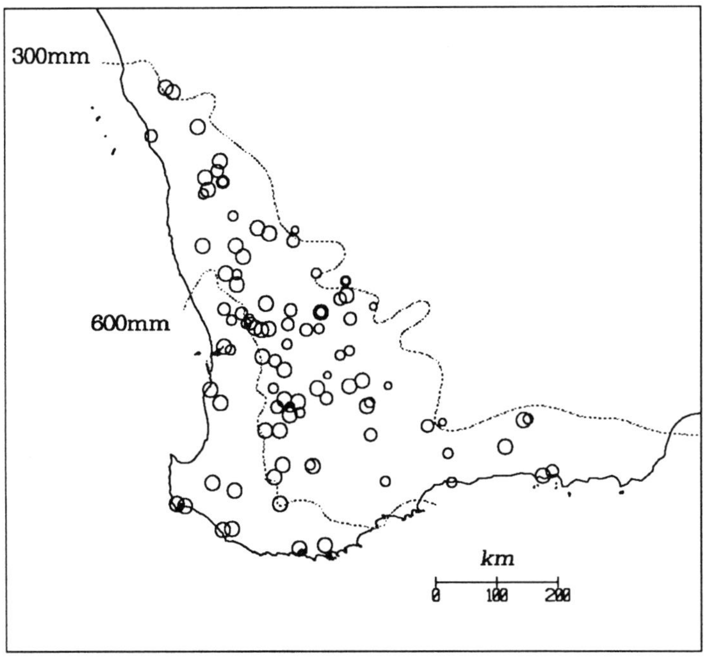

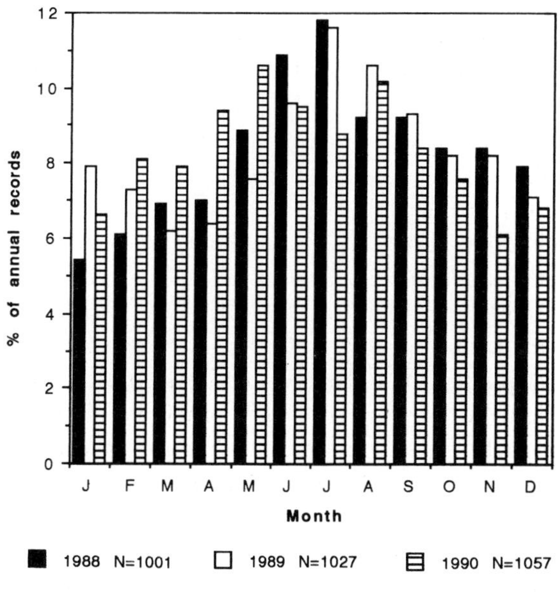

Grey teal

Anas gibberifrons

The grey teal occurs on aquatic habitats, including farm dams in the wheatbelt (Fig. 32). It is a common resident in the higher rainfall districts and an uncommon nomad in the remainder, usually being present during the wetter periods (May to September). It feeds on seed and other plant material, freshwater mussels, snails, dragonfly nymphs, bugs, water boatmen, beetles, weevils, fly larvae, ostracods, yabbies, spiders, and fish. The grey teal breeds at any time of year if conditions are suitable. It nests on the ground or in a tree hollow and lays 6–10 eggs.

The seasonality of occurrence of the grey teal is shown by the percentage of annual recordings for each month of the three complete years of this survey (see Fig. below). In 1988 and 1989 the grey teal was more common in the period May to September, however, in 1990 recording rates were higher from February to September. This is the same pattern of occurrence shown by the black duck and wood duck and was probably influenced by widespread rains in January and February (Fig. 7). The seasonal change in distribution of the grey teal is shown in the distribution maps (Fig. 32) which show that it is much more widely distributed in winter and early spring than it is in summer and early autumn.

There is no evidence to suggest any change in abundance over the last 50 years.

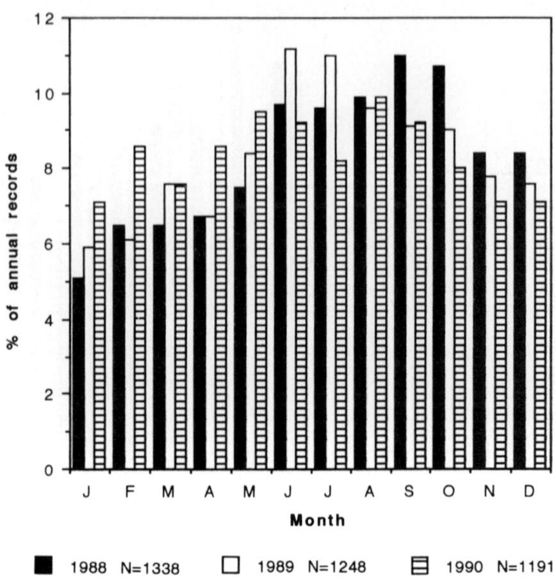

Fig. 32. Distribution 1900–1937

Grey teal

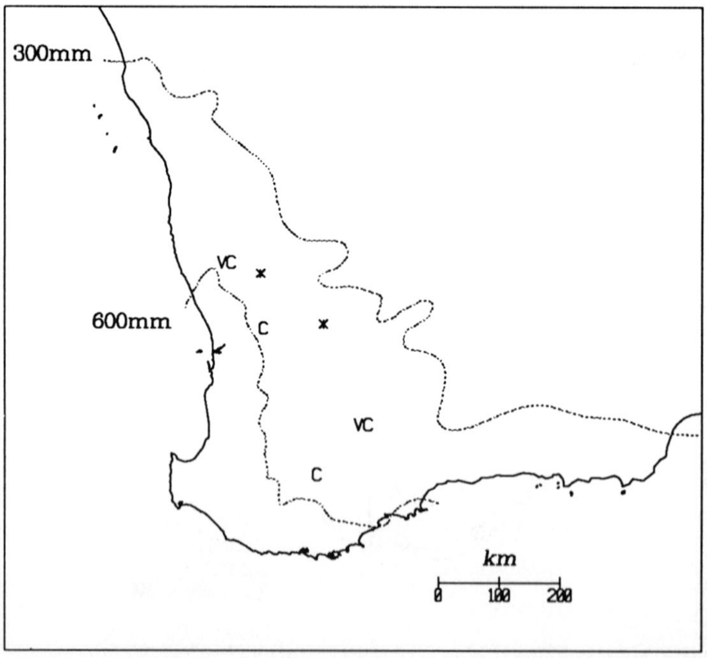

Fig. 32. Distribution 1987–1990

Grey teal

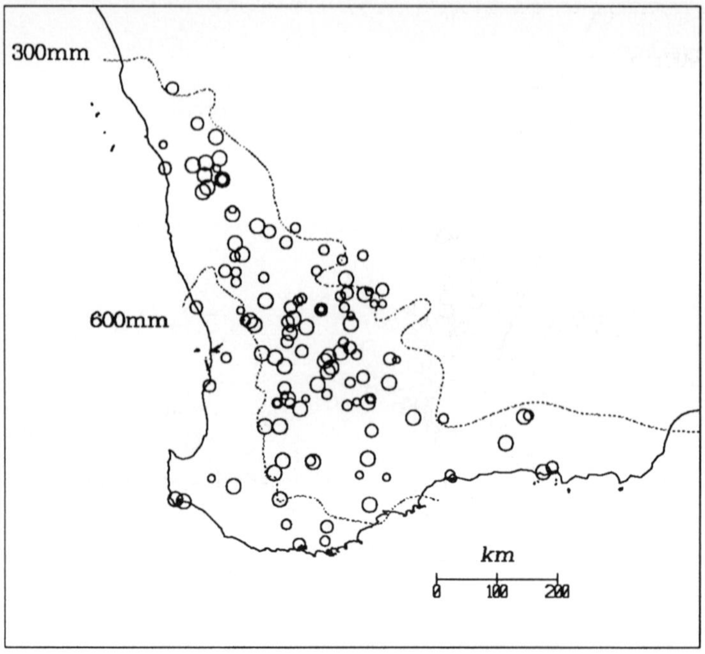

	1900/37	1987	1988	1989	1990
% of observers recording species	60	50	61	56	61
% of recording period species seen	C-VC	19	26	27	28

Fig. 32. Distribution 1987–1990 — January–April

Grey teal

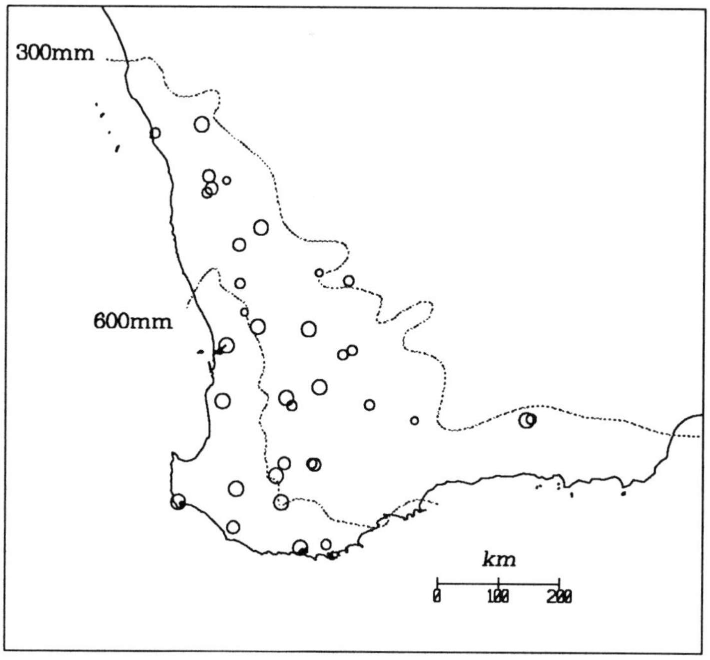

Fig. 32. Distribution 1987–1990 — June–September

Grey teal

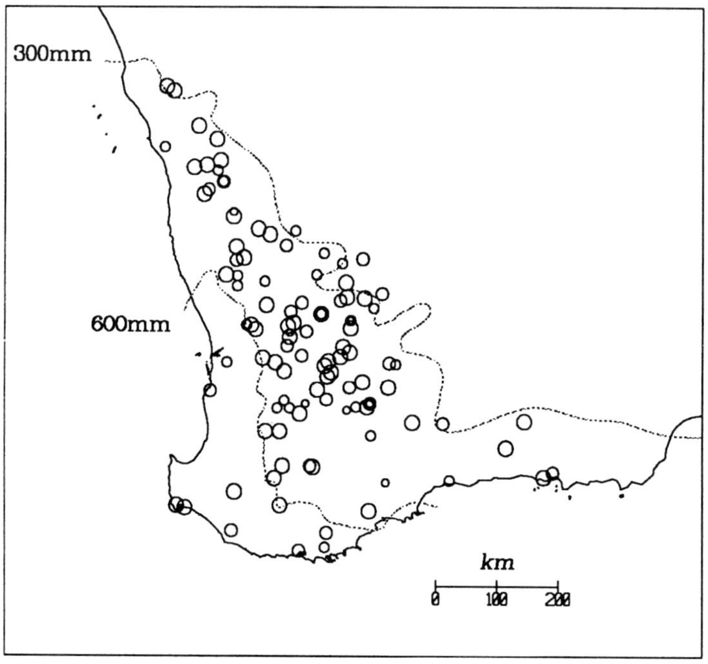

Chestnut teal

Anas castanea

The chestnut teal is an uncommon nomad on aquatic habitats in the wheatbelt (Fig. 33). It feeds on seed and other plant material, and beetles. The chestnut teal breeds in spring and nests on the ground or in a tree hollow near water. It lays 6–10 eggs.

There is no evidence to suggest any change in abundance over the last 50 years.

Fig. 33. Distribution 1900–1937

Chestnut teal

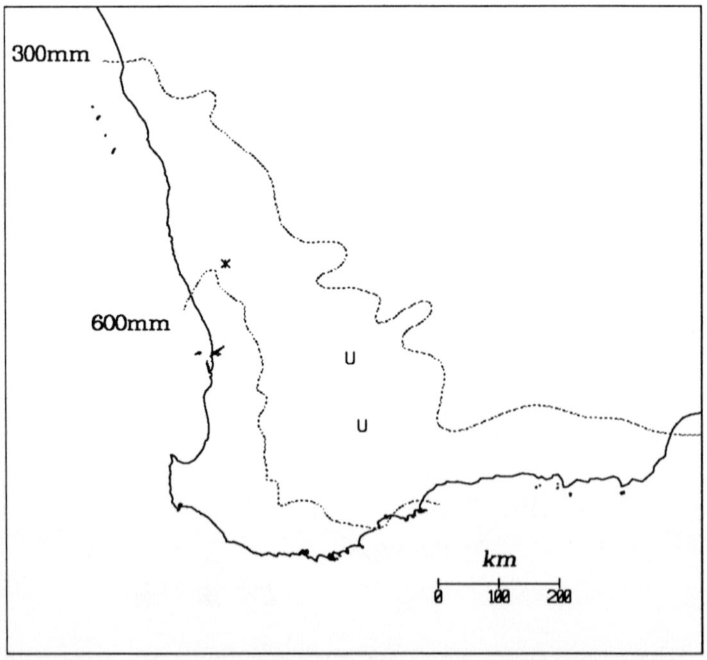

Fig. 33. Distribution 1987–1990

Chestnut teal

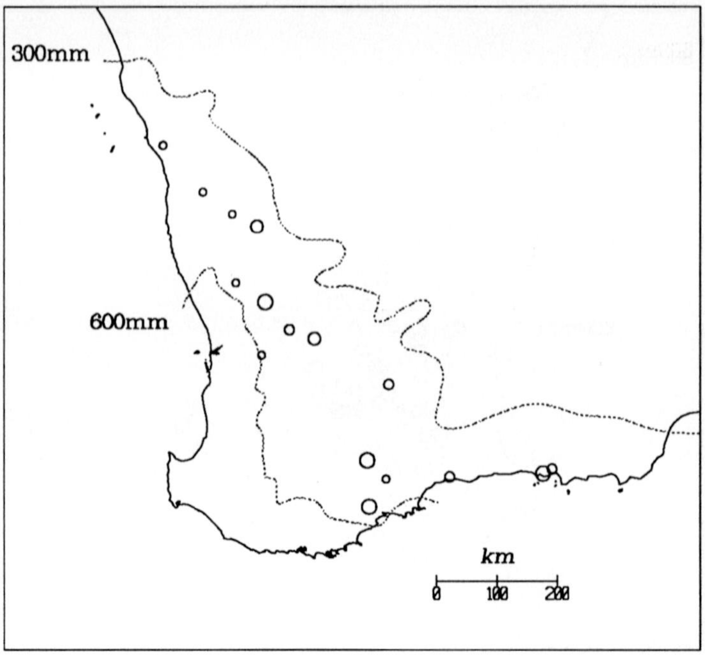

	1900/37	1987	1988	1989	1990
% of observers recording species	30	5.5	8.5	4.3	6.7
% of recording period species seen	U	0.7	3.1	2.5	2.7

Shoveller

Anas rhynchotis

OTHER NAMES:

Australasian shoveller, blue-winged shoveler

The shoveller is an uncommon nomad on aquatic habitats in the wheatbelt (Fig. 34). It feeds on seed and other plant material, molluscs, dragonfly larvae, bugs, backswimmers, water boatmen, beetles, caddis fly larvae, water fleas and ostracods. It breeds in spring or early summer and nests on the ground. It lays 8–10 eggs.

There is no evidence to suggest any change in abundance over the last 50 years.

Fig. 34. Distribution 1987–1990

Shoveller

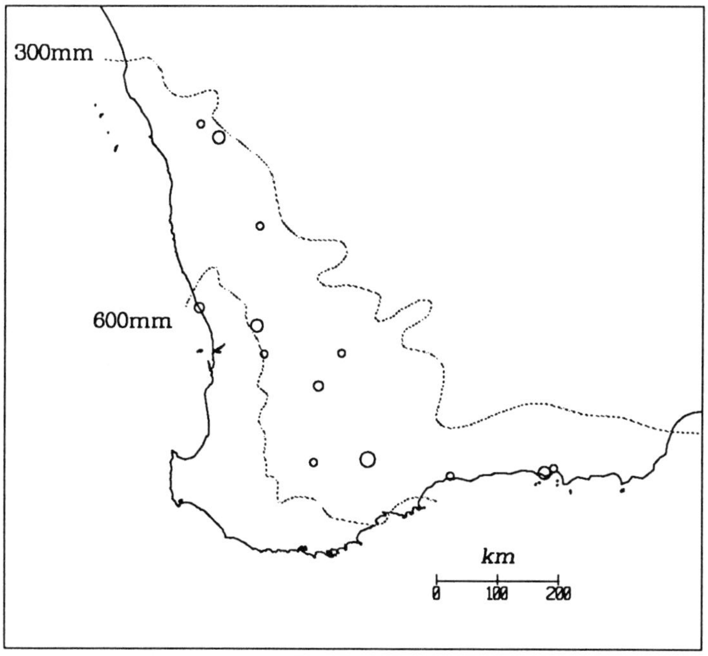

	1900/37	1987	1988	1989	1990
% of observers recording species	0	3.1	6.3	3.4	4.8
% of recording period species seen	NR	1.0	1.2	1.4	1.7

Pink-eared duck

Malacorhynchus membranaceus

The pink-eared duck is an uncommon nomad on aquatic habitats in the wheatbelt (Fig. 35). During 1988 it was recorded more in the period July to November, however, in the following two years there was no seasonal pattern to its occurrence (see Fig. below). The pink-eared duck feeds on seed and other plant material, nematode worms, freshwater molluscs, beetles, fly larvae, water fleas, ostracods, copepods and water mites. It nests in spring and summer on the ground, or in a shrub and lays 5–8 eggs.

There is no evidence to suggest any change in abundance over the last 50 years.

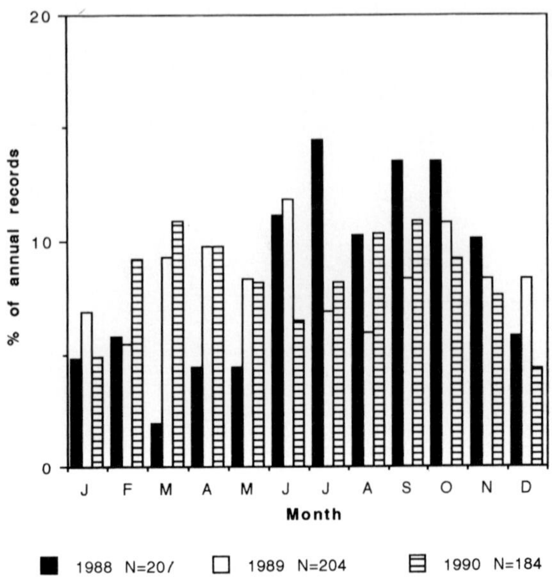

Fig. 35. Distribution 1900–1937

Pink-eared duck

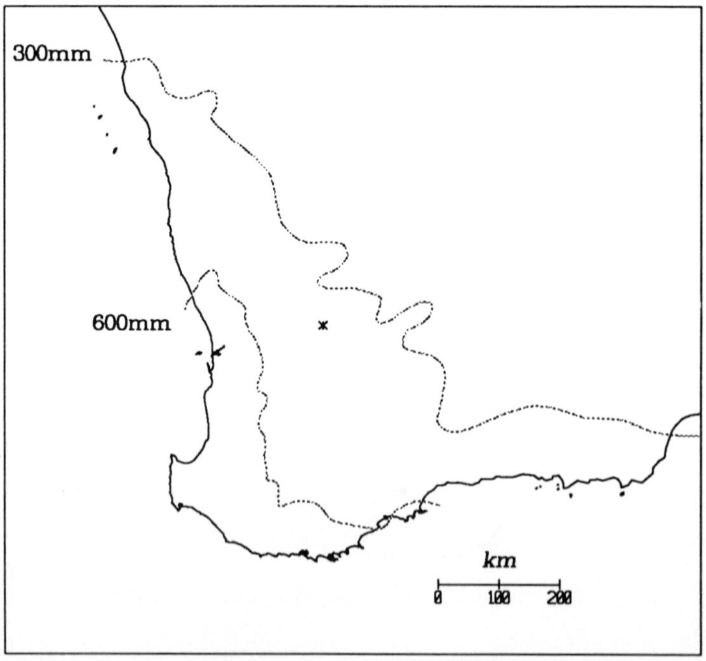

Fig. 35. Distribution 1987–1990

Pink-eared duck

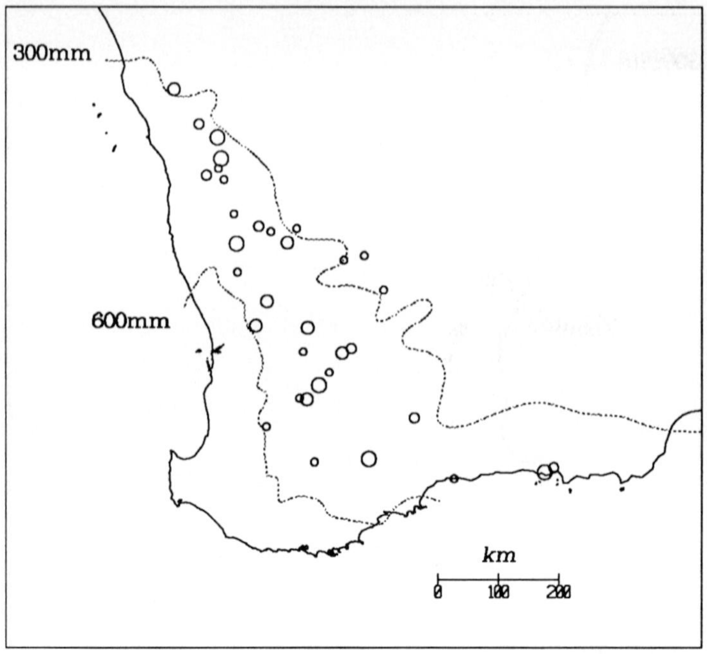

	1900/37	1987	1988	1989	1990
% of observers recording species	10	7.1	15	15	16
% of recording period species seen	—	1.6	4.0	4.5	4.3

Fig. 36. Distribution 1900–1937

White-eyed duck

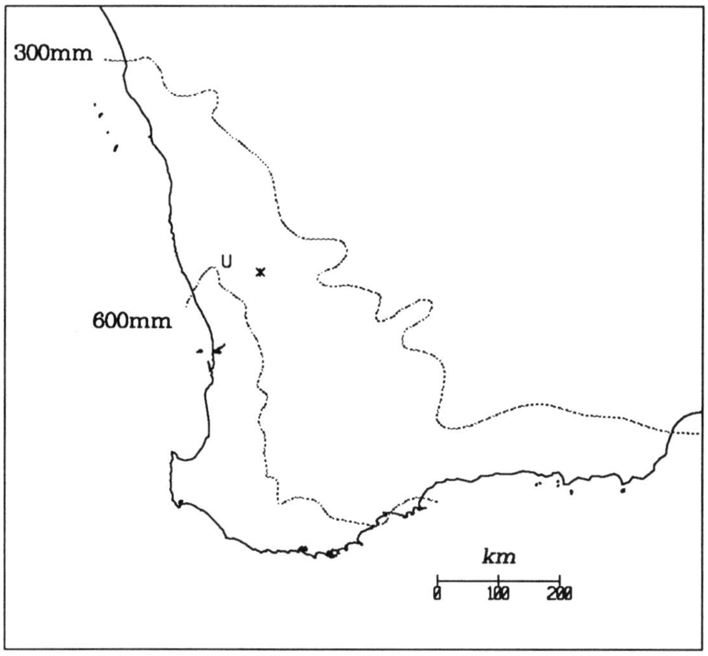

Fig. 36. Distribution 1987–1990

White-eyed duck

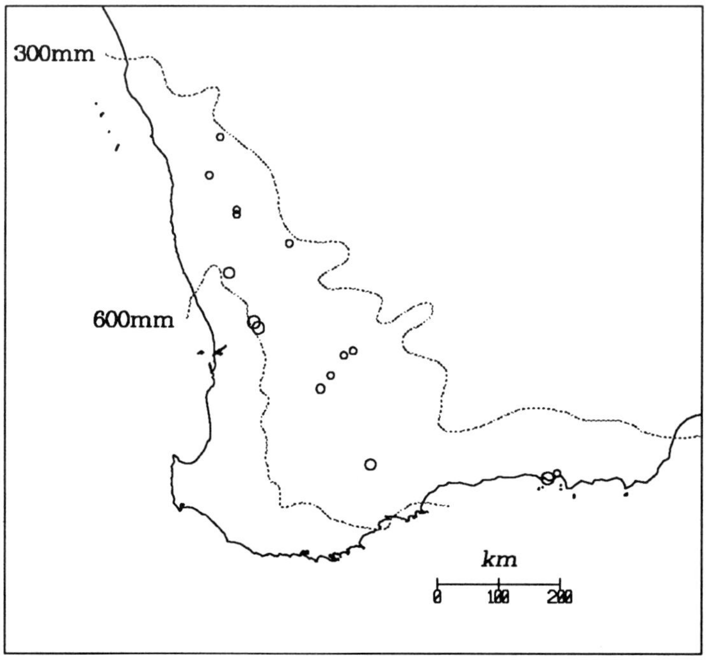

	1900/37	1987	1988	1989	1990
% of observers recording species	20	3.1	7.0	4.3	4.8
% of recording period species seen	U	2.1	3.0	1.0	1.7

White-eyed duck

Aythya australis

OTHER NAME:

Hardhead

The white-eyed duck is an uncommon nomad on aquatic habitats in the wheatbelt (Fig. 36). It may be resident in the higher rainfall districts. It feeds on seed and other plant material, fresh-water molluscs, larvae of damselflies and dragonflies, bugs, midges, caddis fly larvae, ostracods, copepods, shrimp and yabbies. The white-eyed duck breeds in spring on a nest of sticks in a tree or shrub. It lays 8–10 eggs.

There is no evidence to suggest any change in abundance over the last 50 years. However, J. Masters (in litt.) noted that the white-eyed duck has become a regular breeding bird on the Avon River in the last 20 years. He believes that numbers have increased, but they do fluctuate considerably between years.

Wood duck

Chenonetta jubata

OTHER NAMES:

Maned duck, maned goose

The wood duck is a common, seasonal breeding migrant on aquatic habitats including farm dams, and on farmland throughout the wheatbelt (Fig. 37). However, it may be resident on suitable habitats. It feeds on seed and other plant material, molluscs, bugs and beetles. The wood duck breeds in spring in a hollow in a eucalypt and lays 9–12 eggs.

The wood duck is much more widely distributed in winter and early spring than it is in summer and early autumn (see Fig. on facing page). In 1988 and 1989 it was more common in the period May to November. However, in 1990 recording rates were higher from February to November. This may reflect the widespread summer rains in 1990 (Fig. 7).

The provision of farm dams and improved pasture have provided more suitable habitat for this species. It has become more widespread and abundant over the last 80 years. For example, the wood duck was not observed by early visitors to the Stirling Range National Park, however, Sedgwick (1964) noted that it had been seen on the edges of the Park and Burbidge and Rose (1993) noted that it bred in small numbers in the Park. Around the Kellerberrin district, Ford and Stone (1957) classified the wood duck as a rare visitor, yet it is a common breeding migrant in the same area now (Saunders and Curry 1990). In the Northam district, Masters and Milhinch (1974) recorded that the wood duck first appeared along the Avon River in 1950, and was common and still increasing in parts of the district. Sedgwick (1986) noted that the wood duck first appeared on a newly-cleared farm in the eastern wheatbelt within 8 years of the first clearing.

Fig. 37. Distribution 1900–1937

Wood duck

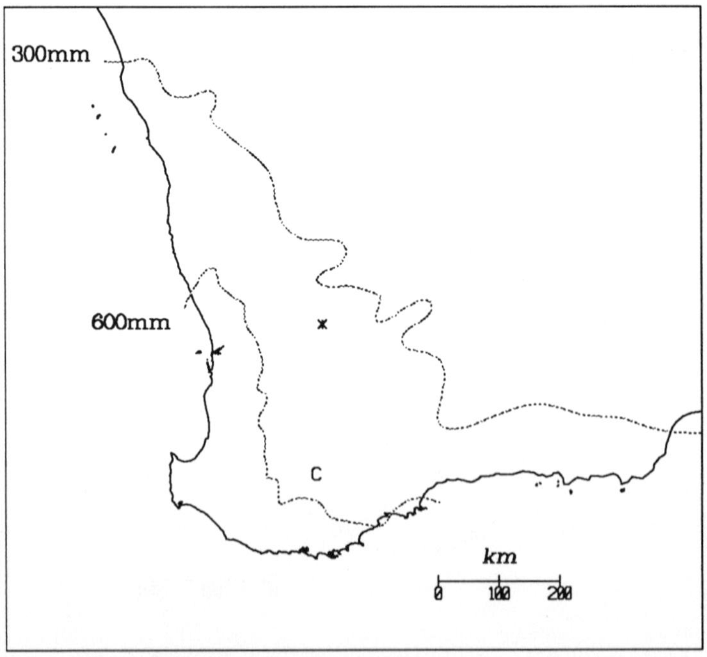

Fig. 37. Distribution 1987–1990

Wood duck

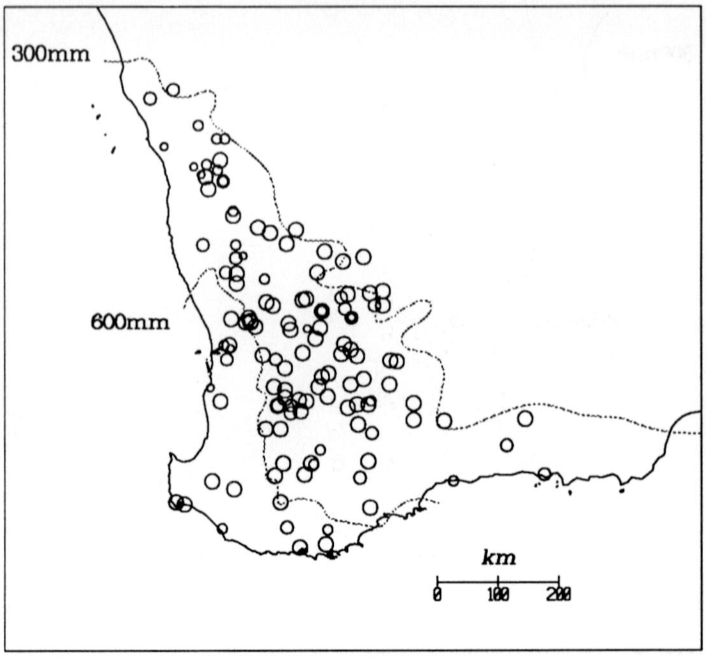

	1900/37	1987	1988	1989	1990
% of observers recording species	20	66	71	76	73
% of recording period species seen	C	38	45	48	52

Fig. 37. Distribution 1987–1990 — January–April

Wood duck

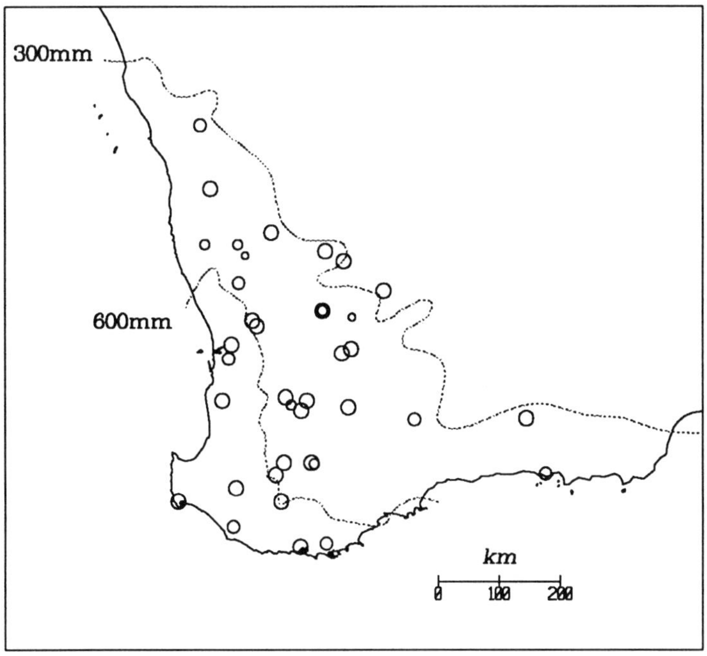

Fig. 37. Distribution 1987–1990 — July–October

Wood duck

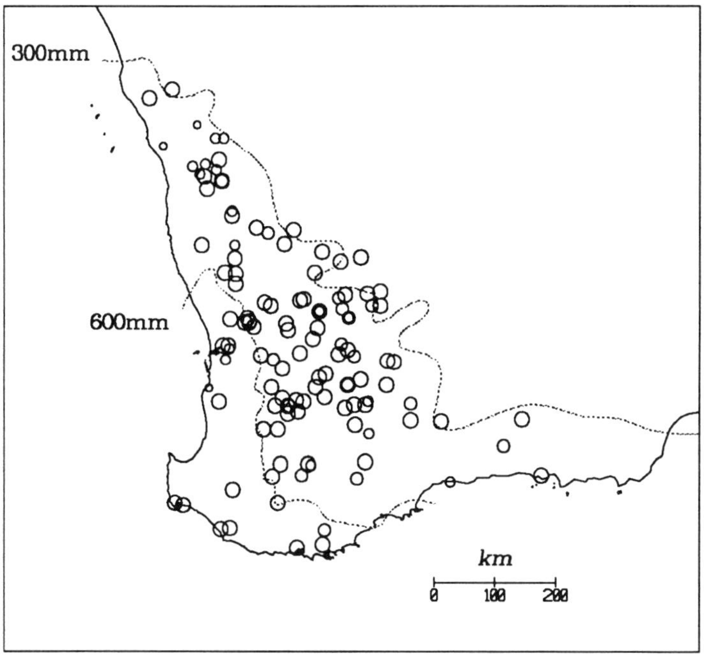

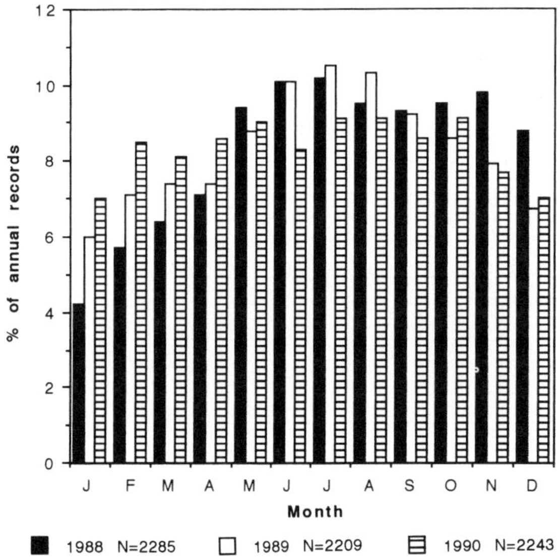

Blue-billed duck

Oxyura australis

The blue-billed duck is a rare nomad on aquatic habitats in the higher rainfall districts of the wheatbelt (Fig. 38). It feeds on seed and other plant material, molluscs, mayflies, larvae of damselflies and dragonflies, beetles, fly larvae, water fleas, copepods and water mites. It breeds in spring and summer, nests on the ground and lays 3–6 eggs.

There is no evidence to suggest any change in abundance over the last 50 years.

Fig. 38. Distribution 1900–1937

Blue-billed duck

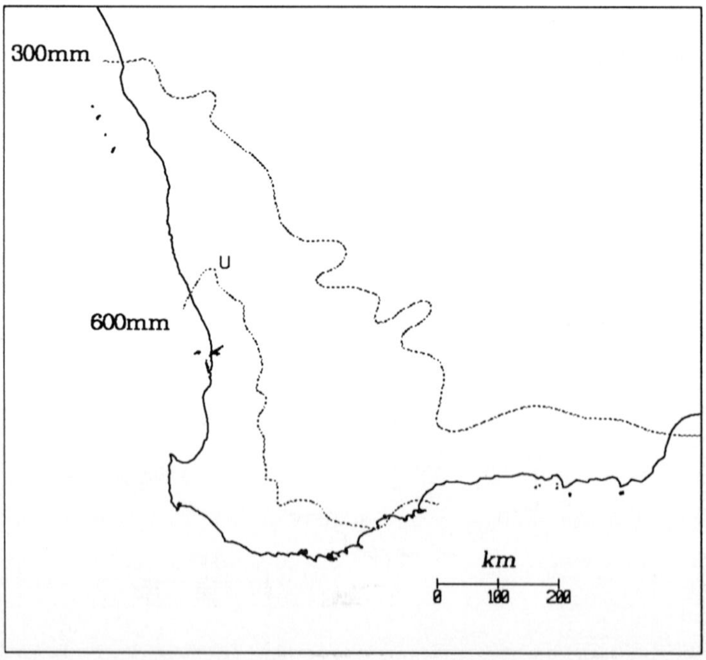

Fig. 38. Distribution 1987–1990

Blue-billed duck

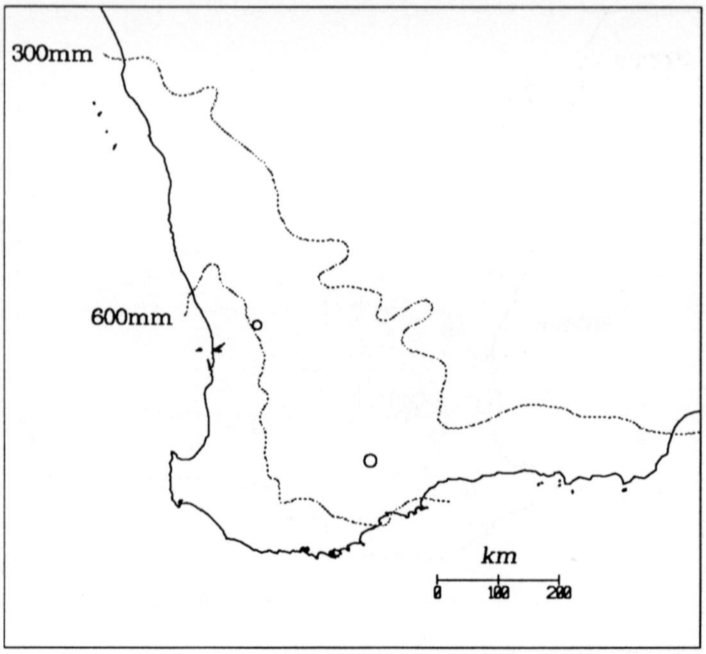

	1900/37	1987	1988	1989	1990
% of observers recording species	10	0.8	1.4	1.7	1.0
% of recording period species seen	U	0.3	0.6	0.9	0.3

Fig. 39. Distribution 1900–1937

Musk duck

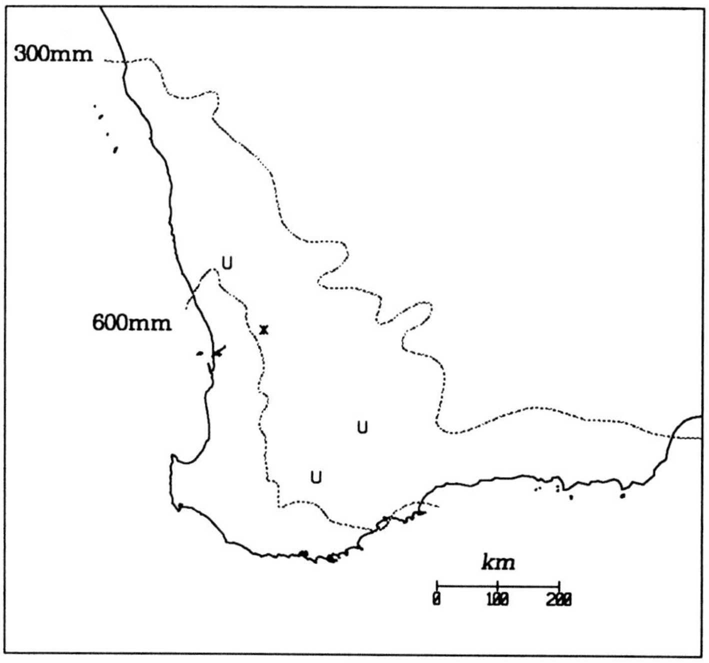

Fig. 39. Distribution 1987–1990

Musk duck

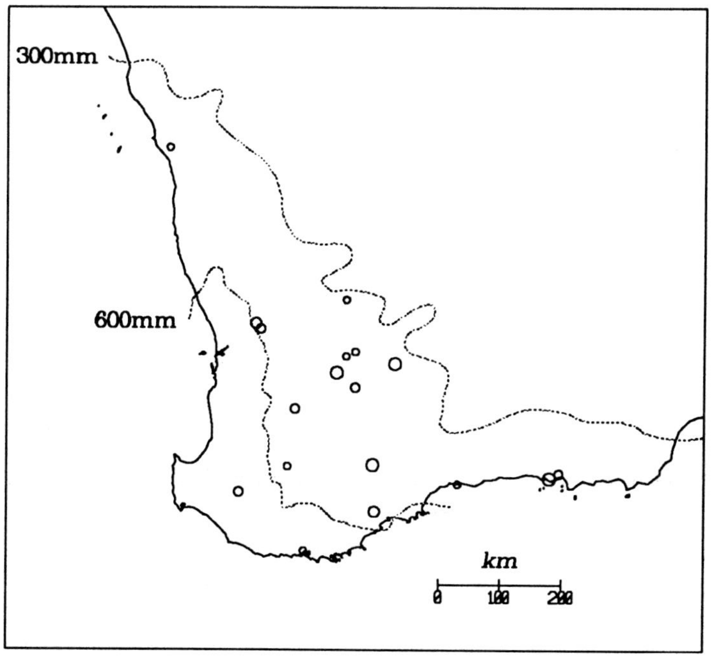

	1900/37	1987	1988	1989	1990
% of observers recording species	40	4.7	7.7	5.1	8.7
% of recording period species seen	U	1.0	3.5	1.6	2.5

Musk duck

Biziura lobata

The musk duck is an uncommon nomad on aquatic habitats in the wheatbelt (Fig. 39). It feeds on seed and other plant material, freshwater mussels and snails, leeches, damselflies, the larvae of mayflies and dragonflies, bugs, backswimmers, water boatmen, beetles, fly larvae, ants, water fleas, ostracods, copepods, shrimp, yabbies, spiders, fish, frogs and tadpoles, and ducklings. It breeds in spring and summer, nests on the ground and lays 2–3 eggs.

There is no evidence to suggest any change in abundance over the last 50 years.

Black-shouldered kite

Elanus notatus

The black-shouldered kite is an uncommon resident on farmland and woodland throughout the wheatbelt (Fig. 40). It feeds on grasshoppers, beetles, wasps, reptiles, small birds, and small mammals like the house mouse. It breeds in winter and early spring on a stick nest in a tree. It lays 3–4 eggs.

During the period 1988 to 1990, the black-shouldered kite demonstrated a seasonal pattern being more commonly recorded in the period April to October (see Fig. below). In addition, recording rates increased between 1988 and 1990, with the latter being more than twice the former.

Serventy and Whittell (1976) regarded the black-shouldered kite as a bird of the arid zone, which appeared in the southwest in cycles of fluctuating numbers. It became numerous in the wheatbelt in 1952, and has increased in range and abundance over the last 40 years. It was not recorded by early observers to Stirling Range National Park, but Sedgwick (1964) noted that it was recorded there in 1953. Ford and Stone (1957) noted that in the Kellerberrin/Kwolyin districts, the black-shouldered kite was observed in the spring of 1945 and it was abundant from the spring of 1951, but declined in the latter half of 1952. They stated that a few individuals were still around in early 1956. This raptor is an uncommon resident in the same area now (Saunders and Curry 1990). Masters and Milhinch (1974) noted that the black-shouldered kite was first recorded in the Northam district in 1935 and became fairly common. It is one of the most commonly recorded raptors in the wheatbelt.

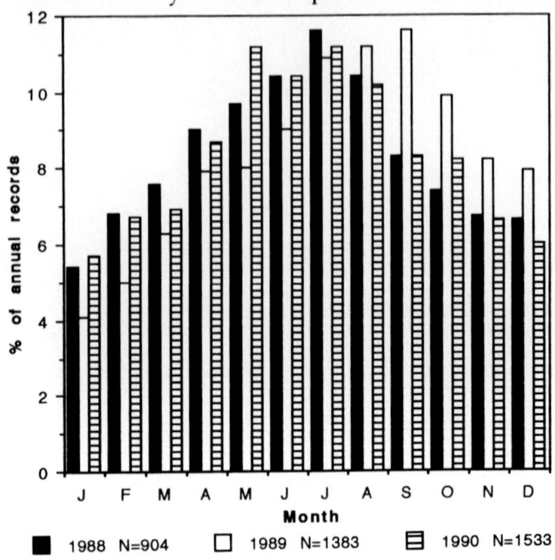

Fig. 40. Distribution 1987–1990

Black-shouldered kite

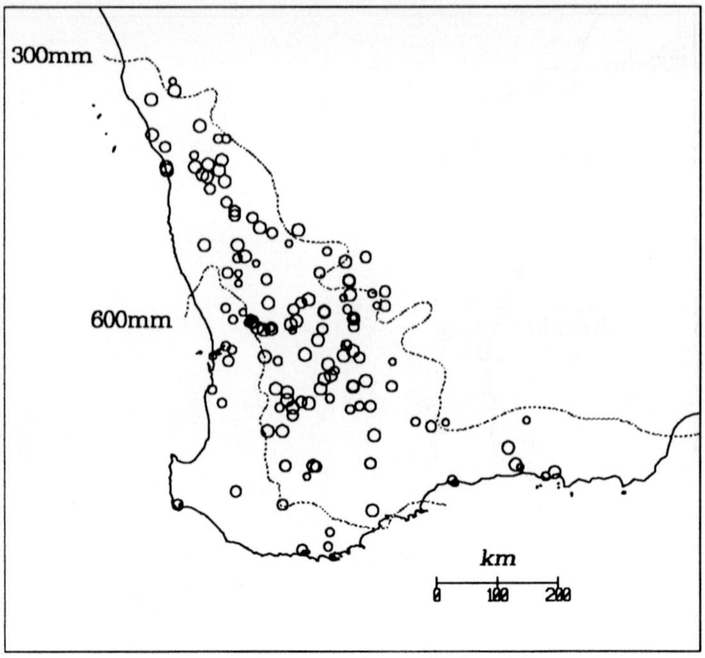

	1900/37	1987	1988	1989	1990
% of observers recording species	0	66	58	76	77
% of recording period species seen	NR	20	17	29	36

Fig. 41. Distribution 1900–1937

Square-tailed kite

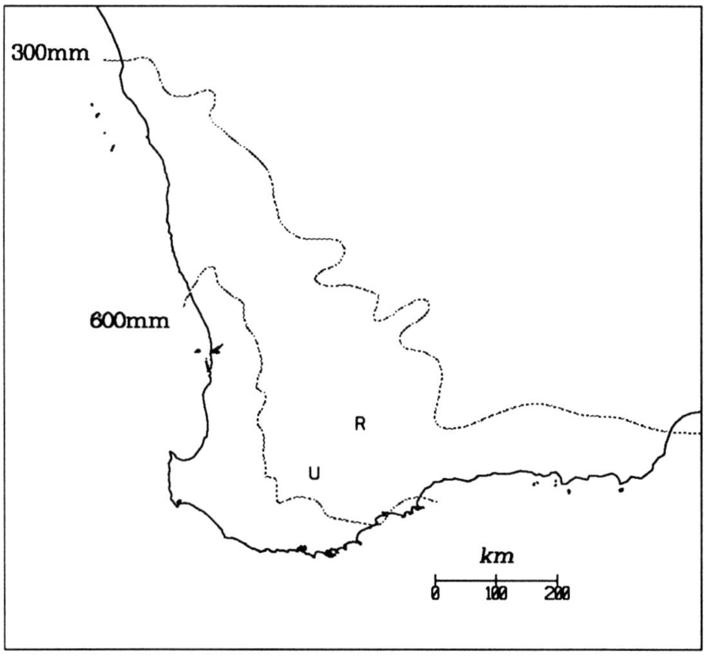

Fig. 41. Distribution 1987–1990

Square-tailed kite

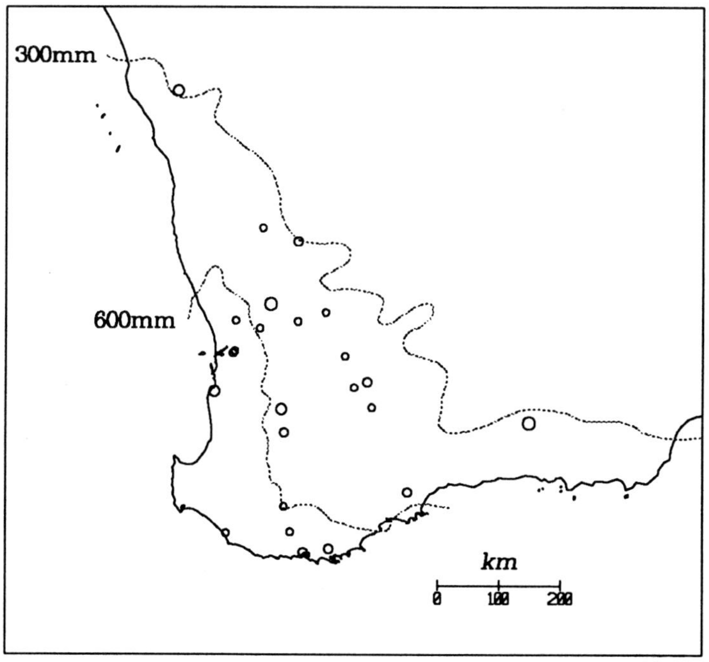

	1900/37	1987	1988	1989	1990
% of observers recording species	20	8.7	10.0	9.4	10.6
% of recording period species seen	R-U	2.4	1.7	2.1	1.6

Square-tailed kite

Lophoictinia isura

The square-tailed kite is an uncommon nomad of woodland, mallee and heath in the wheatbelt (Fig. 41). It feeds on grasshoppers, beetles, ants, lizards, birds and their eggs. It breeds in spring on a stick nest in a tree and lays 2–3 eggs.

The square-tailed kite is regarded as rare (Garnett 1992) and has always been considered uncommon throughout its range. There is no evidence to suggest any change in abundance in the wheatbelt over the last 50 years, although Garnett (1992) believes that because this species is a specialist predator of the canopy of woodland and mallee it may have declined in areas of extensive clearing.

Black kite

Milvus migrans

The black kite is a rare, non-breeding vagrant in woodland, heath and over farmland of the northern half of the wheatbelt (Fig. 42). It feeds on centipedes, cockroaches, termites, mantids, grasshoppers, bugs, beetles, fly larvae, caterpillars, wasps, bees, ants, spiders, fish and fish carrion, frogs, reptiles, birds, mammals including mice and rabbits, and carrion.

In 1952 this species was first recorded in the southwest (Serventy and Whittell 1976) and has been recorded periodically since then. A sighting once every ten years along the Avon River indicates the degree of rarity (J. Masters, in litt.).

Fig. 42. Distribution 1987–1990

Black kite

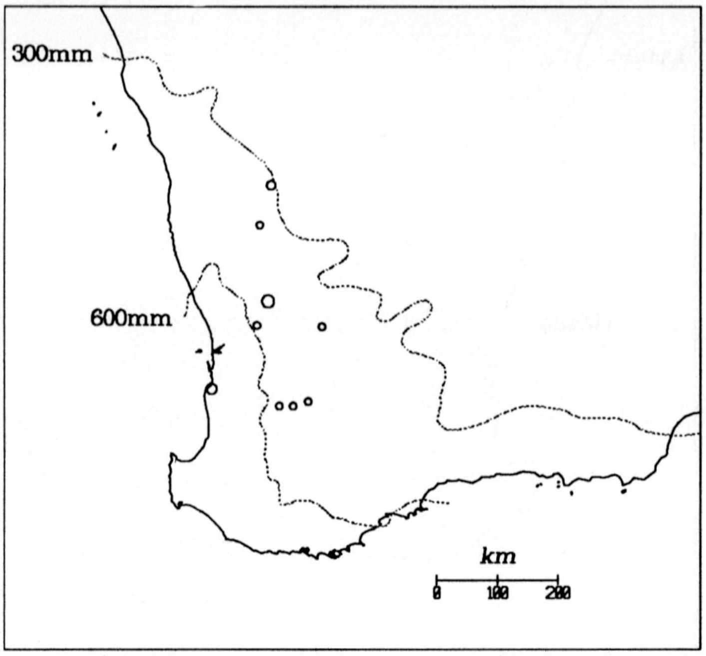

	1900/37	1987	1988	1989	1990
% of observers recording species	0	1.6	2.1	2.6	1.0
% of recording period species seen	NR	0.1	0.3	0.7	0.1

Fig. 43. Distribution 1900–1937

Whistling kite

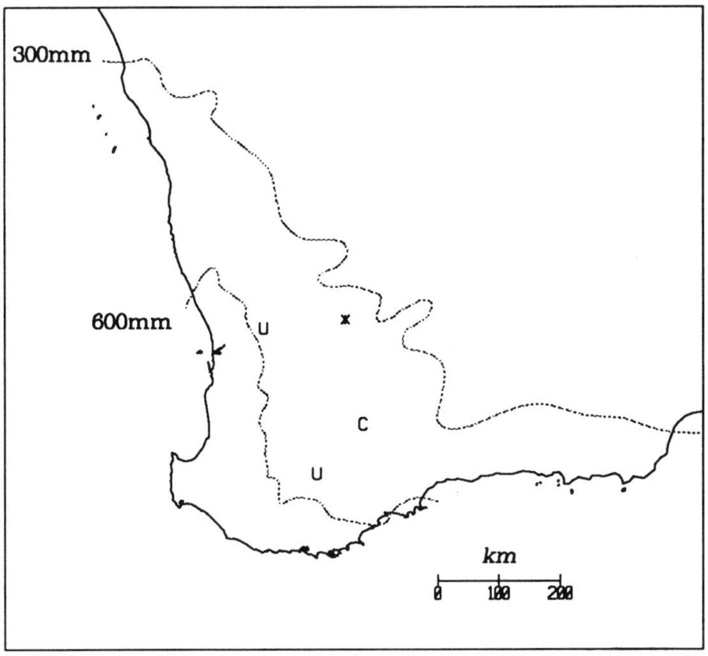

Fig. 43. Distribution 1987–1990

Whistling kite

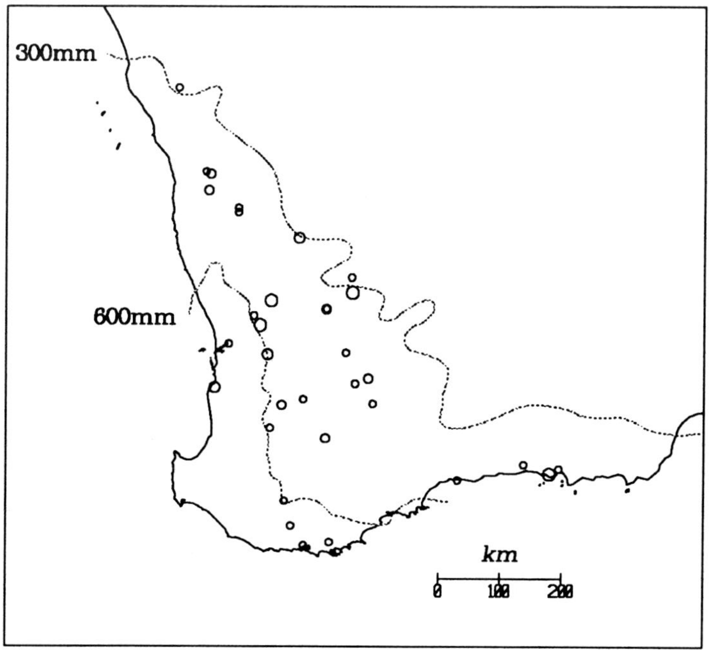

	1900/37	1987	1988	1989	1990
% of observers recording species	40	12	13	7.7	7.7
% of recording period species seen	U-C	3.3	2.1	1.4	1.7

Whistling kite

Haliastur sphenurus

The whistling kite is an uncommon resident around aquatic habitats, and in woodland, heath and over farmland of the wheatbelt (Fig. 43). It is most abundant near swamps, rivers or lakes. It feeds on centipedes, damselflies, dragonflies, grasshoppers, crickets, bugs, beetles, flies, caterpillars, ants, yabbies, fish, reptiles, birds, small mammals and carrion. It breeds in spring on a stick nest in a tree and lays 2–3 eggs.

Although still widespread, this species has decreased in abundance in the wheatbelt over the last 60 years. Carnaby (1933) noted that it was plentiful when the lakes around Lake Grace were full and could be seen in flocks of thirty or forty. In the Northam district, it was classified as fairly common, although it has become greatly reduced in numbers since the 1950s. In the north Dowerin area (G. Evans, in litt.) and around Tammin (J. Masters, in litt.) the whistling kite was regarded as common pre-1940, but it is rare now. Masters and Milhinch (1974) believe this is because rabbits, a major part of its diet, were reduced by myxomatosis. Degradation of its habitat has also contributed to its decline.

Brown goshawk

Accipiter fasciatus

The brown goshawk is an uncommon resident in woodland, shrublands and farmland of the wheatbelt (Fig. 44). It feeds on millipedes, dragonflies, grasshoppers, crickets, locusts, cicadas, beetles, wasps, ants, spiders, frogs, lizards, snakes, birds and small mammals. It breeds in spring and early summer on a stick nest in a tree and lays 2–4 eggs.

Regarded as common in the early part of this century (Orton and Sandland 1913), this species has declined in abundance over the last 80 years (Masters and Milhinch 1974), presumably because of loss of native vegetation. Masters and Milhinch (1974) pointed out that it was common where forest remained.

Fig. 44. Distribution 1900–1937

Brown goshawk

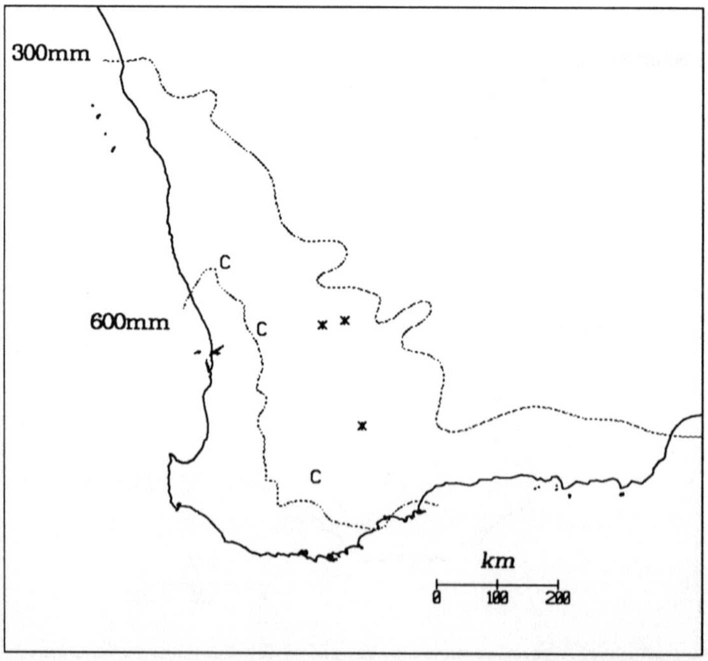

Fig. 44. Distribution 1987–1990

Brown goshawk

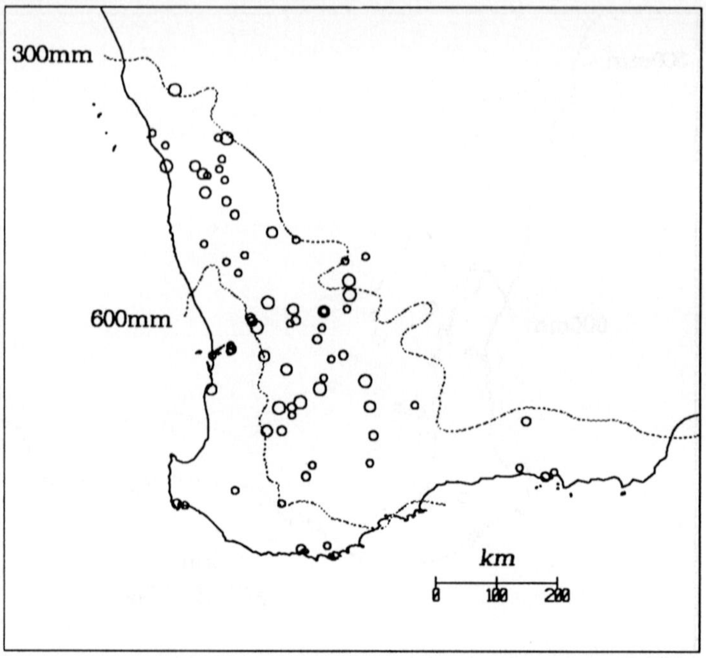

	1900/37	1987	1988	1989	1990
% of observers recording species	60	27	27	31	36
% of recording period species seen	C	7.7	6.3	6.7	6.9

Fig. 45. Distribution 1900–1937

Collared sparrowhawk

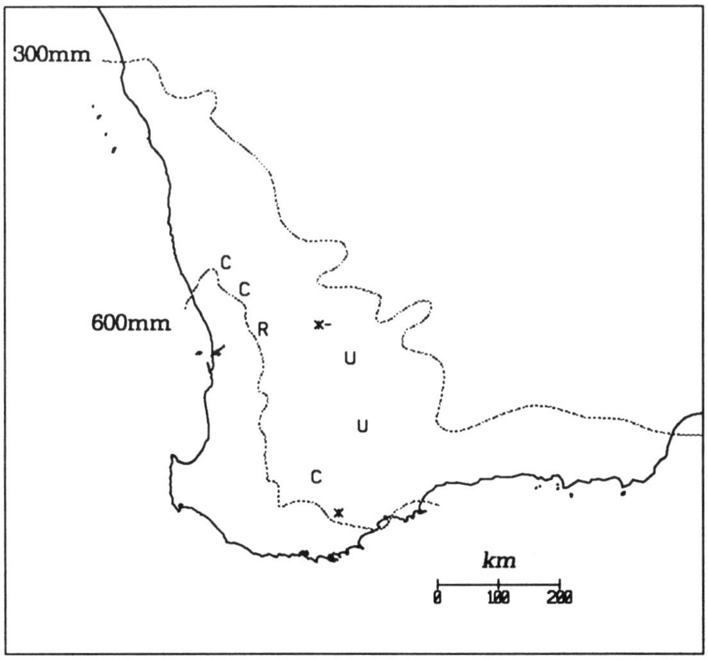

Fig. 45. Distribution 1987–1990

Collared sparrowhawk

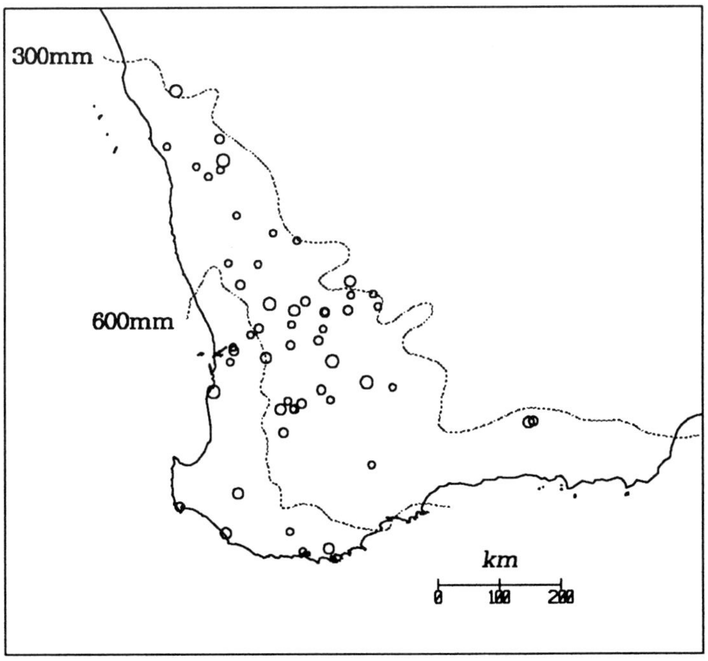

	1900/37	1987	1988	1989	1990
% of observers recording species	80	18	20	22	28
% of recording period species seen	R-C	3.5	4.7	4.6	4.8

Collared sparrowhawk

Accipiter cirrhocephalus

The collared sparrowhawk is an uncommon resident in woodland, shrubland and farmland throughout the wheatbelt (Fig. 45). It feeds on large insects, reptiles, birds and small mammals. It breeds in spring on a stick nest in a tree and lays 2–4 eggs.

This species was regarded as common before 1937 (Orton and Sandland 1913) and has declined in abundance (Masters and Milhinch 1974). Small birds form the bulk of its diet (Readers' Digest 1976). The decline in these as a result of the clearing of native vegetation has undoubtedly contributed to the decline of the collared sparrowhawk.

Wedge-tailed eagle

Aquila audax

The wedge-tailed eagle is the most conspicuous raptor in the wheatbelt. It is an uncommon to common resident in woodland, shrubland, heath and farmland (Fig. 46). It eats carrion and hunts for live prey including reptiles, birds and mammals (Brooker and Ridpath 1980). Brooker and Ridpath estimated a density of one breeding pair per 130 km^2 for the wheatbelt, based on observations from Miling and Newdegate. The wedge-tailed eagle breeds in winter on a large stick nest in a tree and lays 2 eggs.

There is no evidence to suggest any change in abundance over the last 50 years.

Fig. 46. Distribution 1900–1937

Wedge-tailed eagle

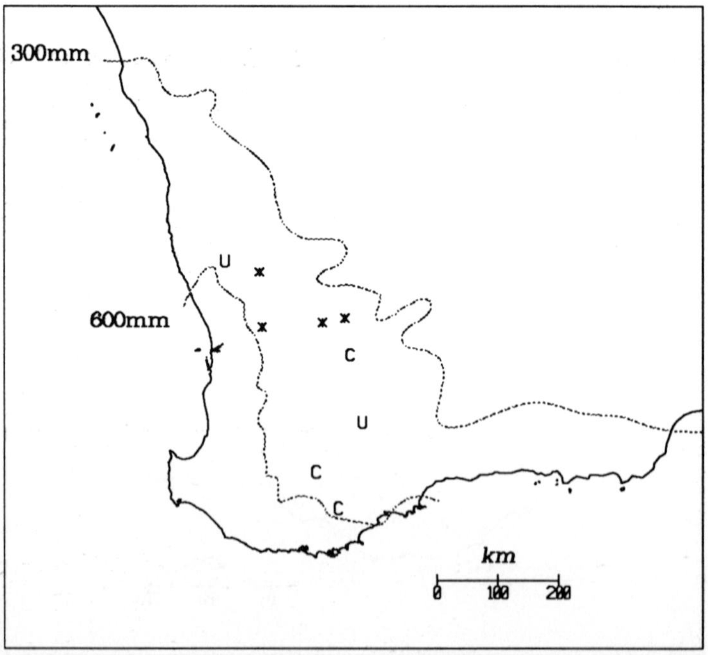

Fig. 46. Distribution 1987–1990

Wedge-tailed eagle

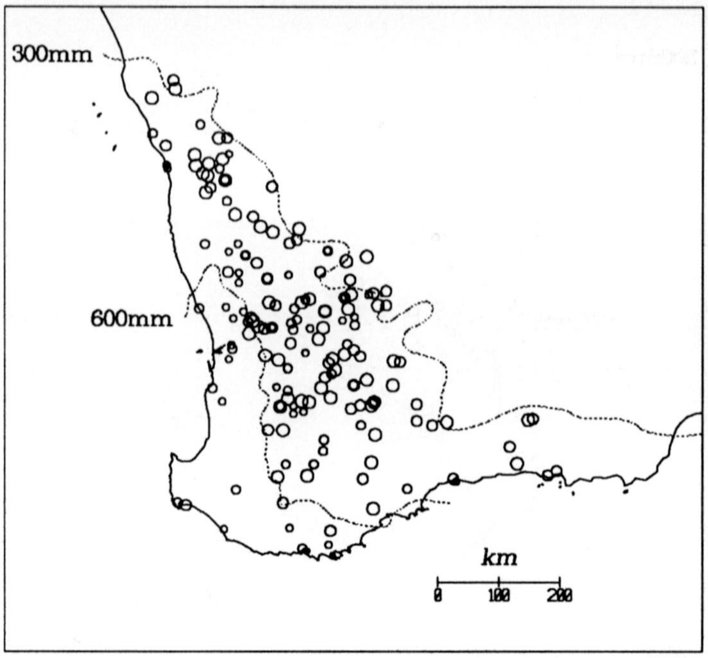

	1900/37	1987	1988	1989	1990
% of observers recording species	90	81	84	79	84
% of recording period species seen	U-C	29	32	27	30

Fig. 47. Distribution 1900–1937

Little eagle

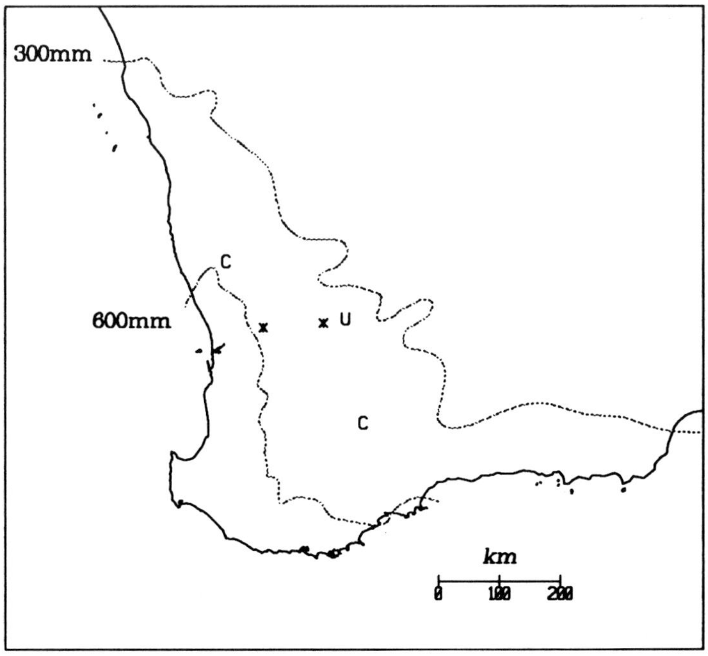

Fig. 47. Distribution 1987–1990

Little eagle

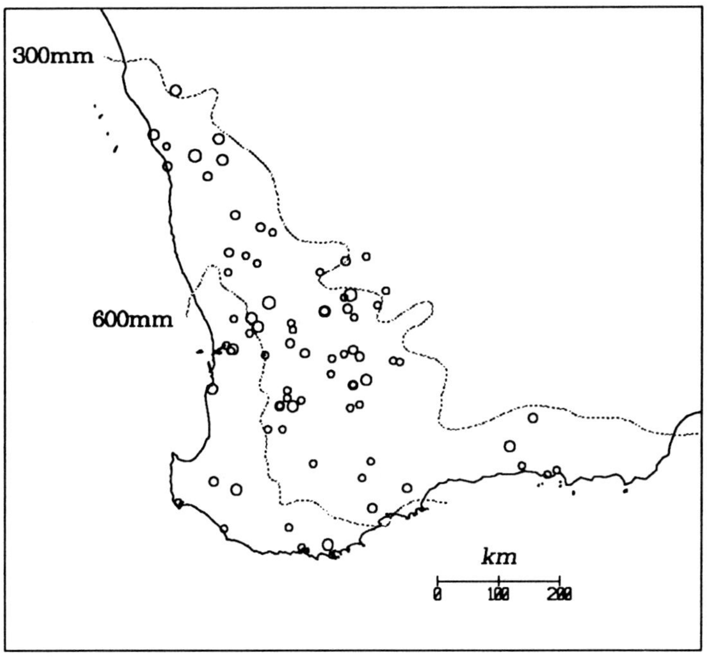

	1900/37	1987	1988	1989	1990
% of observers recording species	50	29	34	25	32
% of recording period species seen	U-C	6.0	4.4	4.6	3.5

Little eagle

Hieraaetus morphnoides

The little eagle is an uncommon resident in woodland, shrubland, heath and farmland of the wheatbelt (Fig. 47). It feeds on cockroaches, beetles, lizards, birds and small mammals, including mice, rabbits and cats. The little eagle breeds in spring on a stick nest in a tree and lays 1–2 eggs.

Regarded as common in some locations in the past, it is no longer common and it is possible that this species has declined in abundance (Storr 1991).

Spotted harrier

Circus assimilis

The spotted harrier is an uncommon resident in shrubland, heath and farmland of the wheatbelt (Fig. 48). It feeds on grasshoppers, locusts, lizards, birds, and small mammals. It breeds in late winter and spring on a stick nest in a tree and lays 2–4 eggs.

The abundance of the spotted harrier does change over time, and it was apparently more common in 1990 than during the previous two years. It was also more commonly recorded in the period August to December (see Fig. below).

There is no evidence to suggest any change in abundance over the last 50 years, although J. Masters (in litt.) believes that the spotted harrier has declined in abundance along the Avon River. He noted that it was a regular summer/autumn visitor, but he has not seen any for the last seven years.

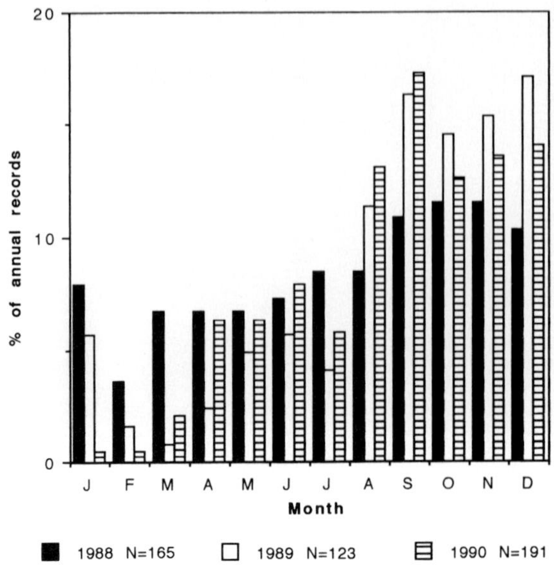

Fig. 48. Distribution 1900–1937

Spotted harrier

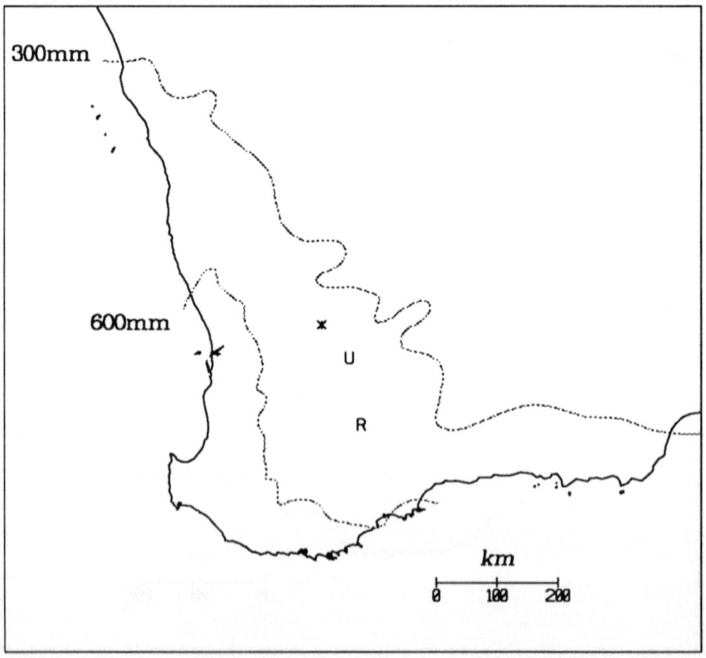

Fig. 48. Distribution 1987–1990

Spotted harrier

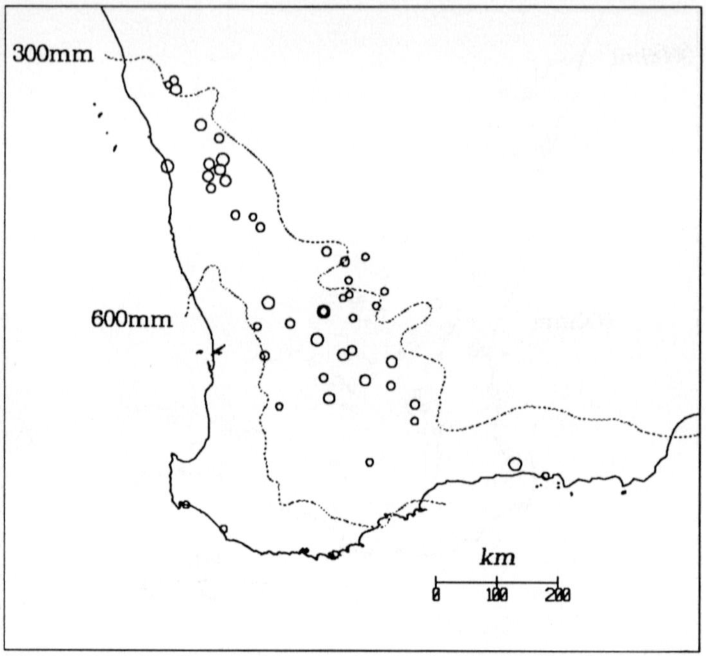

	1900/37	1987	1988	1989	1990
% of observers recording species	30	18	21	15	25
% of recording period species seen	R-U	4.2	3.2	2.5	4.5

Fig. 49. Distribution 1900–1937

Swamp harrier

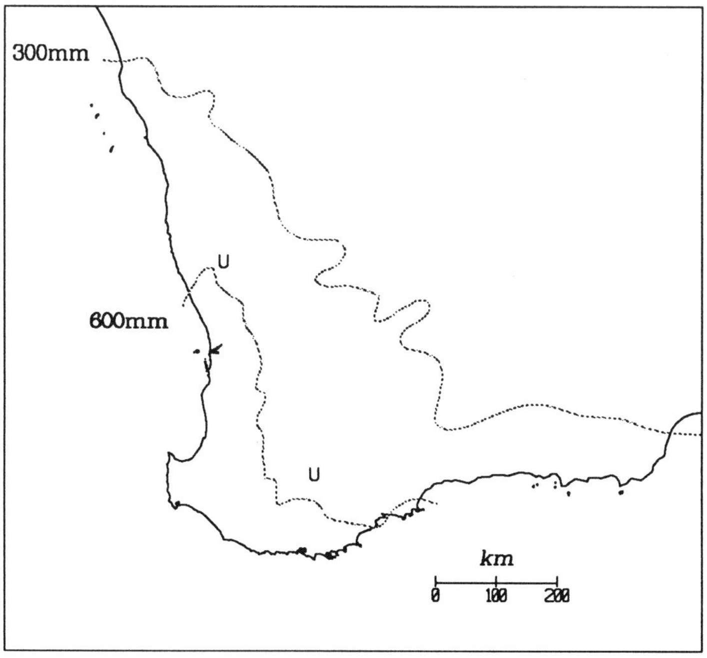

Fig. 49. Distribution 1987–1990

Swamp harrier

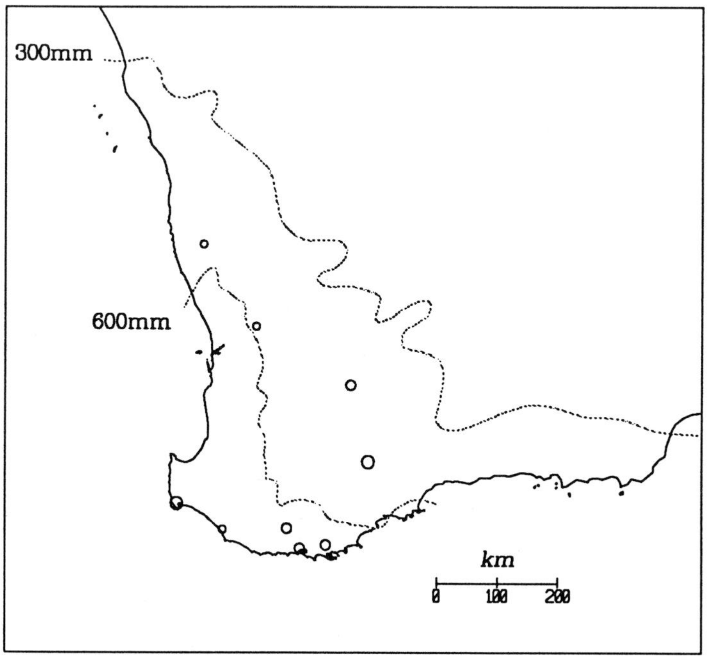

	1900/37	1987	1988	1989	1990
% of observers recording species	20	0	4.9	3.4	6.7
% of recording period species seen	U	0	0.5	0.7	1.4

Swamp harrier

Circus aeruginosus

OTHER NAME:

Marsh harrier

The swamp harrier is a rare resident of open country around swamps of the higher rainfall districts of the wheatbelt (Fig. 49). It feeds on seed, molluscs, grasshoppers, crickets, fish, snakes, birds and their eggs, small mammals such as rats and mice, and carrion. The marsh harrier breeds in spring and early summer. It nests on the ground and lays 3–4 eggs.

This species has probably declined in abundance over the last 50 years as a result of loss of suitable habitat.

Black falcon

Falco subniger

The black falcon is a rare non-breeding vagrant in woodland along watercourses of the northern part of the wheatbelt (Fig. 50). It feeds mainly on birds, but will take grasshoppers and small mammals.

The black falcon was one of the rarest hawks in the lower half of Western Australia (Serventy and Whittell 1976), and this situation has not changed. There is no evidence to suggest any change in abundance over the last 50 years.

Fig. 50. Distribution 1900–1937

Black falcon

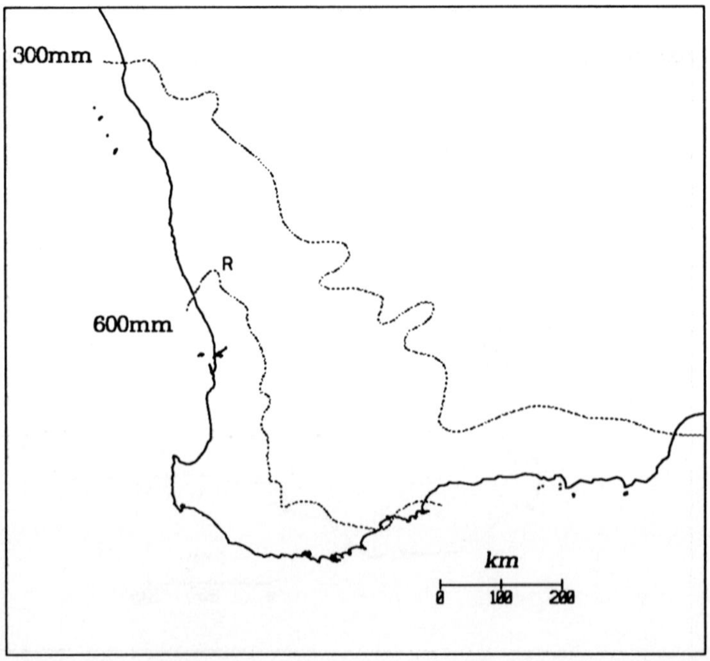

Fig. 50. Distribution 1987–1990

Black falcon

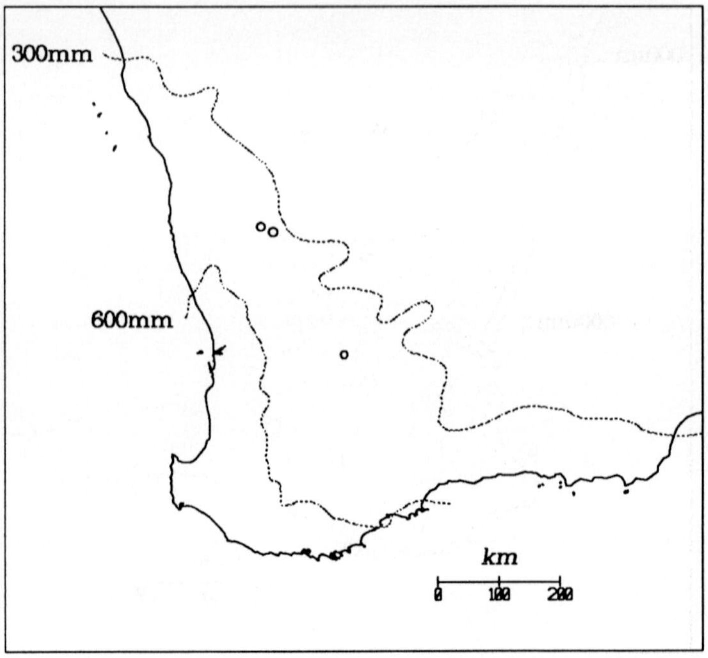

	1900/37	1987	1988	1989	1990
% of observers recording species	10	0	0	1.7	1.0
% of recording period species seen	R	0	0	0.3	0.3

Peregrine falcon

Falco peregrinus

The peregrine falcon is an uncommon resident in woodland, shrubland, heath and farmland throughout the wheatbelt (Fig. 51). It eats birds and small mammals. It breeds in spring in a cave in a cliff, a hollow in a tree or on an abandoned stick nest and lays 3 eggs.

There is no evidence to suggest any change in status over the last 50 years.

Fig. 51. Distribution 1900–1937

Peregrine falcon

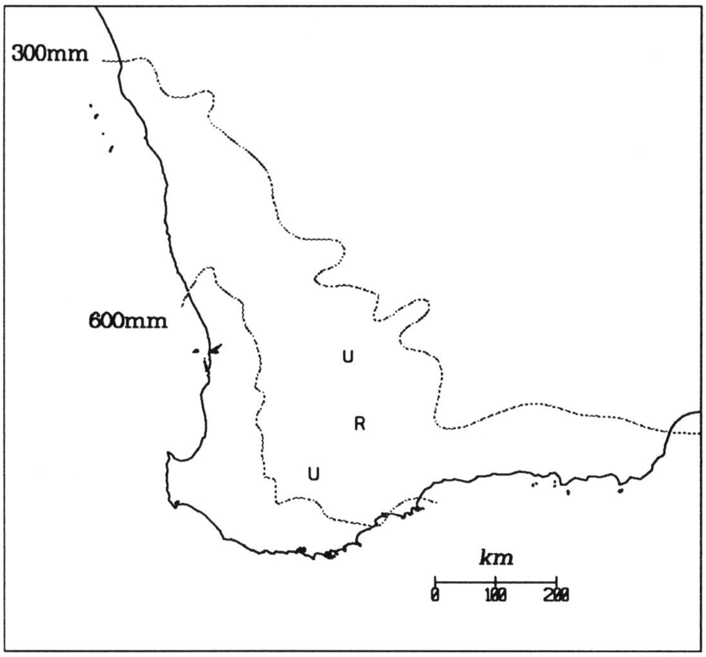

Fig. 51. Distribution 1987–1990

Peregrine falcon

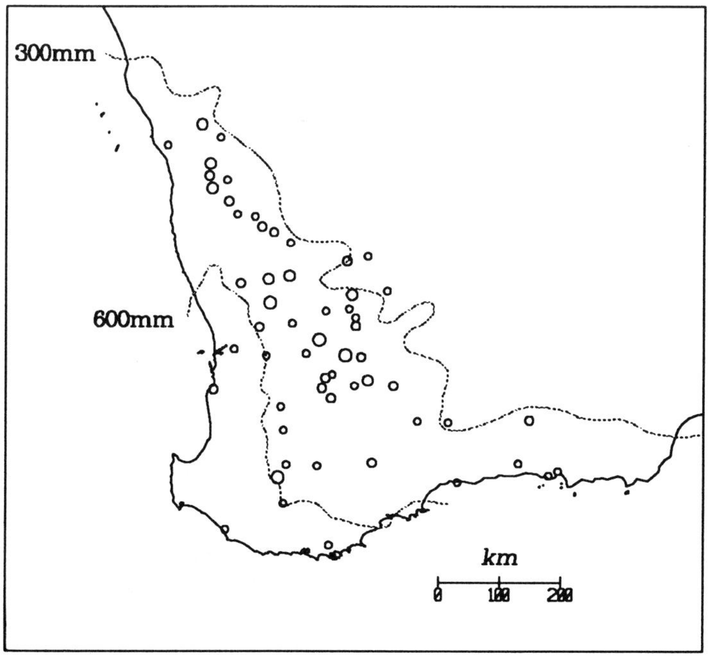

	1900/37	1987	1988	1989	1990
% of observers recording species	30	16	19	23	32
% of recording period species seen	R-U	2.5	3.1	4.5	4.5

Little falcon

Falco longipennis

OTHER NAME:
Australian hobby

The little falcon is an uncommon resident in woodland, shrubland, heath and farmland throughout the wheatbelt (Fig. 52). It feeds on dragonflies, grasshoppers, locusts, cicadas, bugs, beetles, flies, caterpillars, wasps, ants, spiders, as well as small birds and mammals. It breeds in spring on a stick nest in a tree and lays 2–3 eggs.

Serventy and Whittell (1976) regarded the little falcon as a comparatively common species. Before 1937 it was recorded by 6 out of the 10 people who published bird lists for the period. It could not be classed as comparatively common now, and may have declined in abundance.

Fig. 52. Distribution 1900–1937

Little falcon

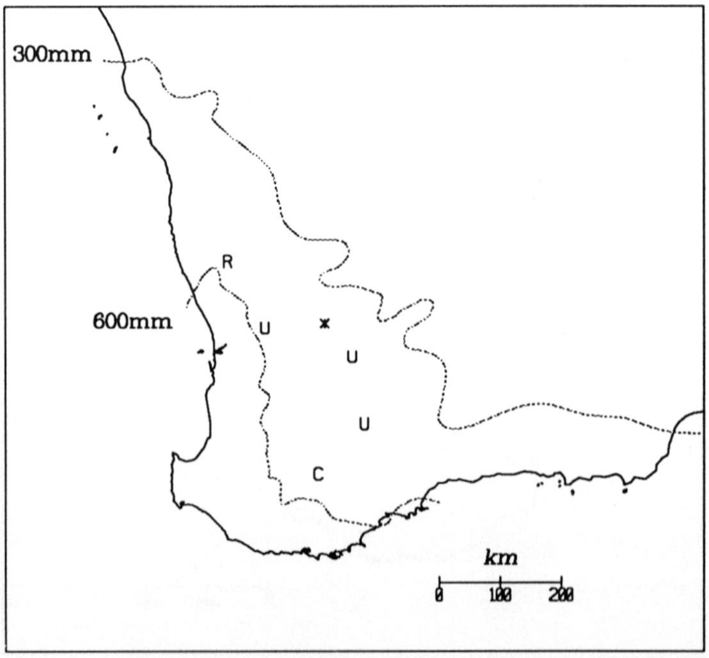

Fig. 52. Distribution 1987–1990

Little falcon

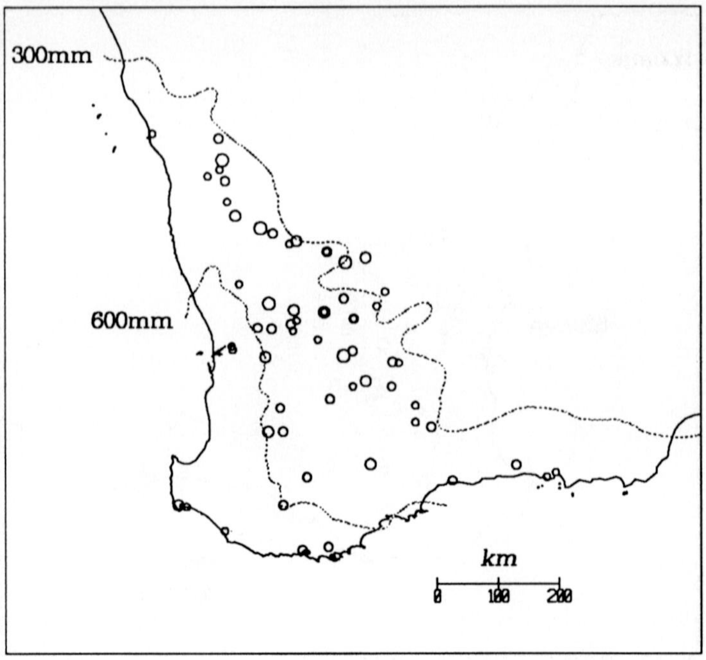

	1900/37	1987	1988	1989	1990
% of observers recording species	60	24	28	26	35
% of recording period species seen	R-C	5.0	4.2	5.6	5.8

Fig. 53. Distribution 1900–1937

Grey falcon

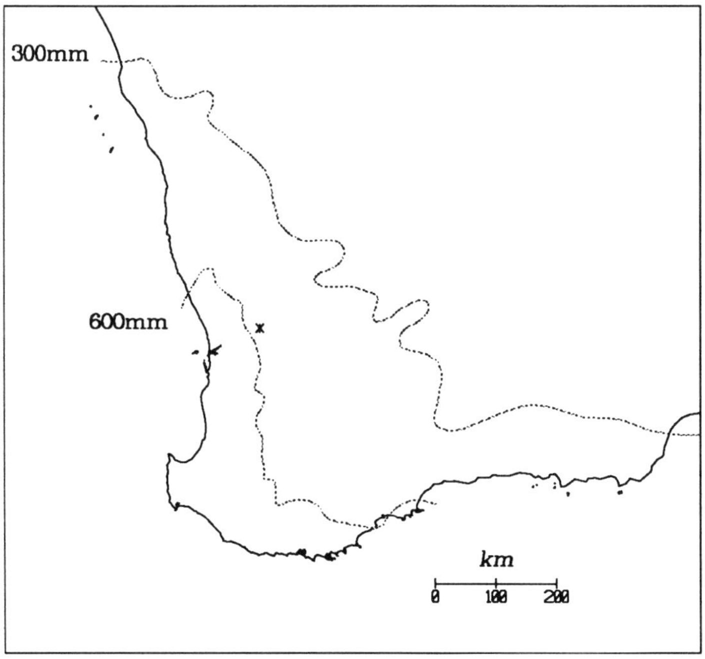

Fig. 53. Distribution 1987–1990

Grey falcon

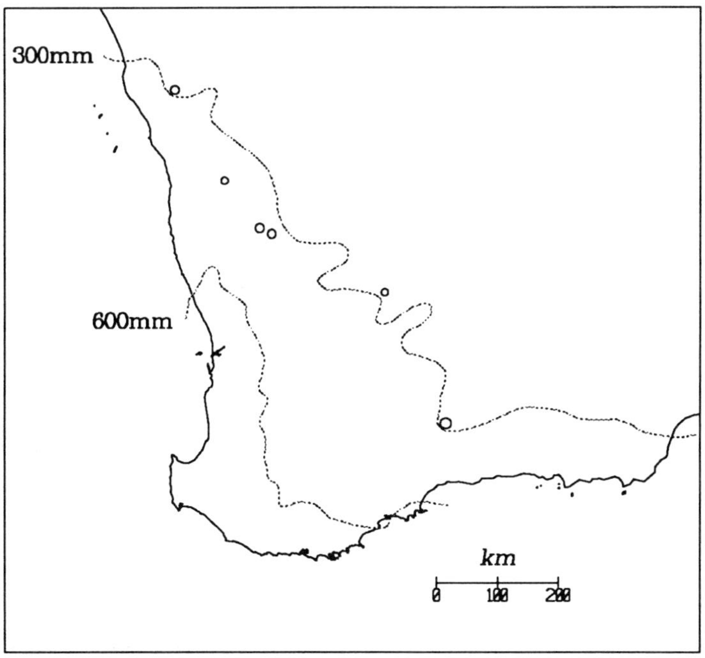

	1900/37	1987	1988	1989	1990
% of observers recording species	10	0.8	1.4	2.3	1.9
% of recording period species seen	—	0.07	0.3	0.4	0.7

Grey falcon

Falco hypoleucos

The grey falcon is regarded as one of Australia's rarest hawks. It occurs in the dry plains of the interior (Readers' Digest 1976). It is a rare non-breeding vagrant in the drier parts of the wheat-belt (Fig. 53) adjacent to the arid zone. Garnett (1992) noted that sightings made outside its normal range are during drought in the inland. It feeds on seed, earthworms, grasshoppers, lizards, birds and small mammals.

There is no evidence to suggest any change in abundance over the last 50 years.

Brown falcon

Falco berigora

The brown falcon is an uncommon resident in woodland, shrubland, heath and farmland throughout the wheatbelt (Fig. 54). It feeds on centipedes, cockroaches, mantids, earwigs, grasshoppers, crickets, locusts, phasmids, cicadas, bugs, beetles, moths, caterpillars, wasps, ants, scorpions, spiders, fish, lizards and snakes, birds and small mammals, as well as carrion. The brown falcon breeds in winter and spring on an abandoned stick nest in a tree and lays 2–3 eggs.

Although widespread, the brown falcon has declined in abundance over the last 50 years. For example, J. Masters (in litt.) noted that it was affected by the reduction in rabbit numbers in the 1950s, with numbers of birds dropping dramatically by 1975.

Fig. 54. Distribution 1900–1937

Brown falcon

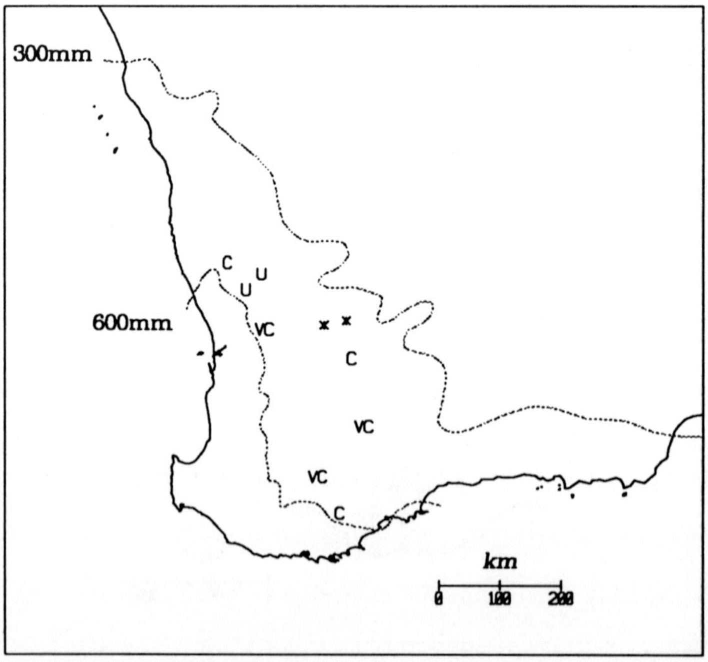

Fig. 54. Distribution 1987–1990

Brown falcon

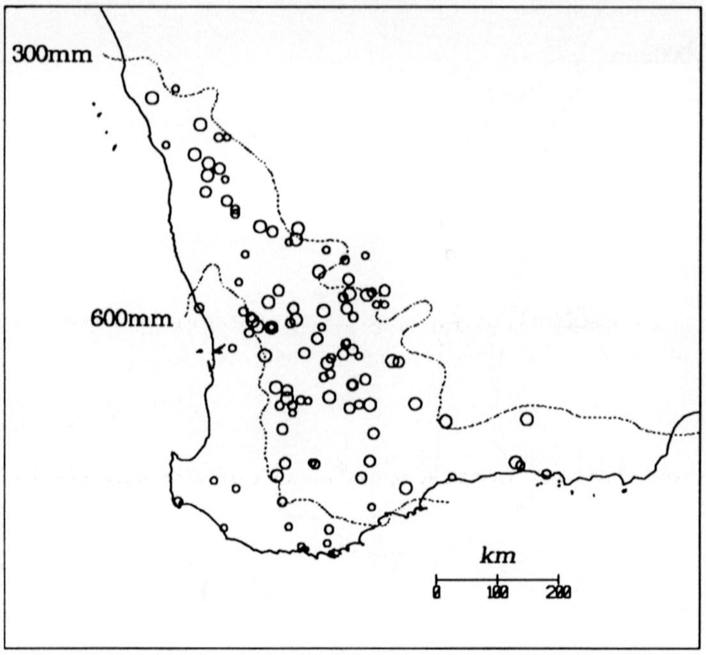

	1900/37	1987	1988	1989	1990
% of observers recording species	100	41	48	47	51
% of recording period species seen	U-VC	17	11	11	13

Fig. 55. Distribution 1900–1937

Nankeen kestrel

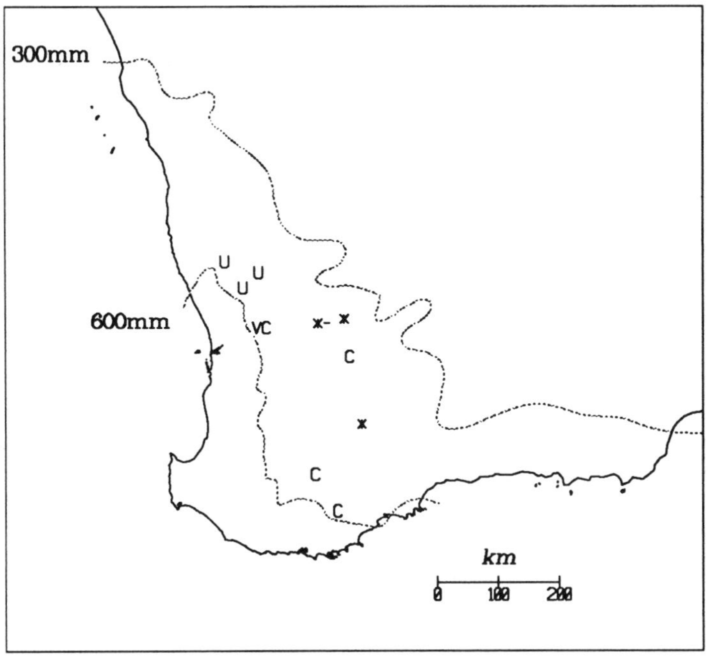

Fig. 55. Distribution 1987–1990

Nankeen kestrel

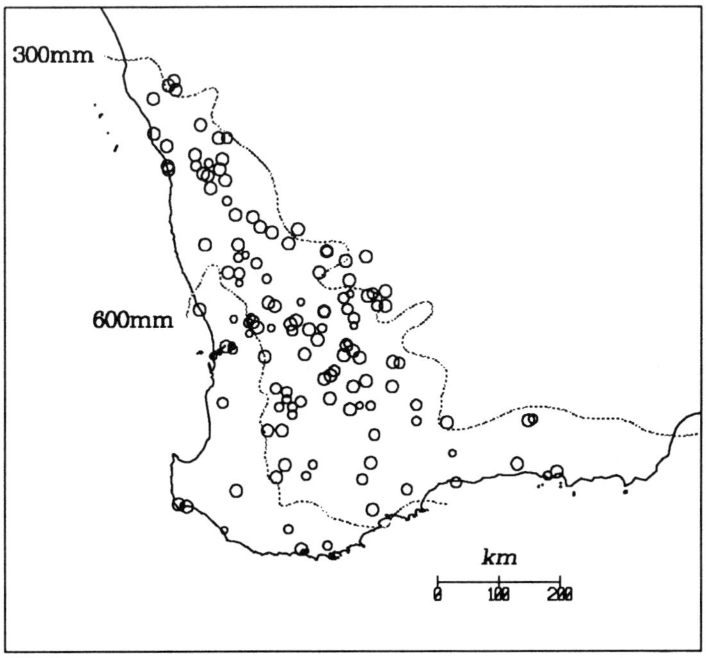

	1900/37	1987	1988	1989	1990
% of observers recording species	100	61	68	68	75
% of recording period species seen	U-VC	31	33	41	45

Nankeen kestrel

Falco cenchroides

OTHER NAME:

Australian kestrel

The nankeen kestrel is a common resident in woodland, shrubland, heath and farmland throughout the wheatbelt (Fig. 55). In fact, it was the raptor most commonly recorded during the survey period 1987 to 1990. It feeds on centipedes, damselflies, cockroaches, mantids, grasshoppers, crickets, locusts, phasmids, bugs, beetles, moths, caterpillars, wasps, ants, spiders, frogs, lizards and small birds and mammals. This species breeds in spring in a tree hollow or on an abandoned stick nest in a tree. It lays 4–5 eggs.

Although common throughout its range, it is possible that it has decreased in abundance (Masters and Milhinch 1974).

Mallee fowl

Leipoa ocellata

Although uncommon, the mallee fowl is resident in some wheatbelt woodland and shrubland patches (Fig. 56). It feeds on seed, fruit, flowers, and insects. The male constructs a large mound of decaying vegetation which he maintains to incubate the 15–24 eggs, which are laid from spring to mid-summer.

In 1842 Gilbert saw the nest and eggs of the mallee fowl in the Wongan Hills, but the bird had disappeared from the area by 1903 (Milligan 1904). Milligan gave drought as the reason for its disappearance. By the early 1970s it was breeding in low numbers in the area (de Rebeira and de Rebeira 1977). Crossman (1909) noted that the mallee fowl was formerly much more plentiful at Cumminin Station (near Bruce Rock), but could give no reason for its decrease. Orton and Sandland (1913) noted that the mallee fowl was more plentiful in the Moora district in the early days, but was rare at the time of writing. Carter (1923) recorded the mallee fowl was plentiful around Broome Hill in 1905, but by 1922 they were rare in the district; land clearing being regarded as the reason for their decline. Ford and Stone (1957) commented that in the Kellerberrin/Kwolyin districts the mallee fowl had been considerably reduced in numbers since the arrival of the fox, and they classed it as rare. In the mid-1970s the mallee fowl was classified as uncommon (Dell 1978) on Durokoppin and Kodj Kodjin Nature Reserves north of Kellerberrin. By the mid-1980s the species had become extinct in the area (Saunders 1989). The mallee fowl occurred in the Goomalling area in the 1930s, but is extinct there (G. Evans, in litt.). In the Stirling Range it was recorded as a breeding species, but has not been seen since 1980 (Burbidge and Rose 1993).

The mallee fowl is vulnerable throughout its range (Garnett 1992) and in need of monitoring to establish trends in population. It is dependent on native vegetation, and extensive loss of habitat and fragmentation of the remainder have resulted in its decline. In addition, grazing on remnant vegetation by domestic stock and rabbits has affected their food supplies. Predation on young birds by foxes and cats have also resulted in the elimination of mallee fowl populations isolated on patches of remnant vegetation. Its future in the wheatbelt is uncertain.

Fig. 56. Distribution 1900–1937

Mallee fowl

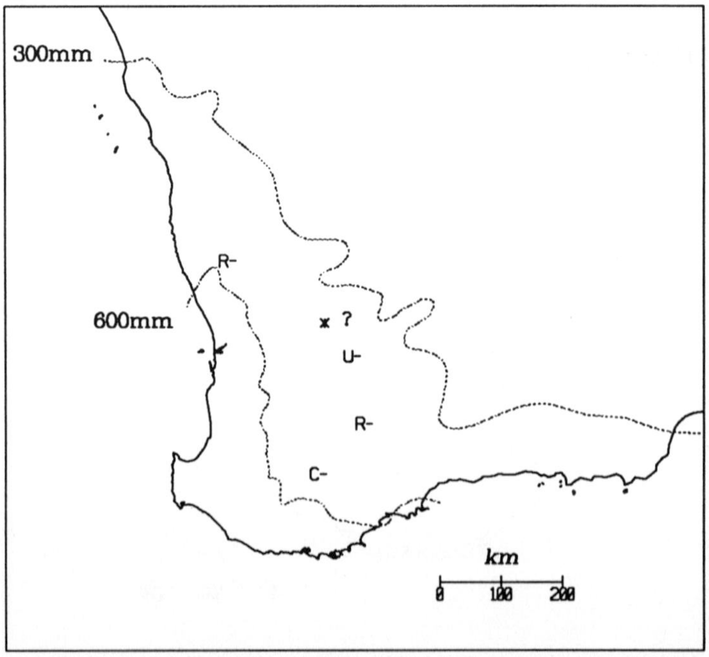

Fig. 56. Distribution 1987–1990

Mallee fowl

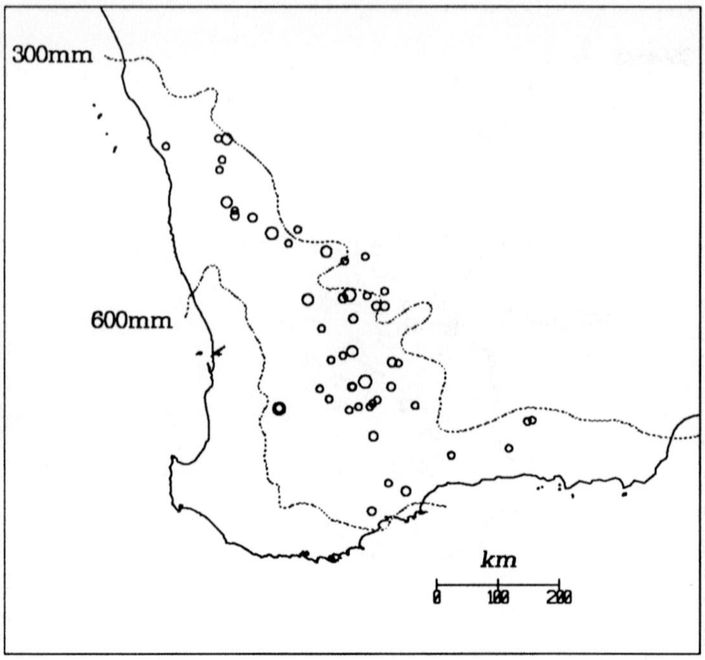

	1900/37	1987	1988	1989	1990
% of observers recording species	60	20	16	21	13
% of recording period species seen	R-C	2.9	2.3	2.5	3.0

Fig. 57. Distribution 1900–1937

Stubble quail

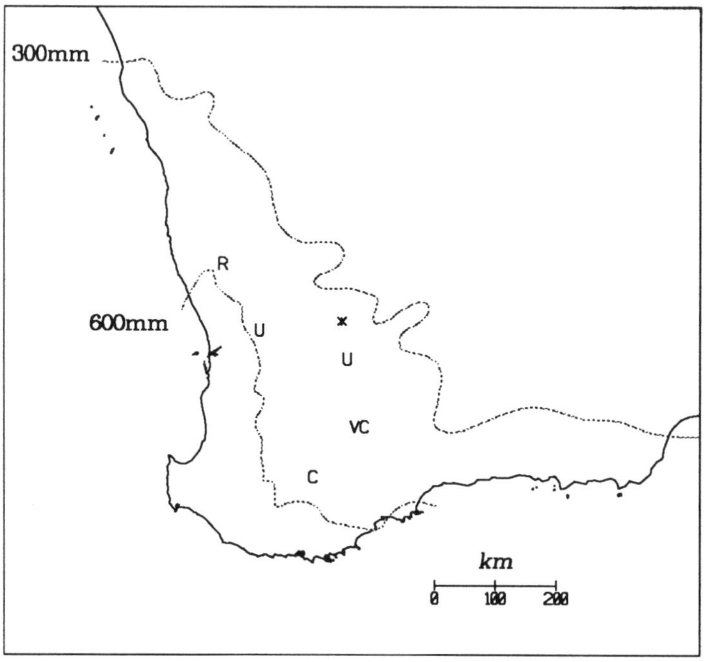

Fig. 57. Distribution 1987–1990

Stubble quail

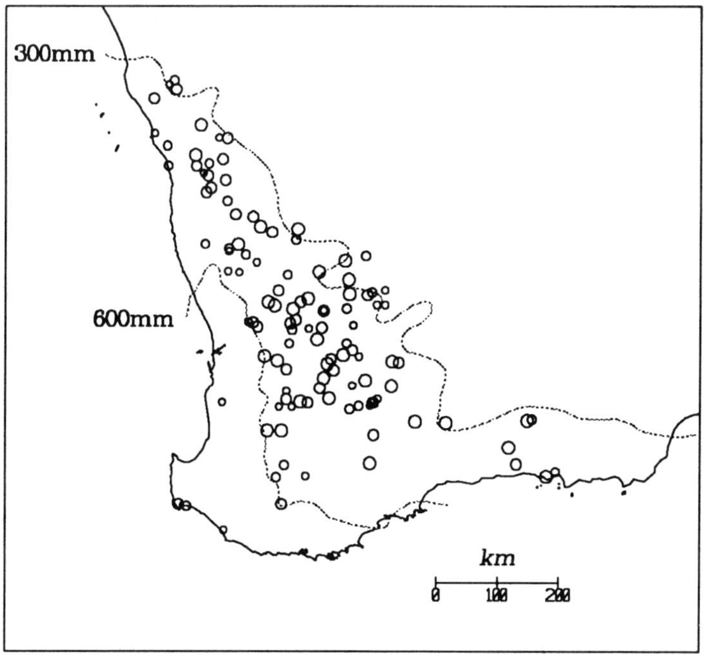

	1900/37	1987	1988	1989	1990
% of observers recording species	60	50	54	56	60
% of recording period species seen	R-VC	20	17	23	23

Stubble quail

Coturnix pectoralis

The stubble quail is an uncommon resident in shrubland, heath and farmland throughout the wheatbelt (Fig. 57). It feeds mainly on seed, and also eats snails, slugs, millipedes, centipedes, springtails, mayflies, cockroaches, termites, mantids, grasshoppers, crickets, bugs, lacewings, beetles, flies, caterpillars, wasps, ants, ostracods, isopods, spiders and frogs. It breeds in spring and summer, nests on the ground and lays 6–8 eggs.

The stubble quail was not recorded in Western Australia before 1901 (Blakers *et al.* 1984). Since then it has invaded most of the southwest as cultivation and improvement of pastures has extended the area of habitat suitable for this species (Serventy and Whittell 1976). Sedgwick (1974) reported that the stubble quail colonized farmland in the eastern wheatbelt within two years of first clearing.

During this survey the stubble quail was more commonly recorded during November and December (see Fig. below). This probably reflects the fact that many of these observations were as a result of the birds being flushed during harvesting (J. Masters, in litt.).

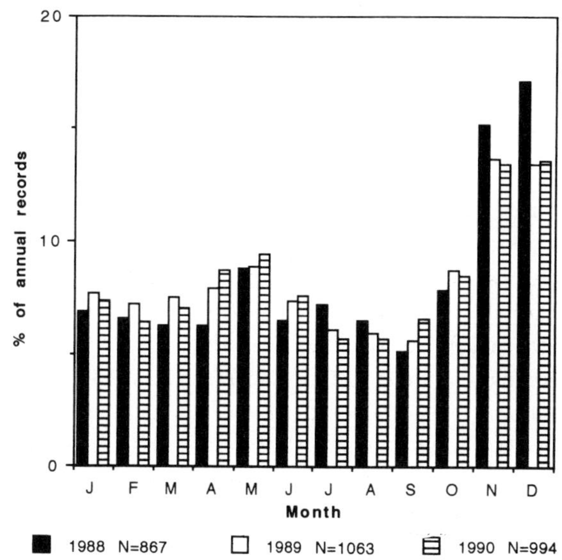

Brown quail

Coturnix australis

The brown quail is an uncommon nomad around creeks and swamps of the wheatbelt (Fig. 58). It feeds on seed and other plant material, and slugs, termites, grasshoppers, bugs, beetles, flies, moths, slaters, and small reptiles like geckoes. The brown quail breeds in spring and summer, nests on the ground and lays 7–11 eggs.

There is no evidence to suggest that this species has changed in status over the last 50 years.

Fig. 58. Distribution 1900–1937

Brown quail

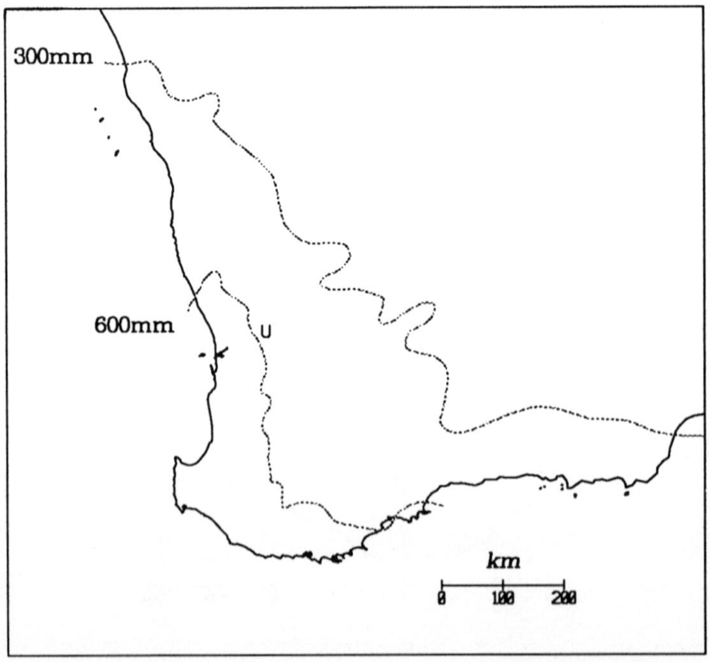

Fig. 58. Distribution 1987–1990

Brown quail

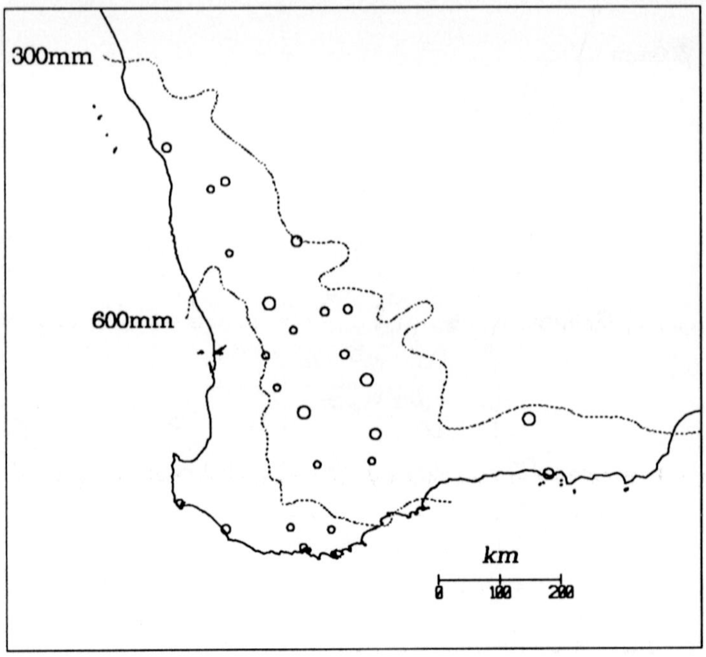

	1900/37	1987	1988	1989	1990
% of observers recording species	10	6.3	8.5	12	14
% of recording period species seen	U	2.2	2.5	2.7	3.5

Fig. 59. Distribution 1900–1937

Painted button quail

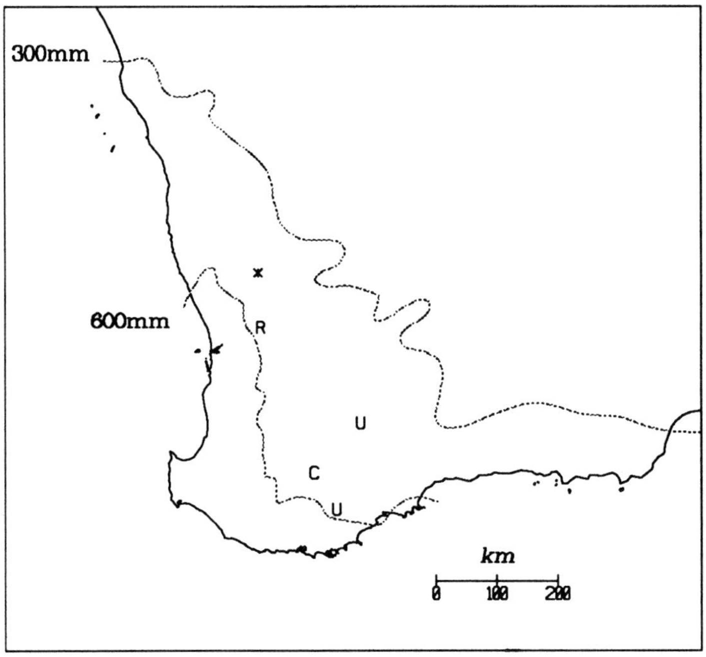

Fig. 59. Distribution 1987–1990

Painted button quail

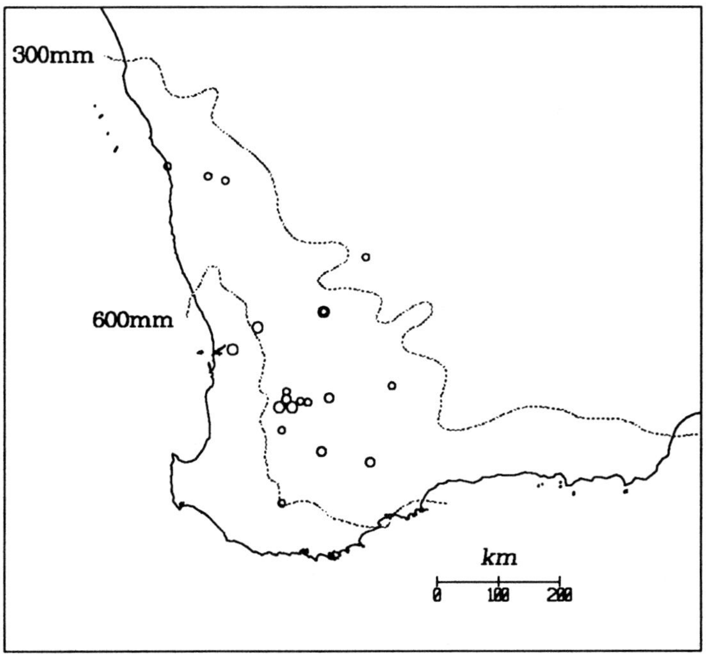

	1900/37	1987	1988	1989	1990
% of observers recording species	50	7.1	7.7	6.0	9.6
% of recording period species seen	R-C	1.1	0.6	1.0	1.1

Painted button quail

Turnix varia

The painted button quail is an uncommon resident in woodland, shrubland and heath in parts of the wheatbelt (Fig. 59). It feeds on seed and fruit, grasshoppers, bugs, beetles, flies, caterpillars and ants. The painted button quail breeds in spring and summer, nests on the ground and lays 4 eggs.

This species is dependent on native vegetation and must have declined in range and abundance over the last 50 years as a result of loss of habitat.

Little button quail

Turnix velox

The little button quail is an uncommon nomad in woodland, shrubland, heath and farmland of the wheatbelt (Fig. 60). It feeds mainly on seed, but also eats crickets, grasshoppers, bugs, beetles, caterpillars and ants. It breeds from spring to mid-summer on the ground and lays 3–4 eggs.

Although there were only 334 recordings of this species during the survey, it is apparent that it is a visitor to the wheatbelt in the spring and early summer (see Fig. below). The majority of observations were made in November and December and these were probably because it is seen most often when flushed during harvesting (J. Masters, in litt.). Leake (1962) noted that the little button quail arrived in the Kellerberrin district in August and left between December and February. J. Masters (in litt.) confirmed this pattern for the Avon Valley. He added "that perhaps about once in ten years a quite dramatic irruption of greater than usual numbers may occur, with much breeding taking place."

There is no evidence to suggest any change in status over the last 50 years.

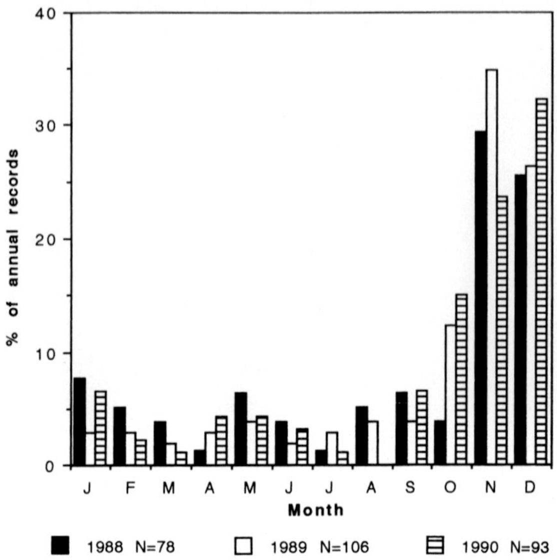

Fig. 60. Distribution 1900–1937

Little button quail

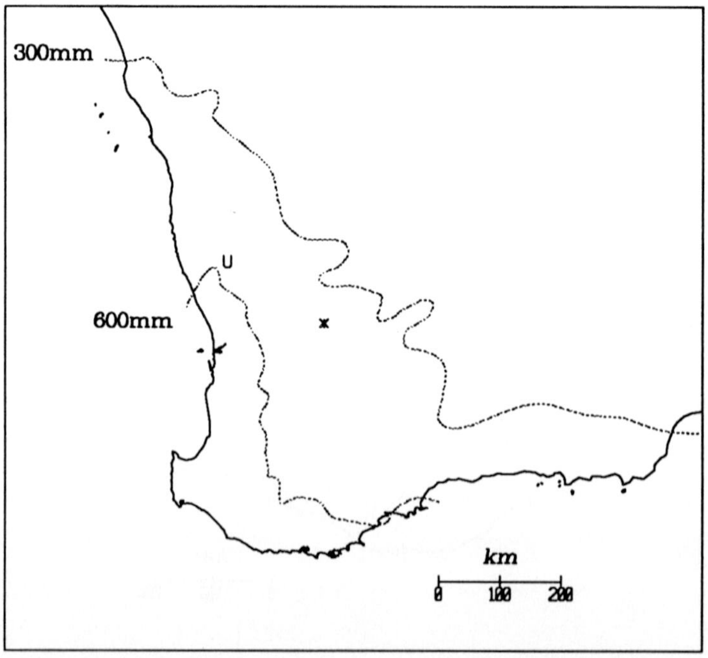

Fig. 60. Distribution 1987–1990

Little button quail

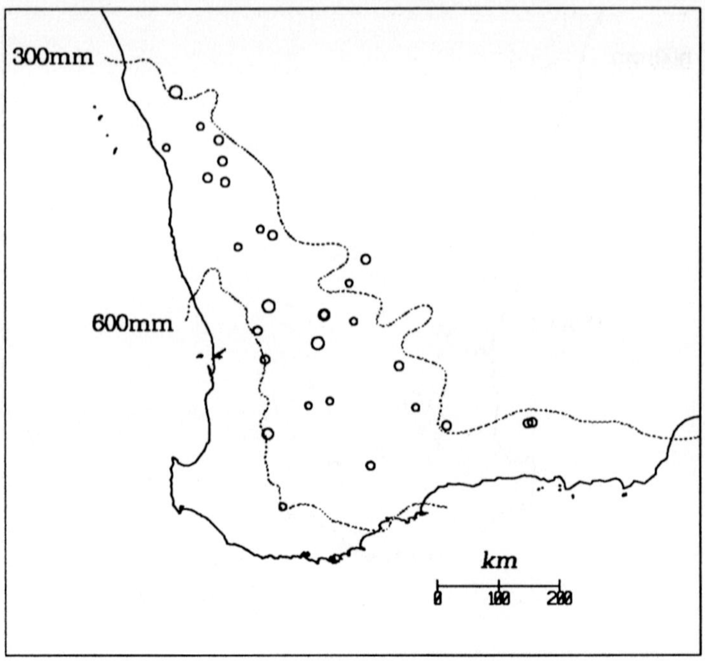

	1900/37	1987	1988	1989	1990
% of observers recording species	20	8.7	11	13	15
% of recording period species seen	U	2.1	1.5	2.2	2.2

Spotted crake

Porzana fluminea

OTHER NAME:

Australian crake

The spotted crake is a rare vagrant in aquatic habitats in the wheatbelt (Fig. 61). The status of this species is unclear as it is hard to see in the habitats it frequents. It feeds on seed, snails, earwigs, grasshoppers, bugs, beetles, flies, caterpillars, wasps, ants, ostracods and spiders.

Fig. 61. Distribution 1990

Spotted crake

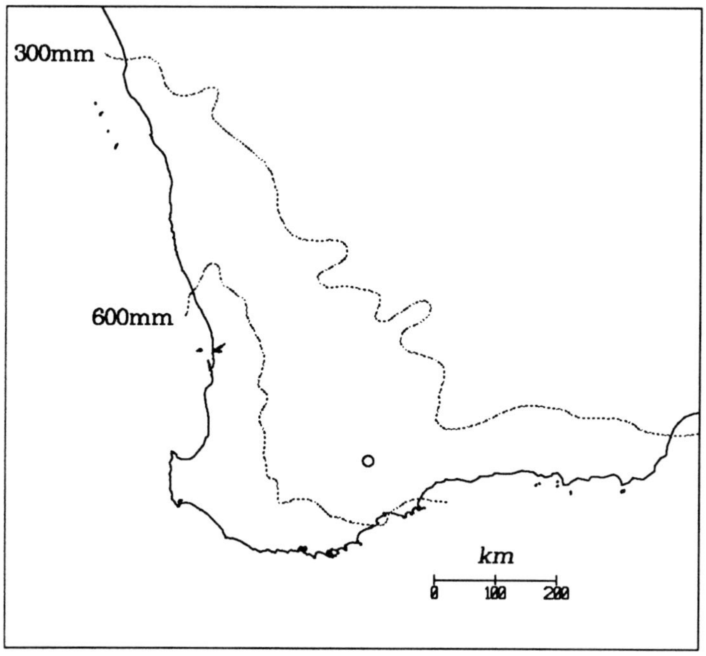

	1900/37	1987	1988	1989	1990
% of observers recording species	0	0	0	0	1.0
% of recording period species seen	NR	0	0	0	0.3

Marsh crake

Porzana pusilla

OTHER NAME:
Baillon's crake

The marsh crake is a rare vagrant on aquatic habitats in the wheatbelt (Fig. 62). The status of this species is unclear as it is hard to see in the habitats it frequents. It feeds on seed, snails, bugs, beetles and copepods.

Fig. 62. Distribution 1990

Marsh crake

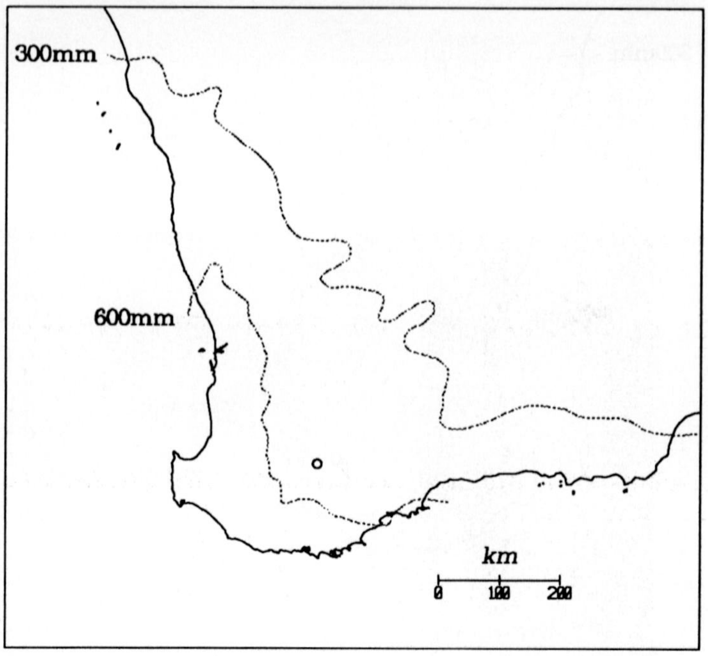

	1900/37	1987	1988	1989	1990
% of observers recording species	0	0	0	0	1.0
% of recording period species seen	NR	0	0	0	0.05

Mallee habitat. Note the lack of understorey (photos by D. A. Saunders).

Black-tailed native hen

Gallinula ventralis

The black-tailed native hen is an uncommon nomad on aquatic habitats, including farm dams, throughout the wheatbelt (Fig. 63). Numbers of this species fluctuate widely over time. It may be rare some years, yet in wet years it may be widespread and abundant, as in 1988 and 1990. It feeds on seed and other plant material, grasshoppers, crickets, beetles, moths and ants. The black-tailed native hen breeds in winter and spring on the ground and lays 5–6 eggs.

This species may arrive suddenly in large numbers if conditions are suitable, breed and depart (Serventy and Whittell 1976) as their habitat dries out and becomes unsuitable. For example, Orton and Sandland (1913) noted that the black-tailed native hen was plentiful in some seasons, as did Carter (1923), and Masters and Milhinch (1974). Ford and Stone (1957) commented that it was a visitor to the Kellerberrin/Kwolyin districts during most of the "wet" years. This species also demonstrates a seasonal pattern of occurrence being more common during the wetter periods of the year, in winter and spring (see Fig. below).

The winter and spring of 1992 was much wetter than average and the black-tailed native hen was common and widespread throughout the wheatbelt. Unfortunately this survey had ceased and no data are available except for personal observations. The wide fluctuation in numbers makes it difficult to assess if the status of the black-tailed native hen has changed over the last 50 years.

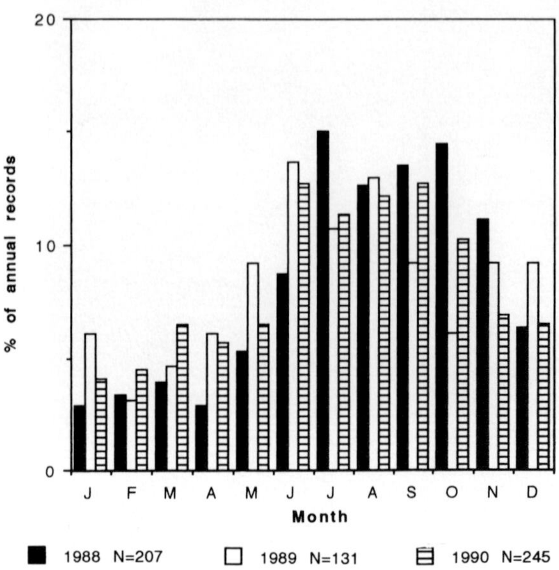

Fig. 63. Distribution 1900–1937

Black-tailed native hen

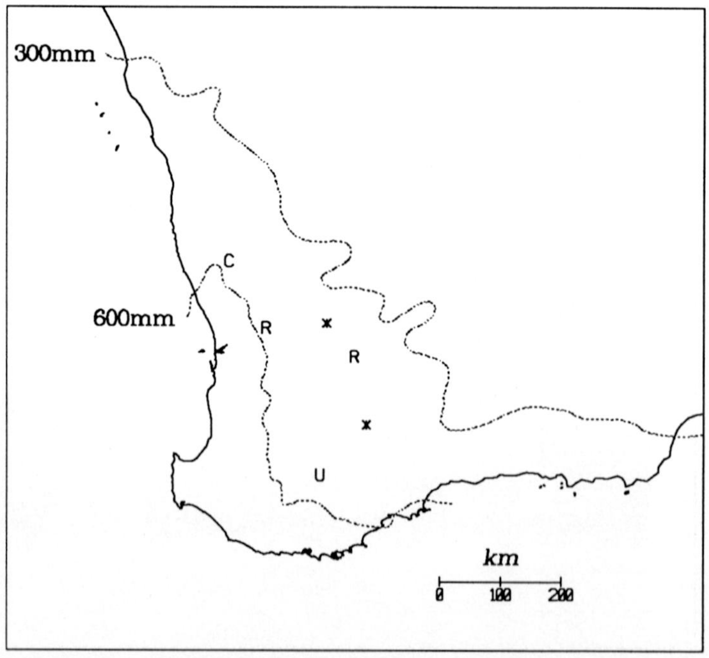

Fig. 63. Distribution 1988

Black-tailed native hen

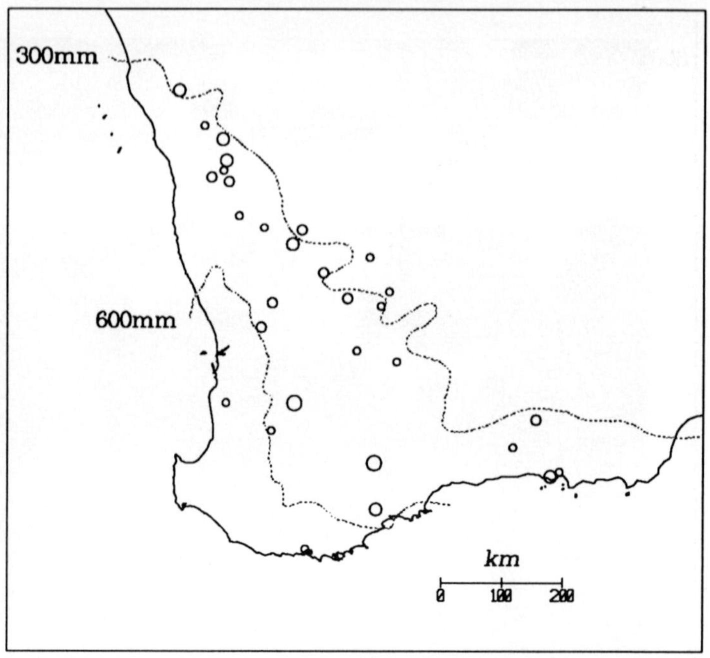

	1900/37	1987	1988	1989	1990
% of observers recording species	60	13	21	8.5	20
% of recording period species seen	R-C	2.4	4.0	2.8	5.9

Fig. 63. Distribution 1989

Black-tailed native hen

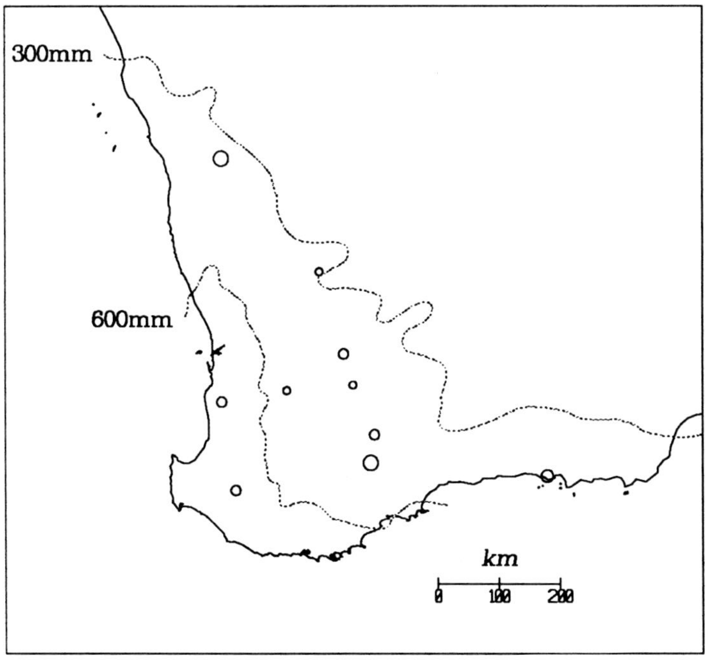

Fig. 63. Distribution 1990

Black-tailed native hen

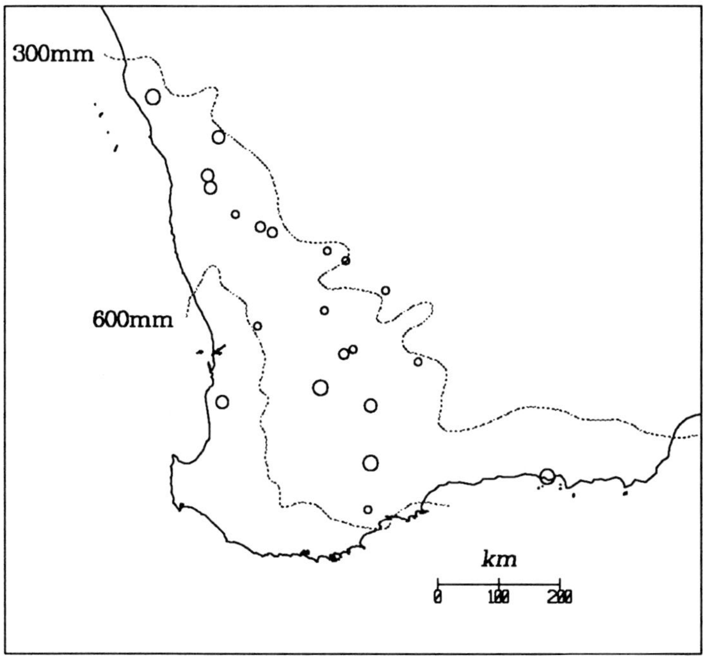

Dusky moorhen

Gallinula tenebrosa

The dusky moorhen is an uncommon nomad on aquatic habitats in the southern half of the wheatbelt (Fig. 64). It may be resident in the higher rainfall districts such as the Avon River. It feeds on seed and other plant material, freshwater mussels, crickets, grasshoppers, bugs, beetles, ants and spiders. It breeds in spring and summer, nests on the ground and lays 6–8 eggs.

There is no evidence to suggest that the status of this species has changed over the last 50 years.

Fig. 64. Distribution 1900–1937

Dusky moorhen

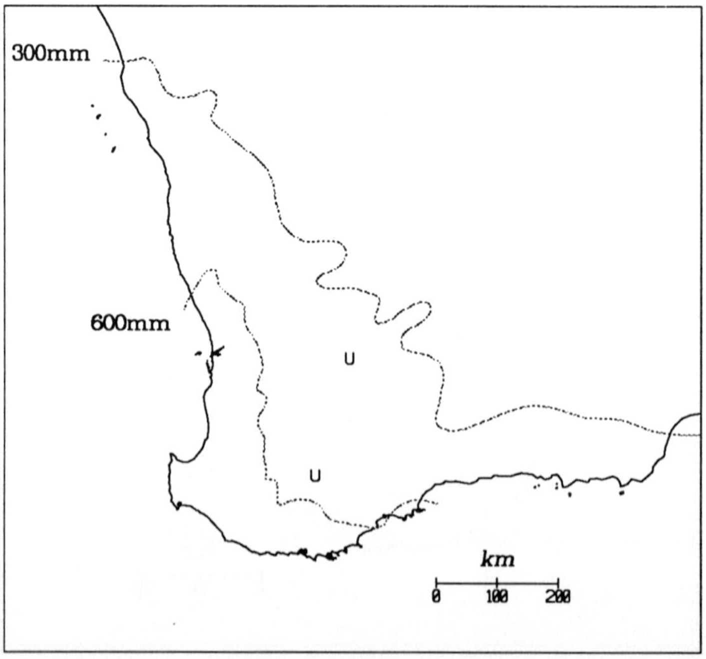

Fig. 64. Distribution 1987–1990

Dusky moorhen

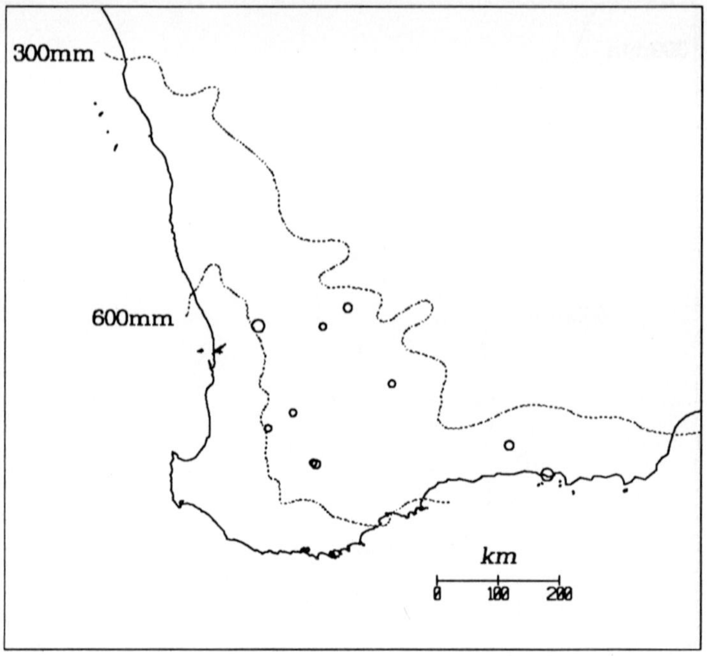

	1900/37	1987	1988	1989	1990
% of observers recording species	20	1.6	4.2	4.3	3.8
% of recording period species seen	U	1.8	1.5	1.6	1.3

Fig. 65. Distribution 1900–1937

Swamp Hen

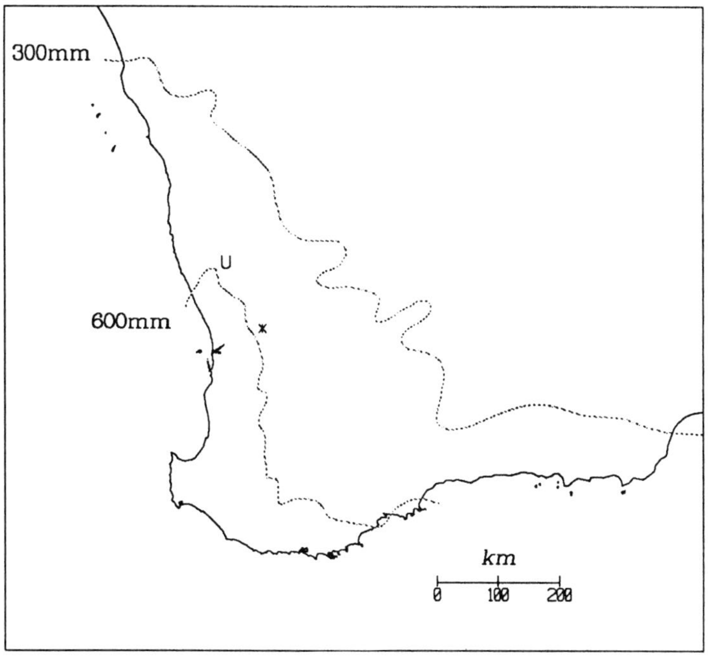

Fig. 65. Distribution 1987–1990

Swamp hen

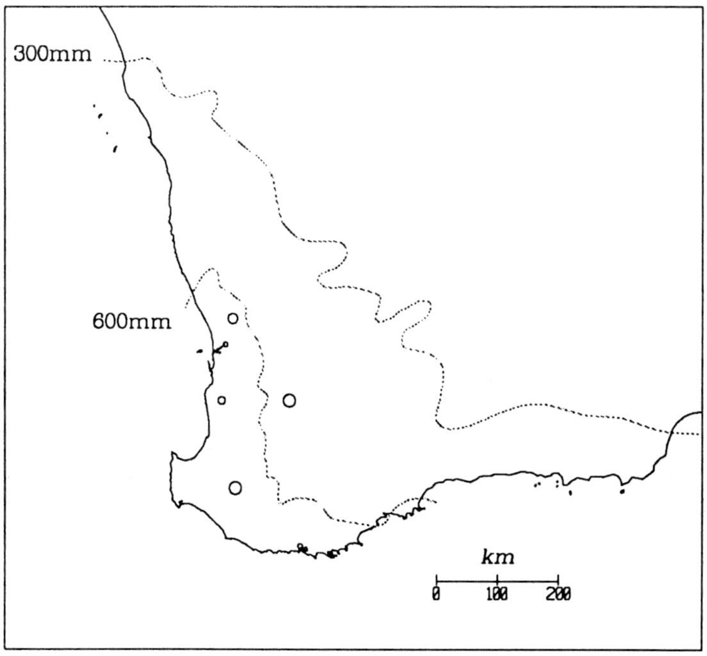

	1900/37	1987	1988	1989	1990
% of observers recording species	20	0	3.5	2.6	3.8
% of recording period species seen	U	0	0.2	0.3	1.4

Swamp hen

Porphyrio porphyrio

OTHER NAME:

Purple swamphen

The swamp hen is a rare nomad on aquatic habitats in the higher rainfall areas of the wheatbelt (Fig. 65). It feeds on seed and other plant material, grasshoppers, water boatmen, beetles, caterpillars, fish and young water birds. The swamp hen breeds in late winter and spring, nests on the ground and lays 4–6 eggs.

This species may have declined as a result of degradation of its habitat. J. Masters (in litt.) recorded that the swamp hen was resident on pools along the Avon River until the 1950s. It disappeared after a considerable rise in salinity. Around Northam, it is resident only on fresh to brackish lakes on sandplains east of the Avon Valley.

Coot

Fulica atra

OTHER NAME:
Eurasian coot

The coot is an uncommon resident on suitable aquatic habitats throughout the wheatbelt (Fig. 66). Although some birds may be resident, many may be nomadic, moving in large numbers to permanent water in the western wheatbelt as lakes dry out over summer (J. Masters, in litt.). For example, a great increase in numbers occurs over autumn on Avon River pools which may be used as a refuge until winter rains refill the ephemeral lakes. The coot feeds on plant material and breeds in spring and summer. It nests on the ground and lays 4–8 eggs.

There is no evidence to suggest that this species has changed in status over the last 50 years.

Fig. 66. Distribution 1900–1937

Coot

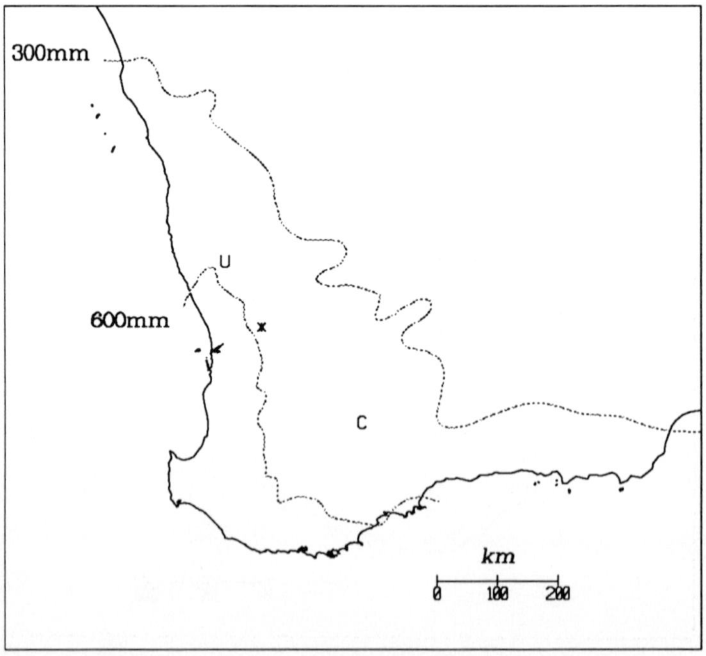

Fig. 66. Distribution 1987–1990

Coot

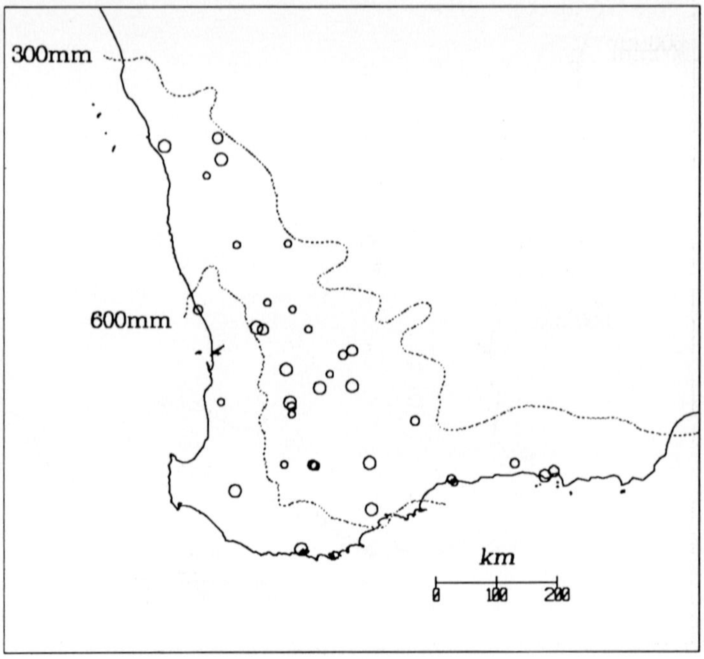

	1900/37	1987	1988	1989	1990
% of observers recording species	30	9.4	15	14	16
% of recording period species seen	U-C	3.6	6.0	6.9	9.0

Fig. 67. Distribution 1900–1937

Bustard

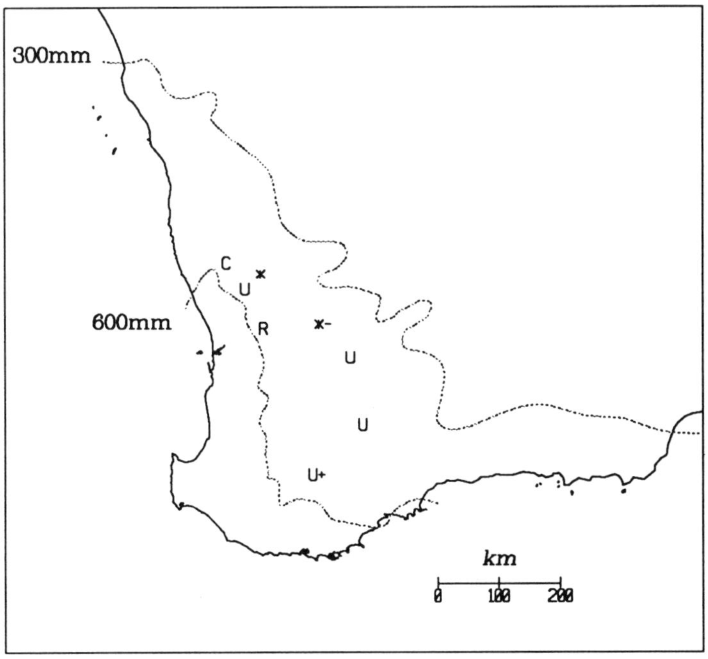

Fig. 67. Distribution 1987–1990

Bustard

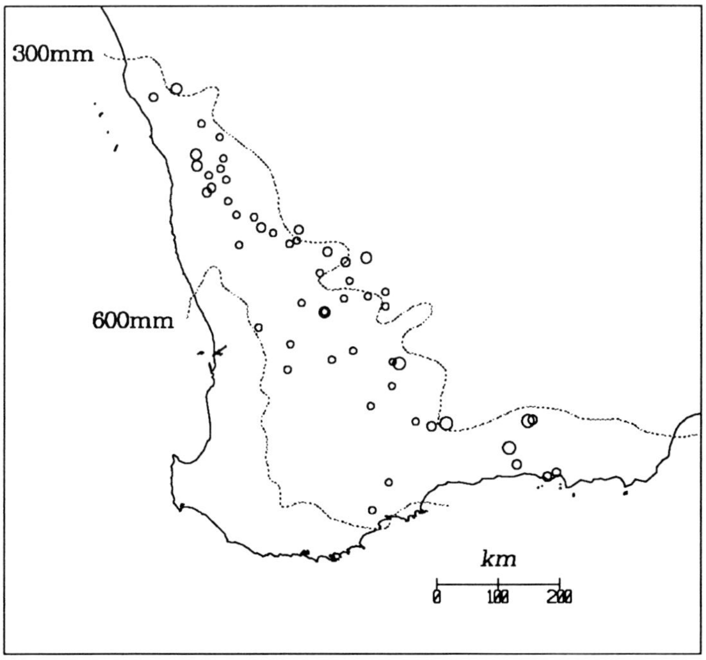

	1900/37	1987	1988	1989	1990
% of observers recording species	80	18	17	13	28
% of recording period species seen	R-C	3.3	2.9	2.9	4.1

Bustard

Ardeotis australis

OTHER NAMES:

Kori bustard, Australian bustard

The bustard is an uncommon nomad in woodland, heath and on farmland throughout the wheatbelt (Fig. 67), but not the forests of the southwest. It may be resident in the drier districts on the wheatbelt's eastern edge. It feeds on seed and other plant material, millipedes, cockroaches, mantids, earwigs, grasshoppers, crickets, phasmids, bugs, cicadas, beetles, caterpillars, wasps, ants, scorpions, spiders, lizards, small ground birds, and small mammals such as the house mouse. The bustard breeds in spring and nests on the ground. It lays one egg.

Serventy and Whittell (1976) noted that the bustard was not really rare except in well-settled districts. It favours open country and may have benefited from clearing of native vegetation (Carter 1923). However, it has declined in range and abundance. For example, the bustard was reported to be very common on the plains around the Wongan Hills (Milligan 1904), but it was not recorded during surveys of the area 70 years later (de Rebeira and de Rebeira 1977). Similarly, it occurred around Goomalling in the 1930s, but is rare there now (F. Boase, in litt.), and it has disappeared from around Merredin (B. Willis, in litt.). Predation on young birds by introduced mammals has undoubtedly been one of the reasons for its decline.

Bush stone curlew

Burhinus magnirostris

OTHER NAME:

Bush thick-knee

The bush stone curlew is an uncommon resident in wheatbelt woodlands (Fig. 68). It feeds on seed and other plant material, snails, centipedes, dragonflies, cockroaches, mantids, earwigs, grasshoppers, crickets, locusts, bugs, beetles, caterpillars, moths, wasps, sawflies, ants, crustacea, spiders, frogs and reptiles. The bush stone curlew breeds in spring and early summer. It nests on the ground and lays 2 eggs.

The bush stone curlew was regarded as common to very common before 1937, although Orton and Sandland (1913) regarded it as rare in the Moora district, noting that it had once been plentiful. In the early 1900s it was frequently heard in the Wongan Hills (Milligan 1904), but was seen there only once during four years of survey in the 1970s (de Rebeira and de Rebeira 1977). Burbidge and Rose (1993) noted that it was plentiful in the Stirling Range when Milligan visited in 1902, but it has not been recorded from there since the 1950s. The bush stone curlew was formerly common in the Northam district (Masters and Milhinch 1974), but has decreased since the 1930s and decreased alarmingly since about 1960. It is extinct north-east of Brookton (E. Fleay, in litt.).

Fig. 68. Distribution 1900–1937

Bush stone curlew

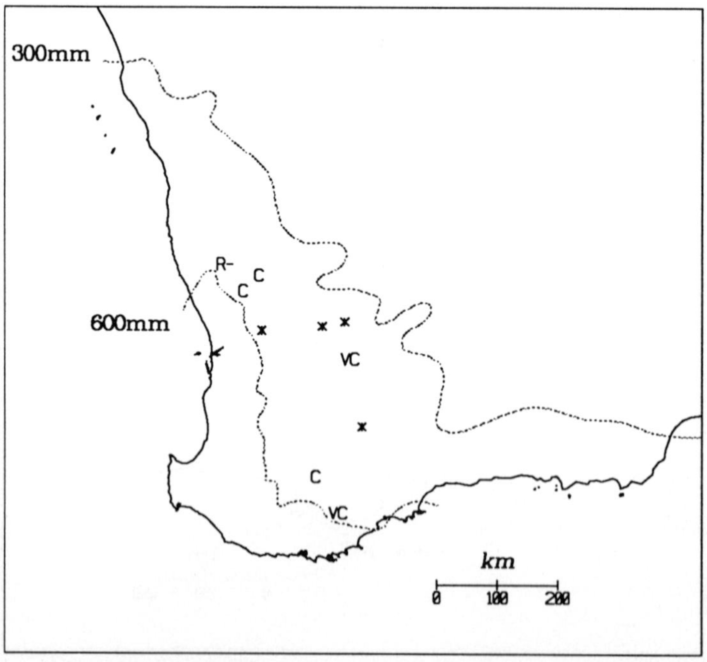

Fig. 68. Distribution 1987–1990

Bush stone curlew

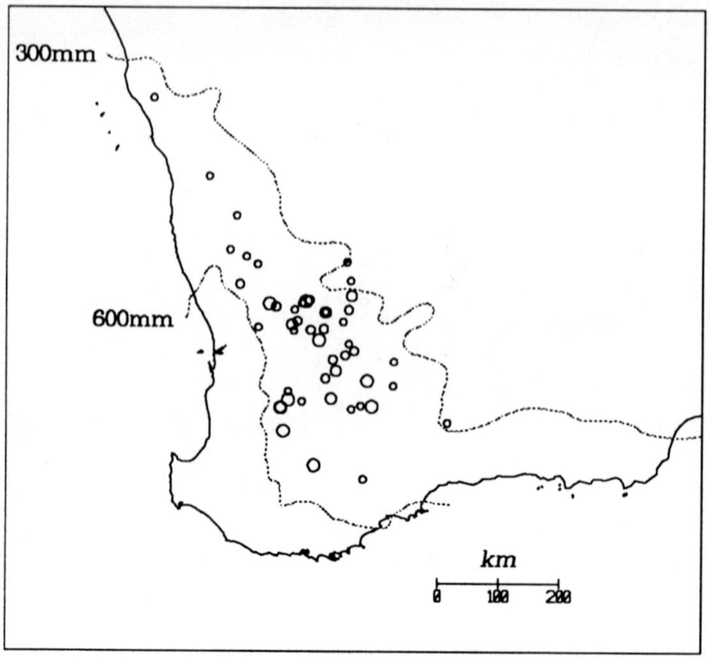

	1900/37	1987	1988	1989	1990
% of observers recording species	100	24	18	18	19
% of recording period species seen	R-VC	7.1	5.7	6.3	5.2

Banded plover

Vanellus tricolor

OTHER NAME:

Banded lapwing

The banded plover is common on farmland with short pasture throughout the wheatbelt (Fig. 69). It feeds on seed and other plant material, snails, cockroaches, earwigs, grasshoppers, crickets, locusts, bugs, beetles, moths, ants and spiders. It breeds in winter and spring, nests on the ground and lays 4 eggs.

During the survey, recording rates for the banded plover were highest during May to July (see Fig. below), indicating some movement. In the Northam district the banded plover is nomadic and most leave the district in summer, returning with the first rains (Masters and Milhinch 1974).

This species favours open areas with short vegetation, and development for agriculture has favoured it (Serventy and Whittell 1976). The banded plover colonizes suitable habitat shortly after it is created (Orton and Sandland 1913; Carter 1923; Leake 1962; Sedgwick 1974). However, Masters and Milhinch (1974) noted that it was decreasing in some areas because of the development of improved pastures with longer grass which is unsuitable habitat.

Fig. 69. Distribution 1900–1937

Banded plover

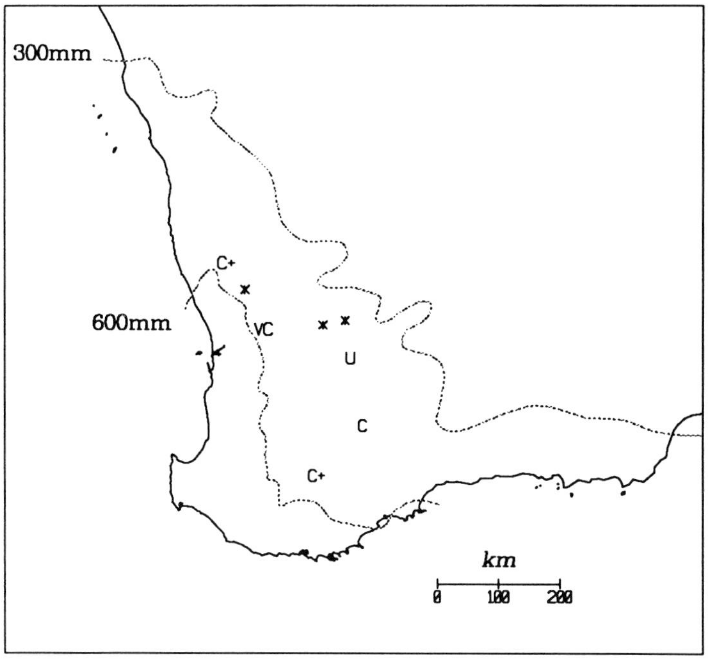

Fig. 69. Distribution 1987–1990

Banded plover

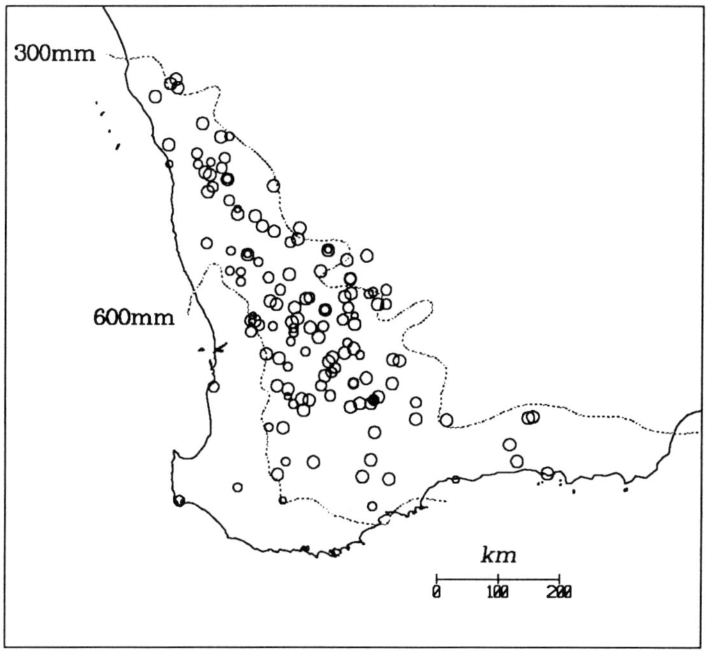

	1900/37	1987	1988	1989	1990
% of observers recording species	80	69	68	71	69
% of recording period species seen	U-VC	31	33	33	36

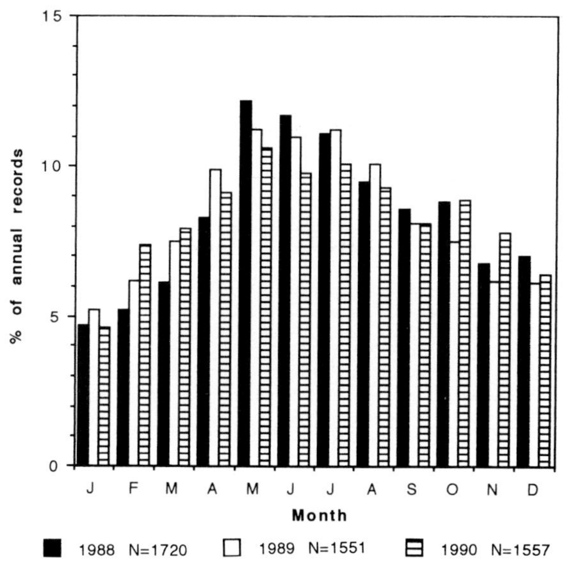

Red-kneed dotterel

Erythrogonys cinctus

OTHER NAME:

Red-kneed plover

The red-kneed dotterel is an uncommon nomad, with no seasonal pattern of occurrence, on aquatic habitats throughout the wheatbelt (Fig. 70). It feeds on seed, snails, mayfly nymphs, earwigs, bugs, water boatmen, beetles, fly larvae, ants, small crustacea and spiders. The red-kneed dotterel breeds in spring, nests on the ground and lays 4 eggs.

There is no evidence to suggest that this species has changed in distribution and abundance over the last 50 years.

Fig. 70. Distribution 1900–1937

Red-kneed dotterel

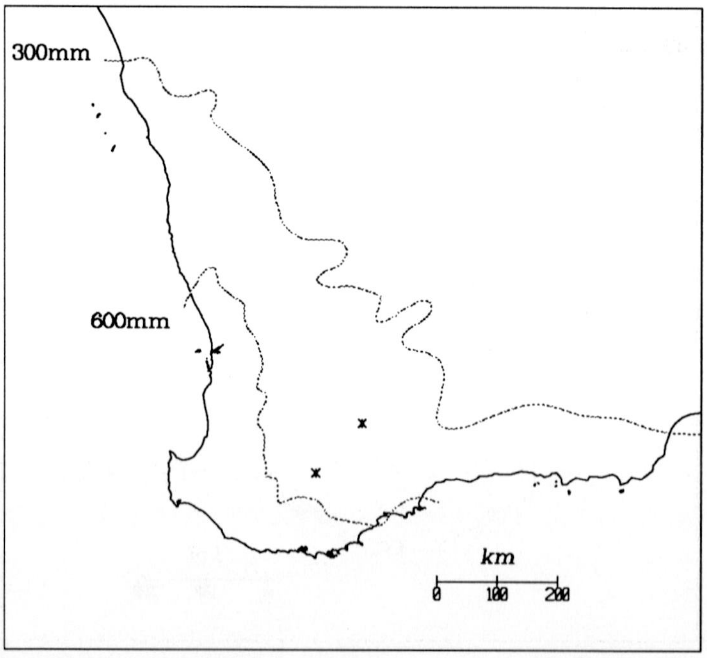

Fig. 70. Distribution 1987–1990

Red-kneed dotterel

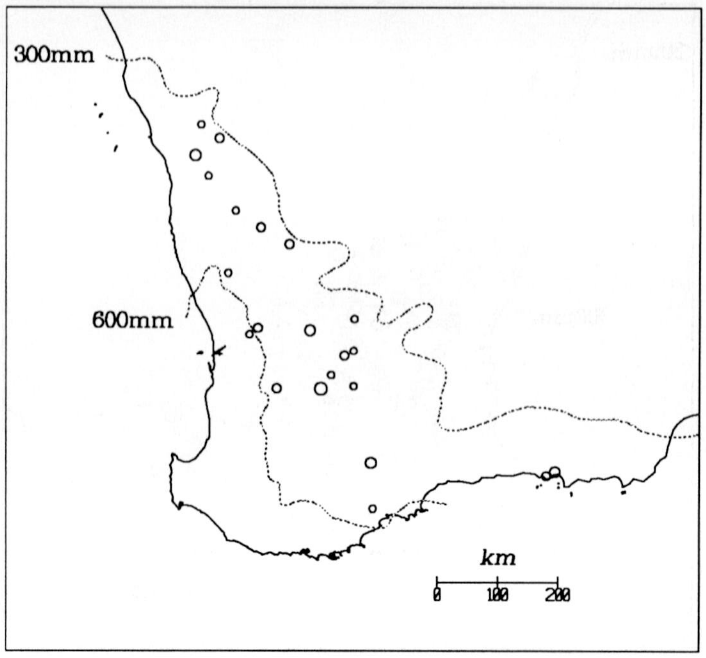

	1900/37	1987	1988	1989	1990
% of observers recording species	20	6.3	10	4.3	8.7
% of recording period species seen	—	1.2	1.1	0.9	1.9

Fig. 71. Distribution 1900–1937

Hooded dotterel

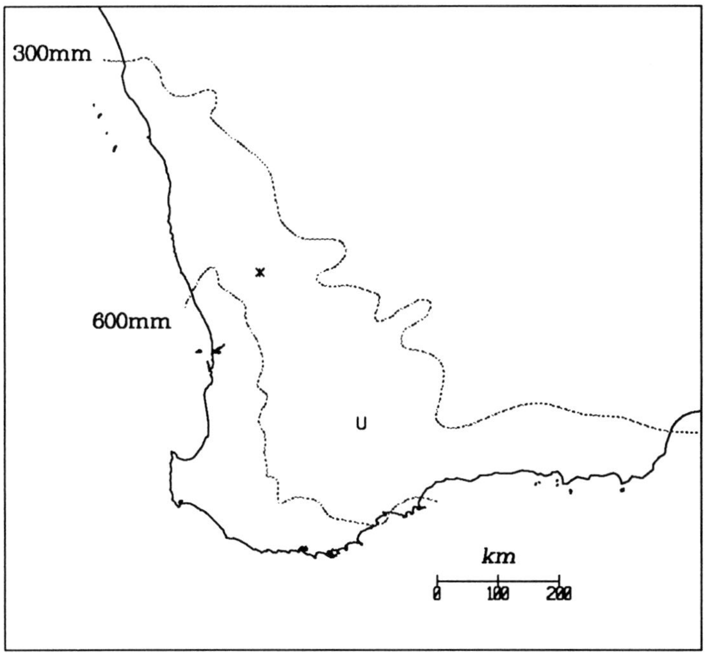

Fig. 71. Distribution 1987–1990

Hooded dotterel

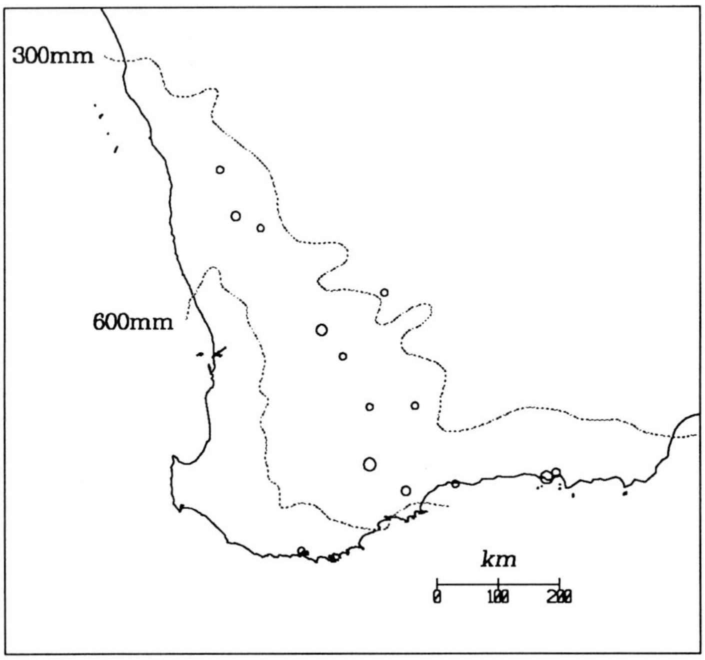

	1900/37	1987	1988	1989	1990
% of observers recording species	20	3.1	7.0	2.6	4.8
% of recording period species seen	U	0.3	1.2	1.2	1.1

Hooded dotterel

Charadrius rubricollis

OTHER NAME:

Hooded plover

The hooded dotterel is an uncommon nomad on salt lakes of the wheatbelt (Fig. 71). It feeds on molluscs, dragonflies and beetles. It breeds in spring and early summer, nests on the ground and lays 2–3 eggs.

There is no evidence to suggest that this species has changed in distribution and abundance over the last 50 years. However, Garnett (1992) listed the hooded dotterel as rare and stated that the species is possibly declining in areas where human disturbance is increasing and where stock destroy nests.

Red-capped dotterel

Charadrius ruficapillus

OTHER NAME:

Red-capped plover

The red-capped dotterel is an uncommon resident on lakes and swamps throughout the wheatbelt (Fig. 72). It feeds on plant material, snails, beetles, midges, ants and isopods. The red-capped dotterel breeds in spring and summer, nests on the ground and lays 2 eggs.

There is no evidence to suggest that this species has changed in distribution and abundance over the last 50 years.

Fig. 72. Distribution 1900–1937

Red-capped dotterel

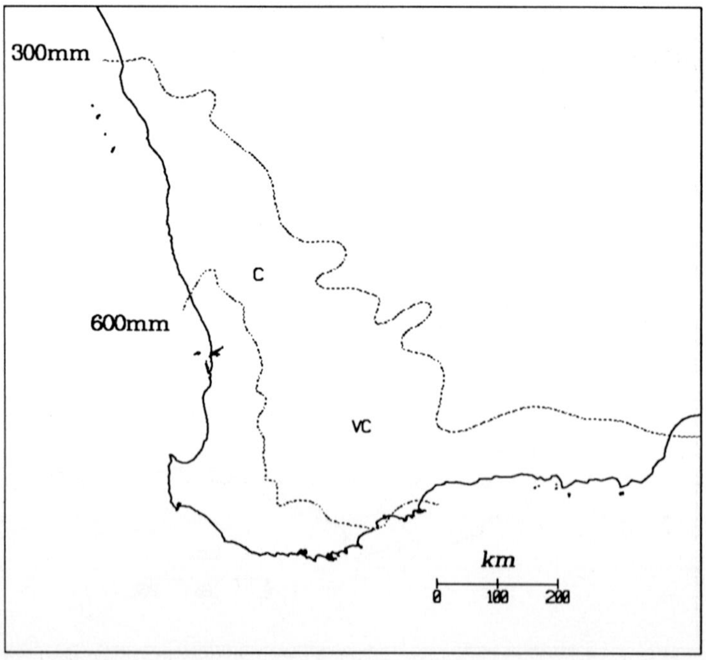

Fig. 72. Distribution 1987–1990

Red-capped dotterel

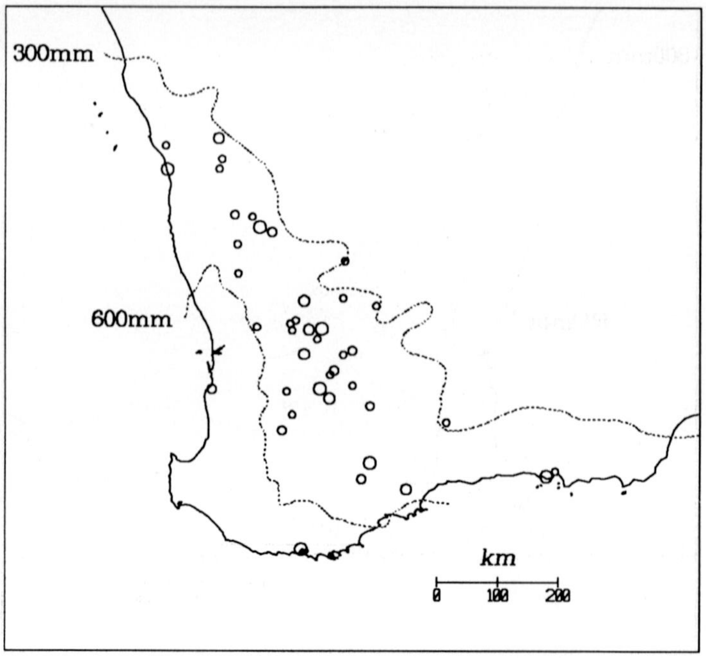

	1900/37	1987	1988	1989	1990
% of observers recording species	20	13	16	17	14
% of recording period species seen	C-VC	3.0	4.3	4.6	6.0

Fig. 73. Distribution 1900–1937

Black-fronted dotterel

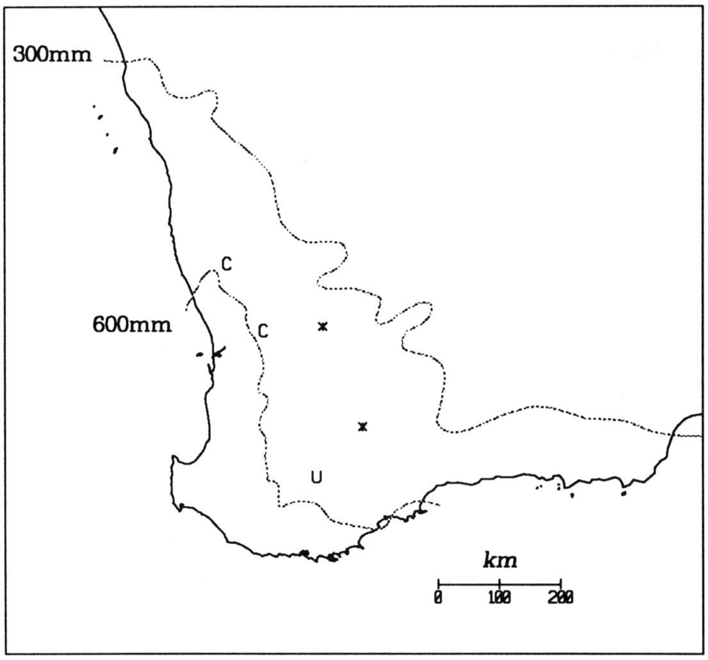

Fig. 73. Distribution 1987–1990

Black-fronted dotterel

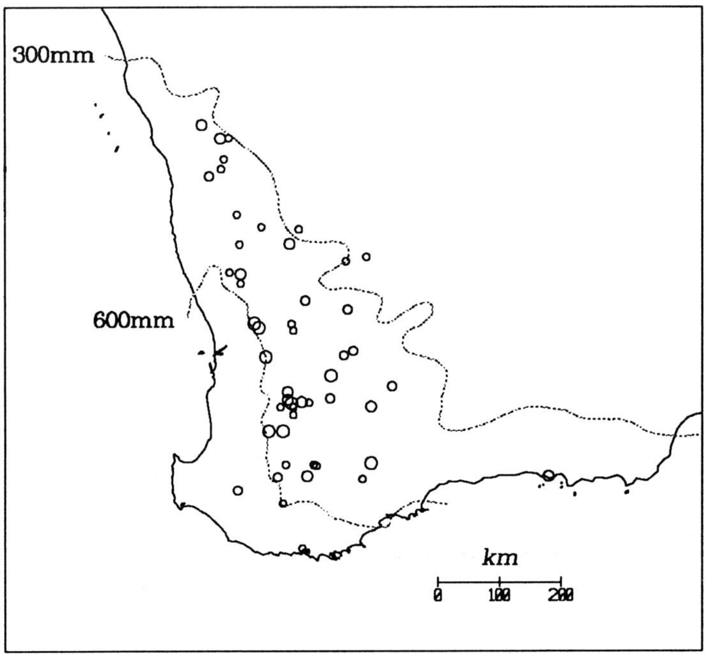

	1900/37	1987	1988	1989	1990
% of observers recording species	50	17	24	21	23
% of recording period species seen	U-C	5.5	7.1	5.1	5.0

Black-fronted dotterel

Charadrius melanops

OTHER NAME:

Black-fronted plover

The black-fronted dotterel is an uncommon resident of fresh water swamps and pools throughout the wheatbelt (Fig. 73). It feeds on seed, freshwater snails, dragonflies, earwigs, grasshoppers, water boatmen, beetles, midge and fly larvae, caterpillars, ants and crustacea. The black-fronted dotterel breeds in spring and summer, nests on the ground and lays 3 eggs.

Serventy and Whittell (1976) noted that this species was widespread and plentiful and that a pair or two were likely to be found on any fresh water swamp, pool or dam. Although still widespread, the species is not plentiful. It is possible that salination of suitable habitat has affected this species and resulted in its decline in the wheatbelt.

Australian dotterel

Peltohyas australis

OTHER NAME:
Inland dotterel

The Australian dotterel is an uncommon resident of open, bare areas including farmland in the wheatbelt (Fig. 74). It is not found in the higher rainfall districts in the southwest corner of the state. It feeds on seed and other plant material, grasshoppers, bugs, backswimmers, beetles, fly larvae, caterpillars, wasps, ants and scorpions. The Australian dotterel breeds in autumn and winter, nests on the ground and lays 3 eggs.

During the survey, this species was more commonly recorded in winter, especially in May and June (see Fig. below).

The Australian dotterel has benefited from clearing for agriculture and has colonized suitable habitat on farmland over the last 80 years (Sedgwick 1974).

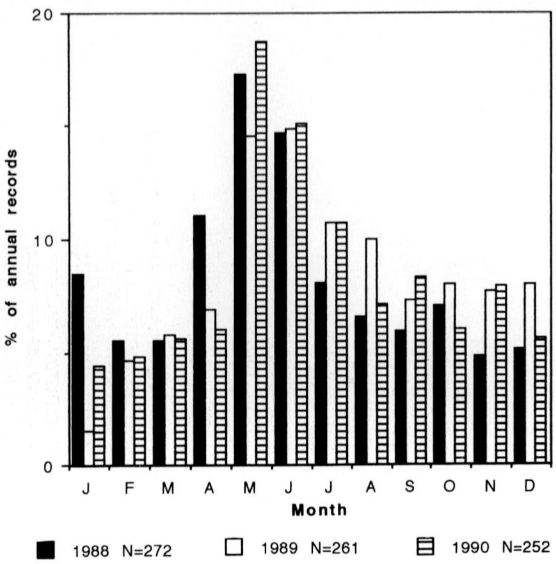

Fig. 74. Distribution 1900–1937

Australian dotterel

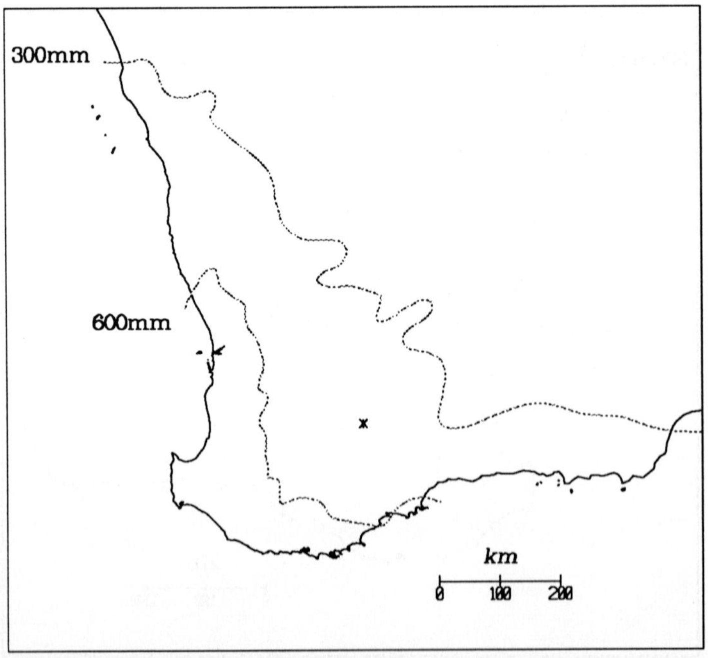

Fig. 74. Distribution 1987–1990

Australian dotterel

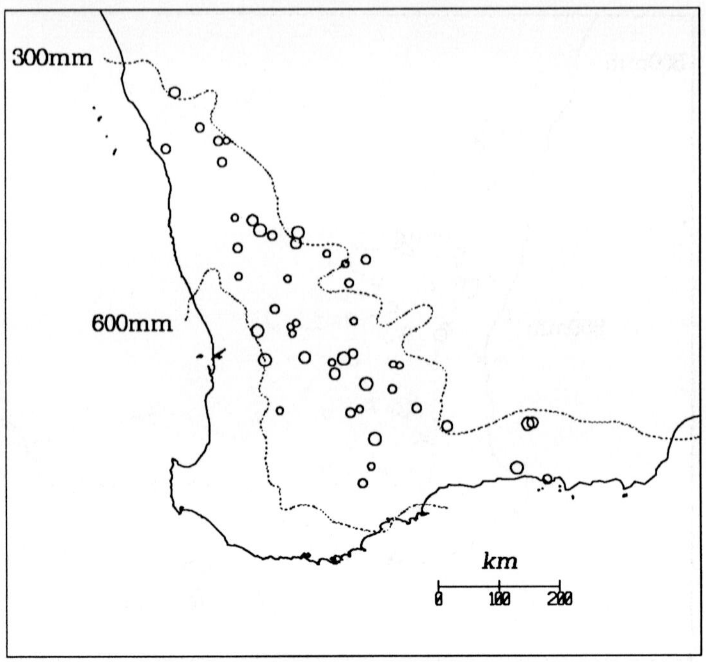

	1900/37	1987	1988	1989	1990
% of observers recording species	10	16	19	21	22
% of recording period species seen	—	3.7	5.3	5.5	5.9

Pied stilt

Himantopus himantopus

OTHER NAME:
Black-winged stilt

The pied stilt is an uncommon nomad on aquatic habitats in most of the wheatbelt (Fig. 75). It may be resident in the higher rainfall districts. It feeds on seed, snails, mayfly and dragonfly larvae, bugs, backswimmers, water boatmen, beetles, caddis fly larvae and ants. The pied stilt breeds in spring, nests on the ground and lays 4 eggs.

During the survey the pied stilt was more commonly recorded during the late winter and spring in 1988 and 1989. This pattern was not demonstrated in 1990 (see Fig. below) when a wet summer (Fig. 7) provided suitable habitat over a wider area than normal. The pied stilt has greatly increased in numbers on the Avon River since 1930 because of an increase in salinity (Masters and Milhinch 1974).

Fig. 75. Distribution 1900–1937

Pied stilt

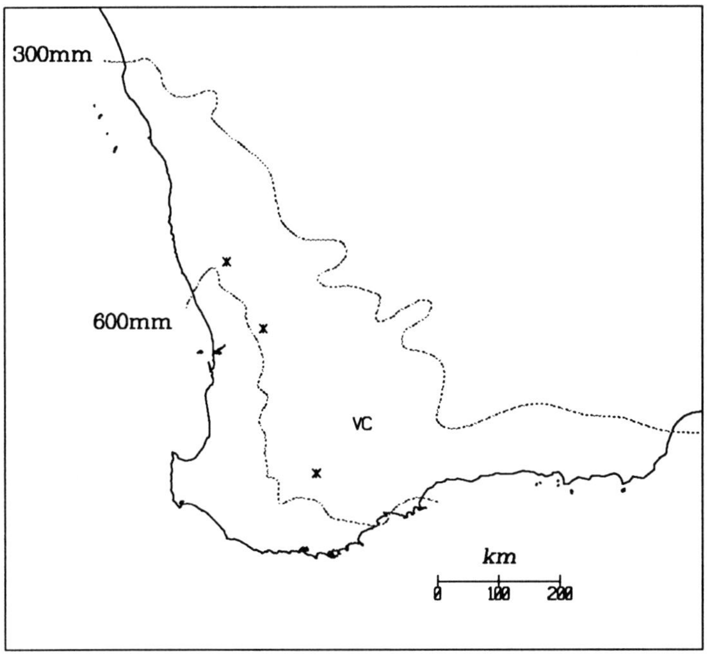

Fig. 75. Distribution 1987–1990

Pied stilt

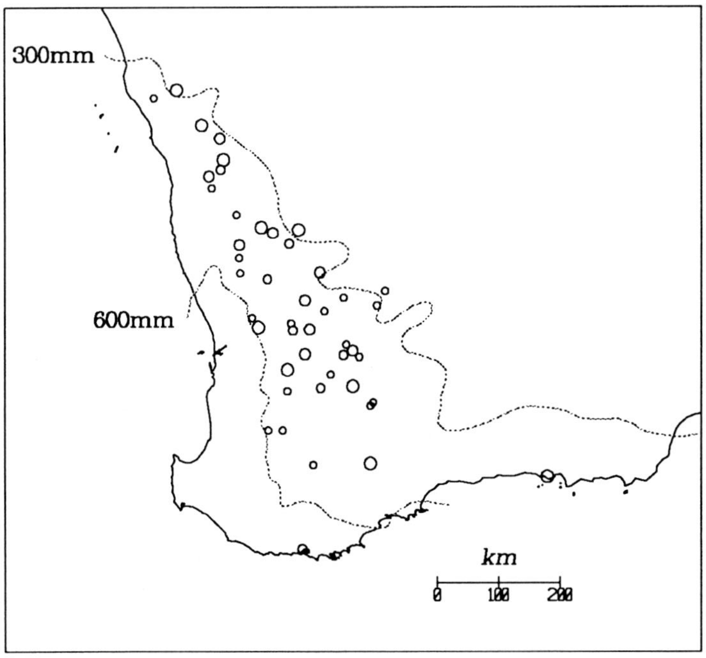

	1900/37	1987	1988	1989	1990
% of observers recording species	40	18	23	18	21
% of recording period species seen	—	5.3	7.2	6.9	5.4

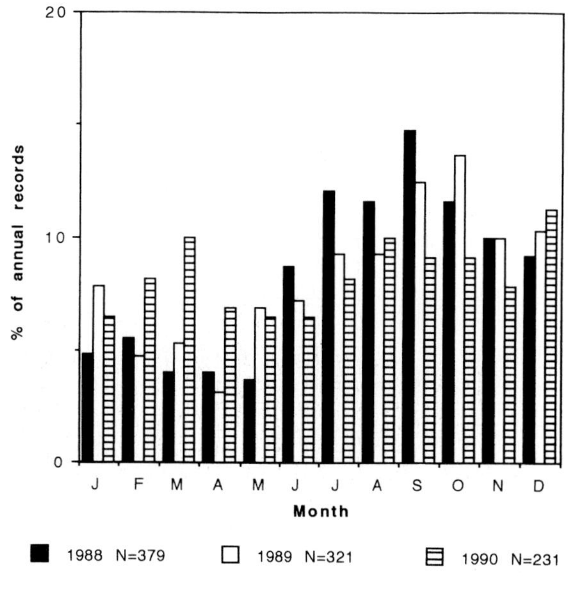

Banded stilt

Cladorhynchus leucocephalus

The banded stilt is an uncommon nomad on salt lakes throughout the wheatbelt (Fig. 76). It feeds on molluscs, water boatmen, beetles, mosquito pupae, shrimps, particularly brine shrimps, water fleas and ostracods. The banded stilt nests on the ground and lays 3–4 eggs. It only breeds in years when there has been extensive flooding of the inland salt lake systems (Serventy and Whittell 1976).

During this survey the banded stilt was more commonly recorded during the period September to April (see Fig. below), a seasonal pattern demonstrated on Rottnest Island just off the coast near Perth (Saunders and de Rebeira 1986).

There is no evidence to suggest that this species has changed in status over the last 50 years.

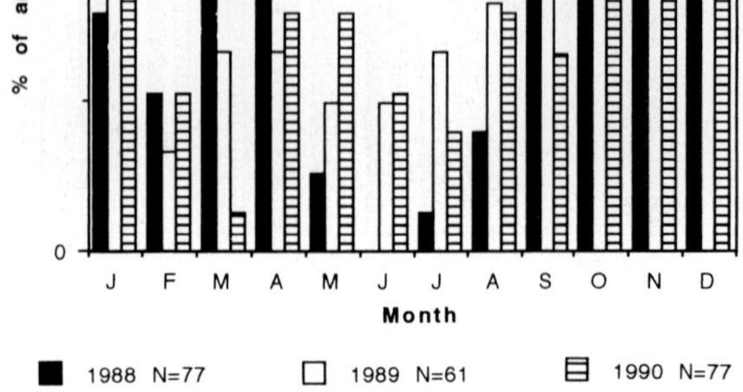

Fig. 76. Distribution 1900–1937

Banded stilt

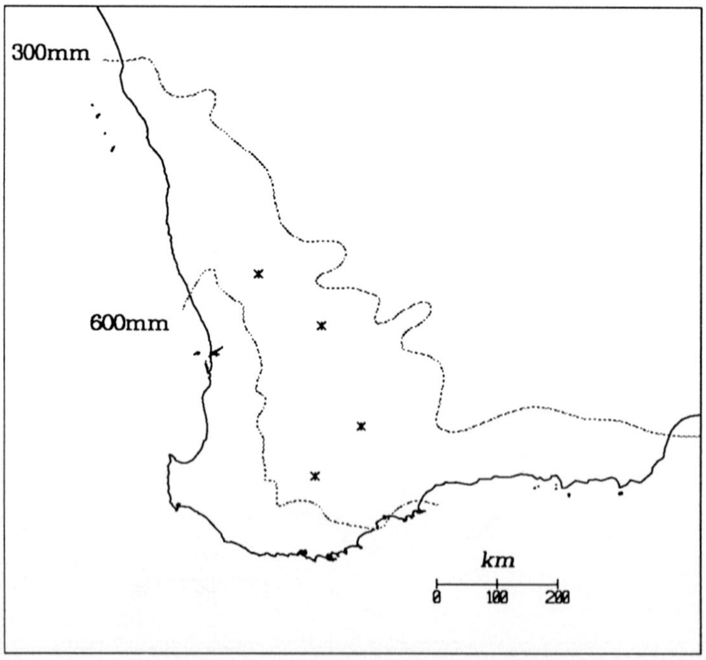

Fig. 76. Distribution 1987–1990

Banded stilt

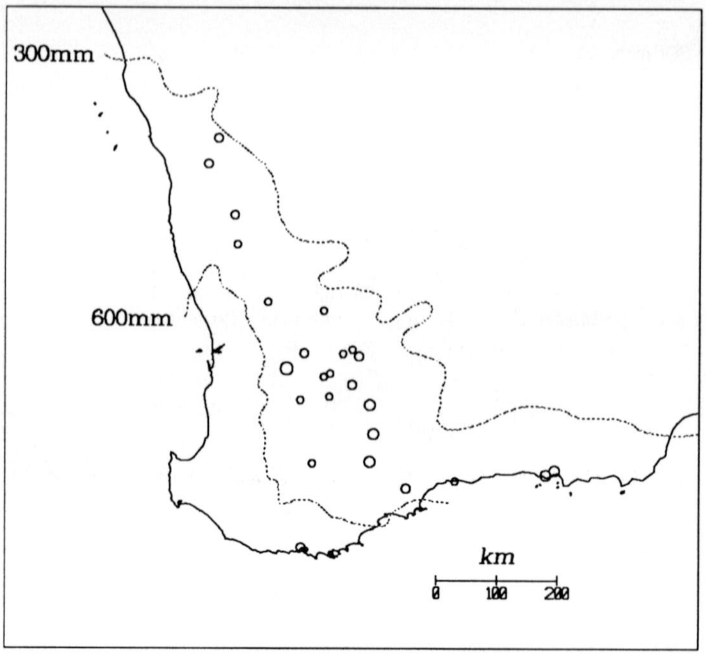

	1900/37	1987	1988	1989	1990
% of observers recording species	40	6.3	6.3	11	9.6
% of recording period species seen	—	1.7	1.5	1.3	1.8

Fig. 77. Distribution 1900–1937

Red-necked avocet

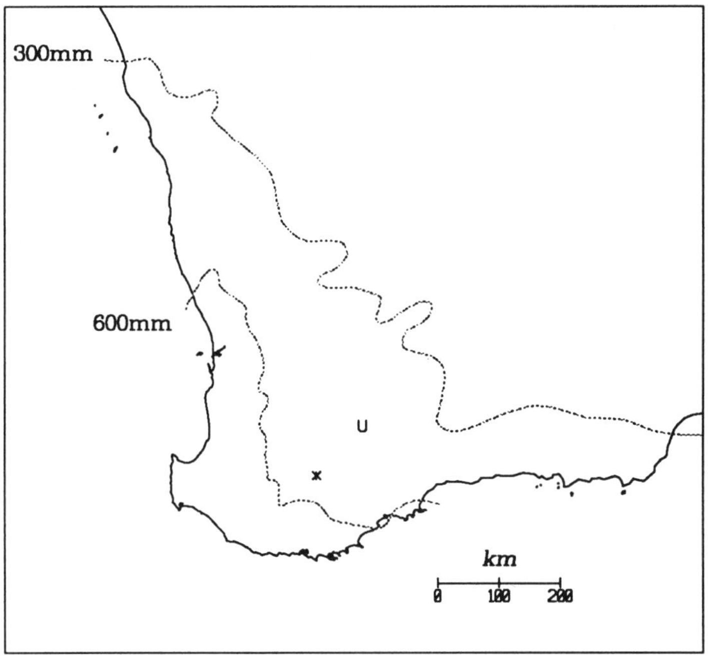

Fig. 77. Distribution 1987–1990

Red-necked avocet

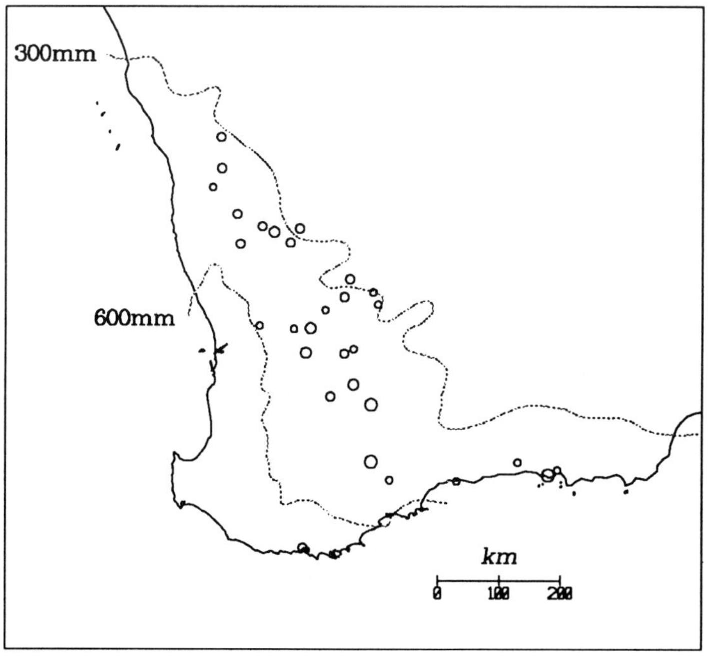

	1900/37	1987	1988	1989	1990
% of observers recording species	20	7.9	13	15	13
% of recording period species seen	U	1.3	3.4	2.7	2.2

Red-necked avocet

Recurvirostra novaehollandiae

The red-necked avocet is an uncommon nomad on salt lakes throughout the wheatbelt (Fig. 77). It may be resident in some areas. It feeds on seed and other plant material, insect larvae and beetles. It breeds in winter and spring, nests on the ground and lays 4 eggs.

The pattern of occurrence of the red-necked avocet is variable within and between years. During this survey, the avocet was more commonly recorded during 1988 in the winter and spring. However, in 1989 and 1990 it was recorded uniformly throughout the year (see Fig. below).

There is no evidence to suggest that this species has changed in status over the last 50 years.

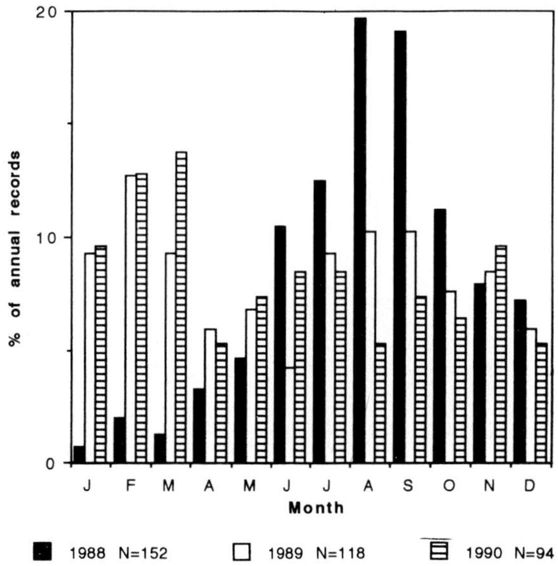

Wood sandpiper

Tringa glareola

The wood sandpiper breeds in the Northern Hemisphere and spends the northern winter in the Southern Hemisphere. It is a rare nomad on fresh water swamps and ponds of the wheatbelt (Fig. 78). It feeds on insects.

The wood sandpiper is one of the least common migratory waders visiting Australia (Readers' Digest 1976). There is no evidence to suggest that this species has changed in status over the last 50 years.

Fig. 78. Distribution 1987–1990

Wood sandpiper

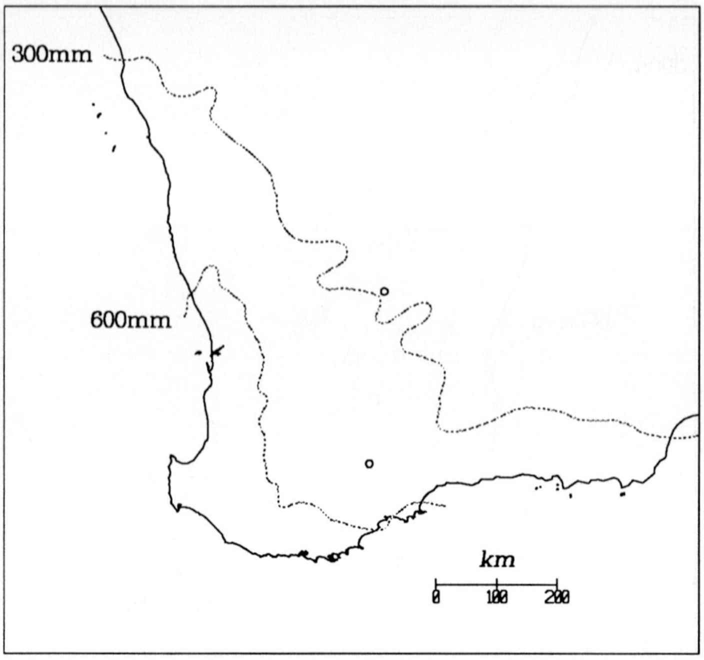

	1900/37	1987	1988	1989	1990
% of observers recording species	0	0	0.7	0.9	1.9
% of recording period species seen	NR	0	0.1	0.1	0.07

Fig. 79. Distribution 1900–1937

Greenshank

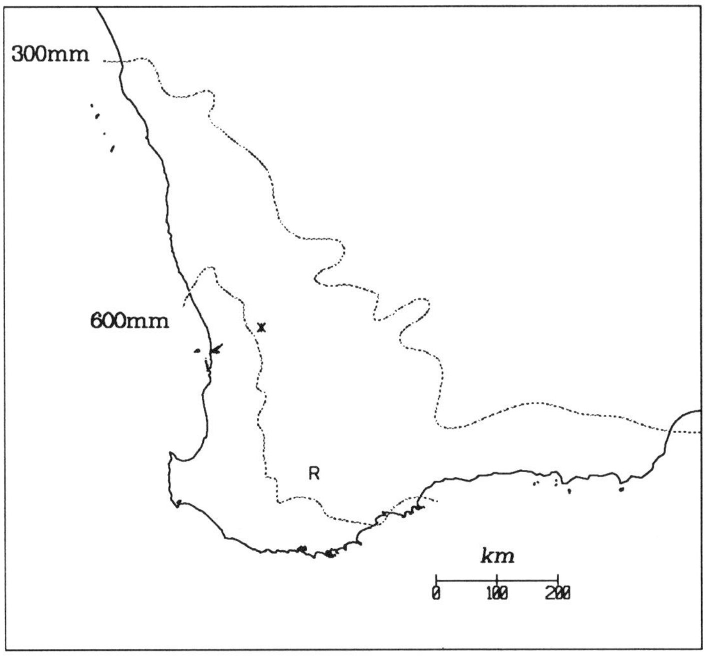

Fig. 79. Distribution 1987–1990

Greenshank

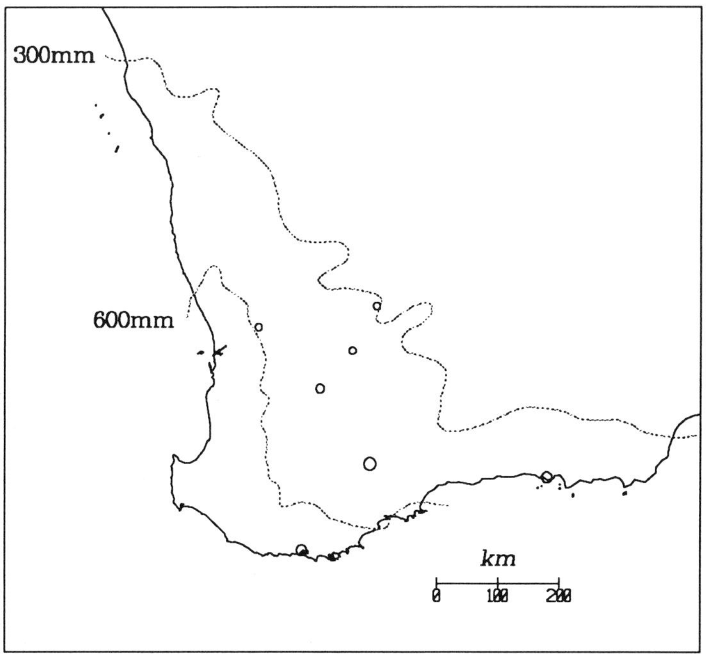

	1900/37	1987	1988	1989	1990
% of observers recording species	20	0.8	2.8	4.3	3.8
% of recording period species seen	R	0.07	1.3	1.5	1.8

Greenshank

Tringa nebularia

The greenshank breeds in the Northern Hemisphere and spends the northern winter in the Southern Hemisphere. It is an uncommon seasonal nomad on fresh water lakes, swamps and dams of the wheatbelt (Fig. 79). The greenshank feeds on mussels, larvae of damsel and dragonflies, crickets, grasshoppers, bugs, backswimmers, water boatmen, beetles, flies, caterpillars, ants, brine shrimps, water fleas, ostracods, shrimps, fish and frogs.

There is no evidence to suggest that this species has changed in status over the last 50 years.

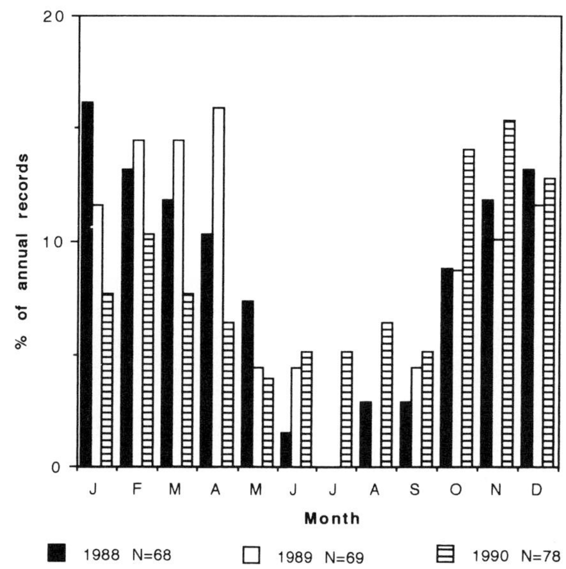

Marsh sandpiper

Tringa stagnatilis

The marsh sandpiper breeds in the Northern Hemisphere and spends the northern winter in the Southern Hemisphere. It is a rare vagrant on fresh water habitats in the wheatbelt (Fig. 80). It feeds on dragonflies, beetles and ants.

There is no evidence to suggest that this species has changed in distribution and abundance over the last 50 years.

Fig. 80. Distribution 1987–1990

Marsh sandpiper

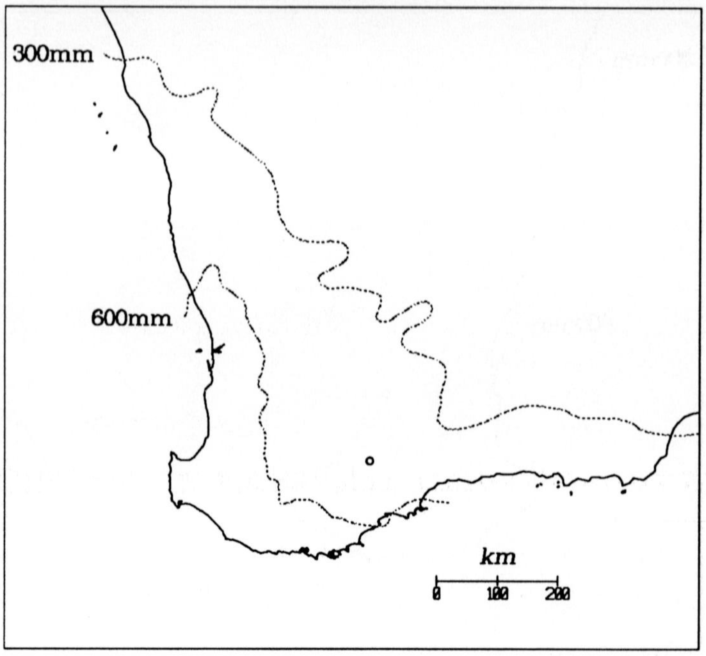

	1900/37	1987	1988	1989	1990
% of observers recording species	0	0	0	0.9	1.0
% of recording period species seen	NR	0	0	0.04	0.1

Fig. 81. Distribution 1900–1937

Common sandpiper

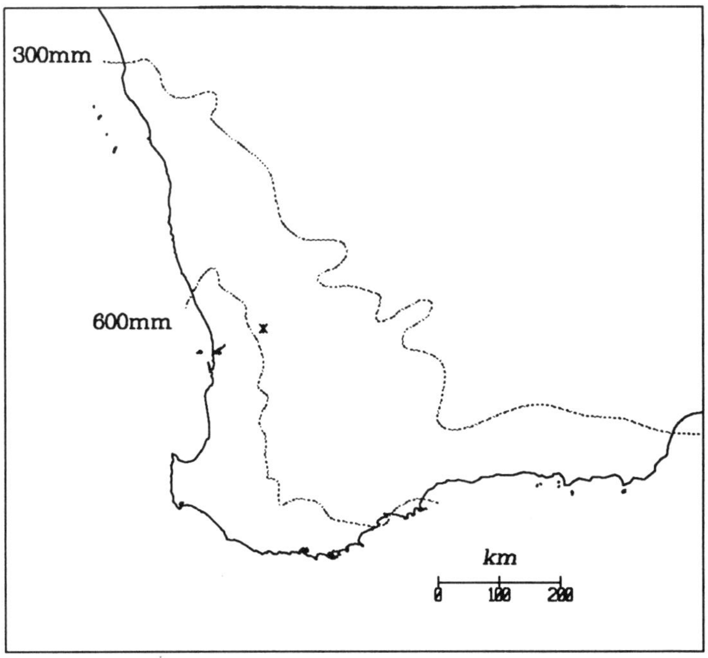

Fig. 81. Distribution 1987–1990

Common sandpiper

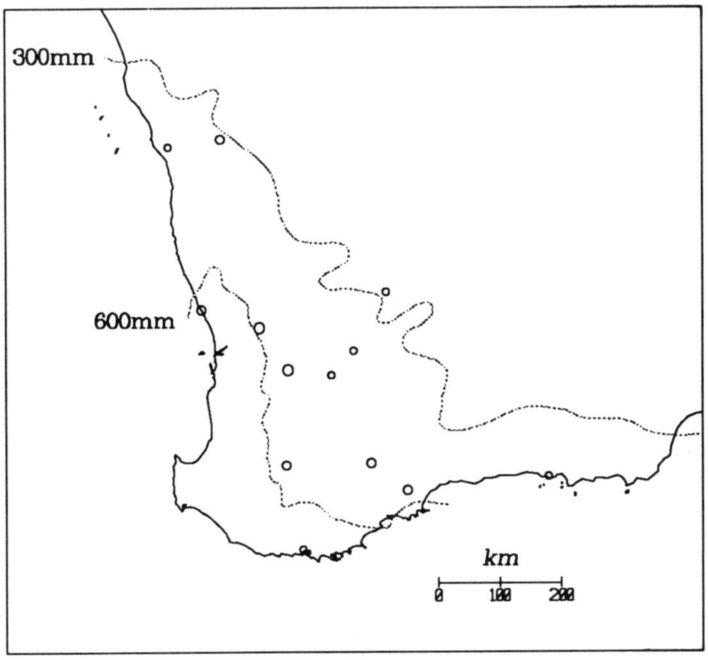

	1900/37	1987	1988	1989	1990
% of observers recording species	10	4.7	6.3	2.3	5.8
% of recording period species seen	—	1.3	1.0	0.9	0.7

Common sandpiper

Tringa hypoleucos

The common sandpiper breeds in the Northern Hemisphere and spends the northern winter in the Southern Hemisphere. It is an uncommon seasonal nomad on fresh water habitats including farm dams in the wheatbelt (Fig. 81). According to J. Masters (in litt.), the common sandpiper is the most regular transequatorial migrant visiting pools along the Avon River, with each substantial pool being visited by a lone bird. It feeds on molluscs, crickets, beetles, ants, crabs and sand hoppers.

There is no evidence to suggest that this species has changed in distribution and abundance over the last 50 years.

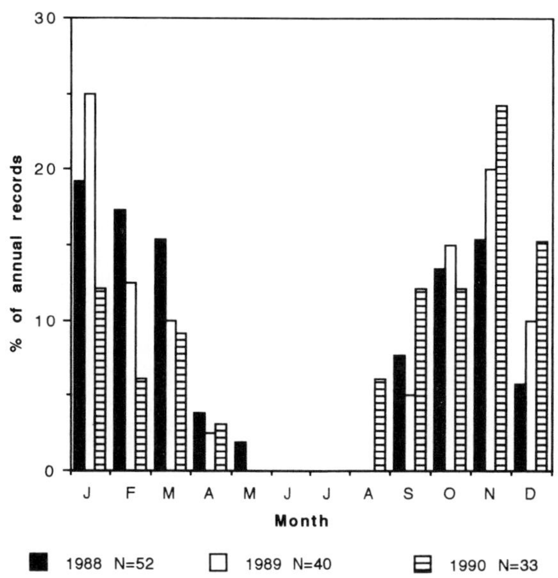

Sharp-tailed sandpiper

Calidris acuminata

The sharp-tailed sandpiper breeds in the Northern Hemisphere and spends the northern winter in the Southern Hemisphere. It is a rare seasonal nomad on fresh water habitats of the wheatbelt (Fig. 82). It feeds on seed, dragonfly larvae, earwigs, grasshoppers, bugs, beetles, fly larvae, caterpillars and ants.

There is no evidence to suggest that this species has changed in distribution and abundance over the last 50 years.

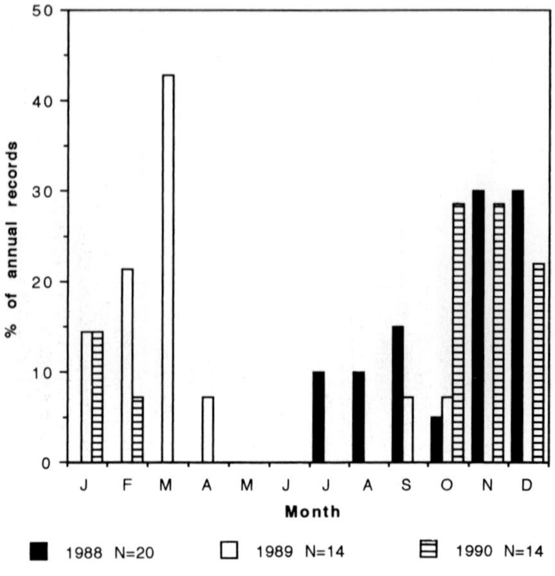

Fig. 82. Distribution 1987–1990

Sharp-tailed sandpiper

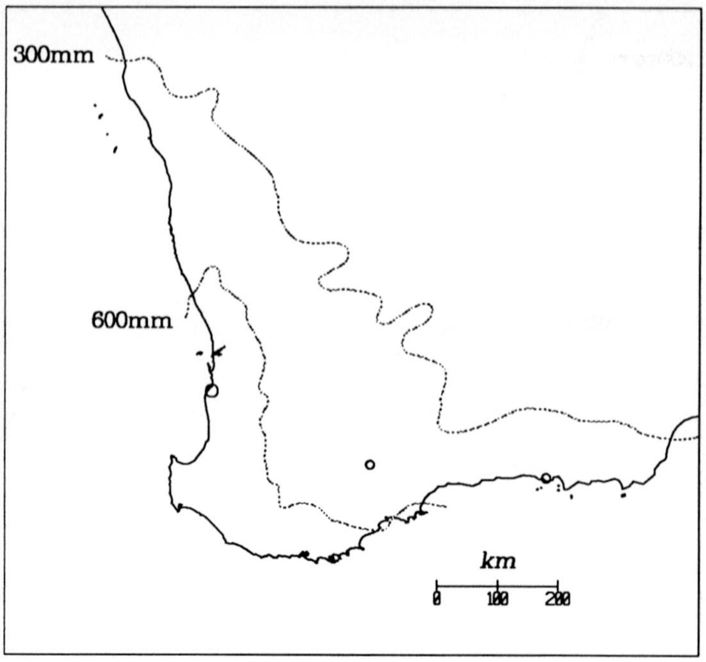

	1900/37	1987	1988	1989	1990
% of observers recording species	0	0	1.4	1.7	1.9
% of recording period species seen	NR	0	0.4	0.3	0.2

Red-necked stint

Calidris ruficollis

The red-necked stint is the most numerous small wader visiting Australia (Readers' Digest 1976). It breeds in the Northern Hemisphere and spends the northern winter in the Southern Hemisphere. It is a rare, seasonal nomad on salt and fresh water habitats of the wheatbelt (Fig. 83). It feeds on seed, gastropods, beetles, midge larvae, caterpillars, ants, crustacea, ostracods and spiders. The red-necked stint is found in the southwest from late winter through to autumn (see Fig. below) although some individuals remain over winter.

There is no evidence to suggest that this species has changed in distribution and abundance over the last 50 years.

Fig. 83. Distribution 1987–1990

Red-necked stint

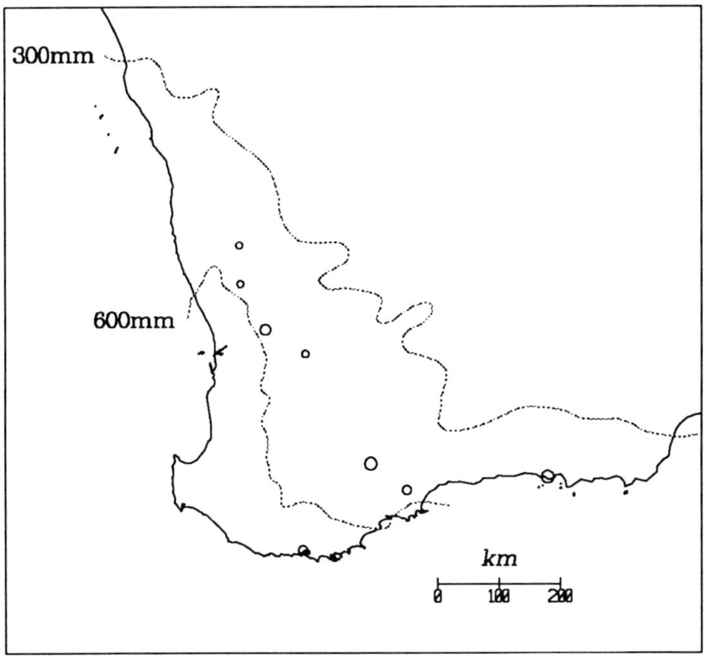

	1900/37	1987	1988	1989	1990
% of observers recording species	0	1.6	2.8	2.3	4.8
% of recording period species seen	NR	0.6	0.9	1.1	0.8

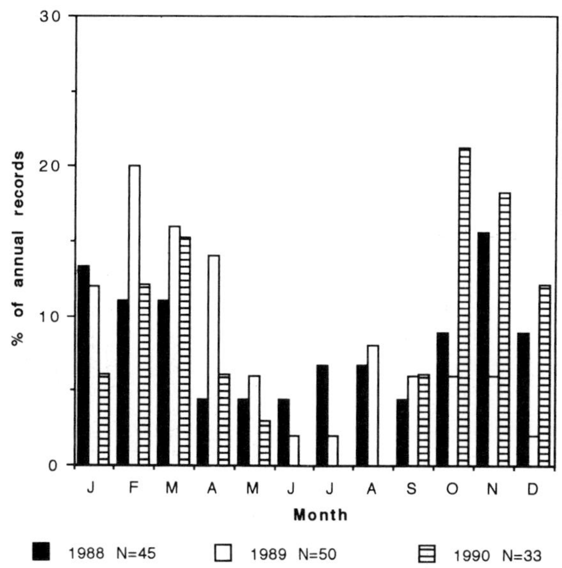

Long-toed stint

Calidris subminuta

The long-toed stint breeds in the Northern Hemisphere and spends the northern winter in the Southern Hemisphere. It is a rare vagrant on aquatic habitats of the wheatbelt (Fig. 84). It feeds on insects.

There is no evidence to suggest that this species has changed in distribution and abundance over the last 50 years.

Fig. 84. Distribution 1989

Long-toed stint

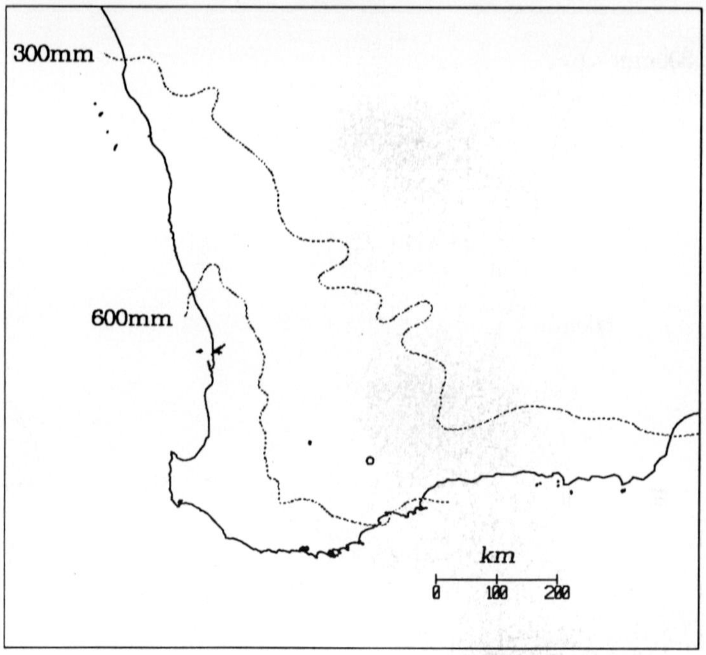

	1900/37	1987	1988	1989	1990
% of observers recording species	0	0	0	0.9	0
% of recording period species seen	NR	0	0	0.04	0

Curlew sandpiper

Calidris ferruginea

The curlew sandpiper breeds in the Northern Hemisphere and spends the northern winter in the Southern Hemisphere. It is a rare seasonal nomad on aquatic habitats in the wheatbelt (Fig. 85). It feeds on seed and other plant material, molluscs, polychaetes, beetles, and crustacea.

There is no evidence to suggest that this species has changed in distribution and abundance over the last 50 years.

Fig. 85. Distribution 1987–1990

Curlew sandpiper

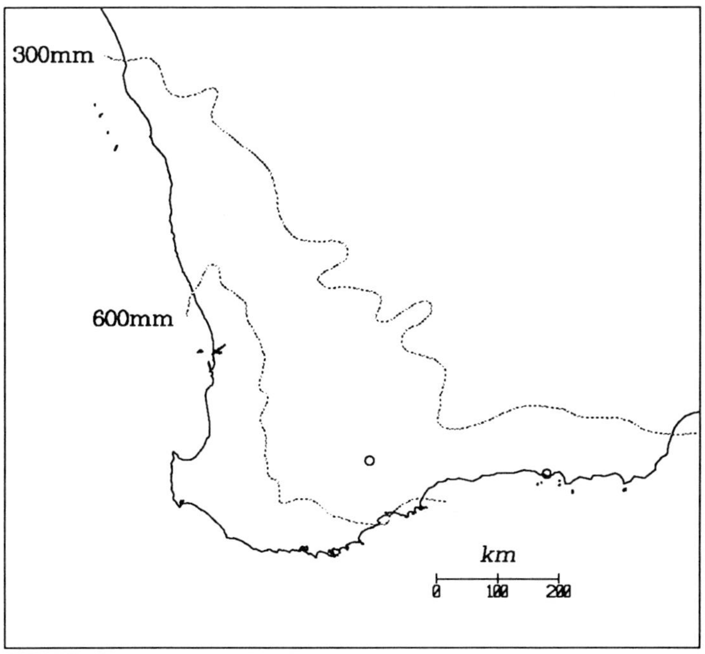

	1900/37	1987	1988	1989	1990
% of observers recording species	0	0	1.4	1.7	1.0
% of recording period species seen	NR	0	0.4	0.2	0.1

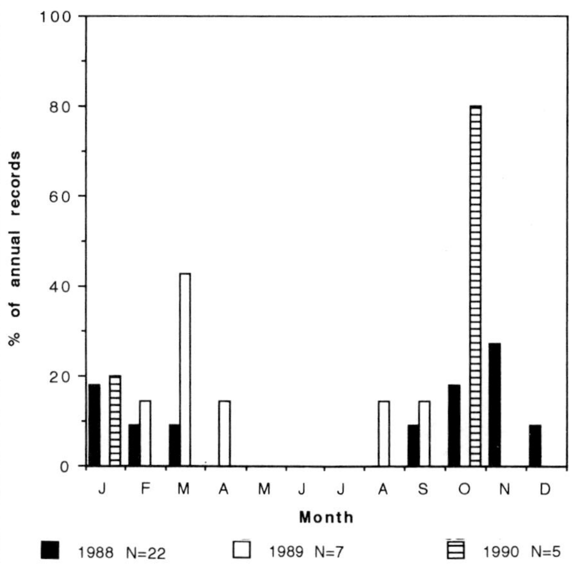

Silver gull

Larus novaehollandiae

OTHER NAME:
Seagull

The silver gull is a coastal species which has increased in range and abundance as a result of human activities. Early this century, this species was not recorded inland other than on Lake Muir (Carter 1923). Now it is an uncommon vagrant on inland water bodies throughout the wheatbelt (Fig. 86). It is a scavenger, frequenting rubbish tips and may breed opportunistically on inland lakes. It nests on the ground and lays 2–4 eggs.

Fig. 86. Distribution 1987–1990

Silver gull

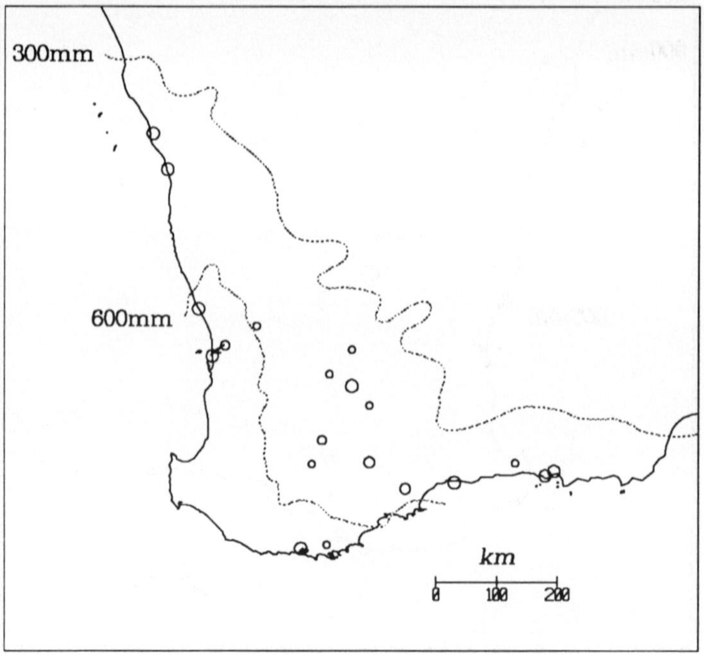

	1900/37	1987	1988	1989	1990
% of observers recording species	0	6.0	7.0	9.0	11
% of recording period species seen	NR	3.0	3.8	5.0	6.6

Whiskered tern

Chlidonias hybrida

The whiskered tern is a rare nomad on inland swamps and lakes in the wheatbelt (Fig. 87). It feeds on centipedes, damselfly larvae, earwigs, grasshoppers, water boatmen, beetles, caterpillars, ants, shrimps, freshwater crayfish, spiders, small fish, frogs and tadpoles, young birds and small mammals. It breeds in spring and summer, nests on vegetation on water and lays 3 eggs.

There is no evidence to suggest that this species has changed in distribution and abundance over the last 50 years.

Fig. 87. Distribution 1987–1990

Whiskered tern

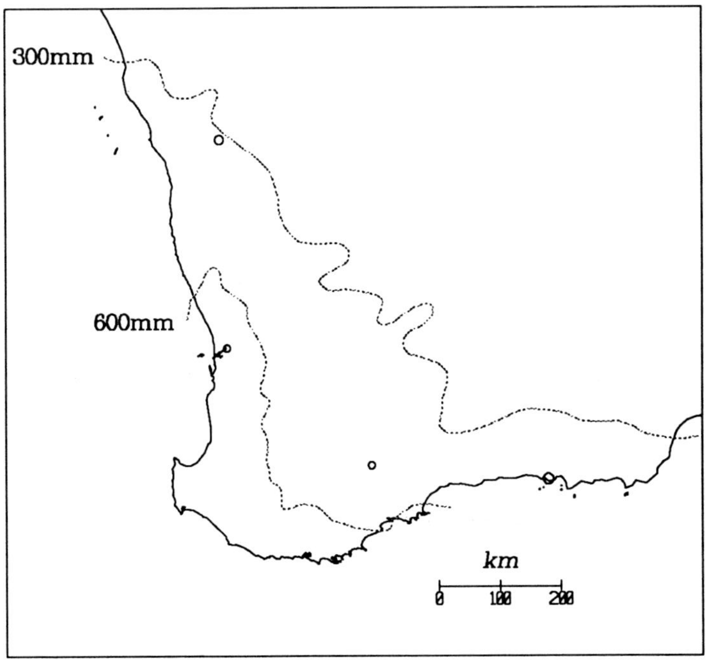

	1900/37	1987	1988	1989	1990
% of observers recording species	0	0.8	2.1	1.7	1.9
% of recording period species seen	NR	0.3	0.2	0.1	0.1

Domestic pigeon

Columba livia

OTHER NAME:

Feral pigeon

Originally brought from overseas and released in Perth, the domestic pigeon has rapidly extended its range through the wheatbelt where it is resident in towns, around farms and in woodland (Fig. 88). It feeds on seed and scraps. It may breed at any time of year and nests on a flat surface or in a tree hollow. It lays 2 eggs.

Fig. 88. Distribution 1987–1990

Domestic pigeon

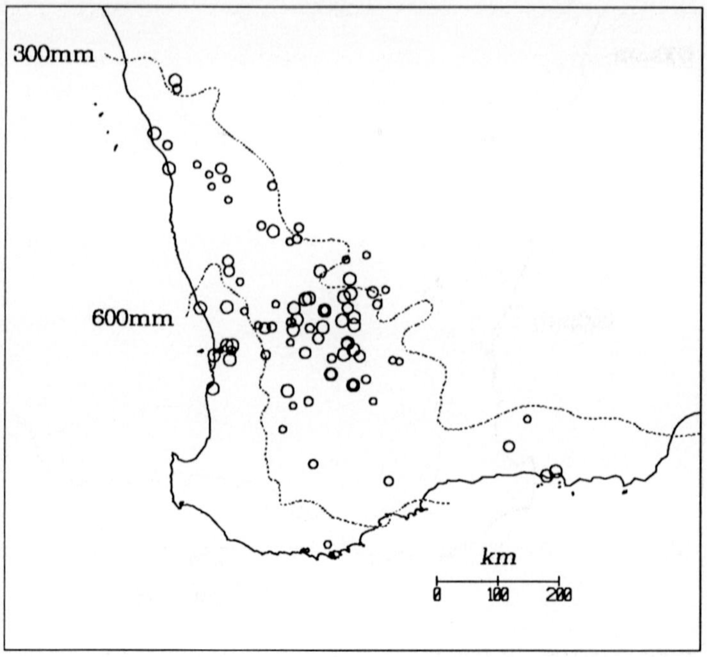

	1900/37	1987	1988	1989	1990
% of observers recording species	0	35	33	38	34
% of recording period species seen	NR	19	14	16	18

Fig. 89. Distribution 1900–1937

Laughing turtledove

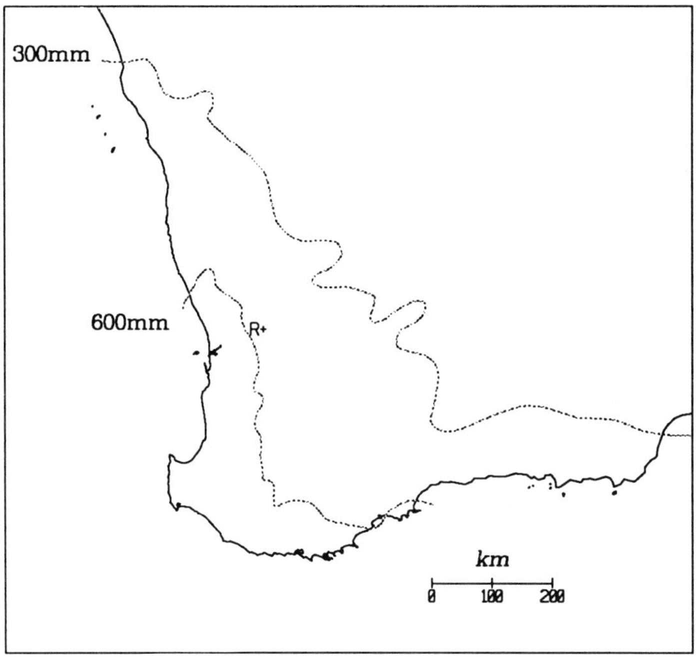

Fig. 89. Distribution 1987–1990

Laughing turtledove

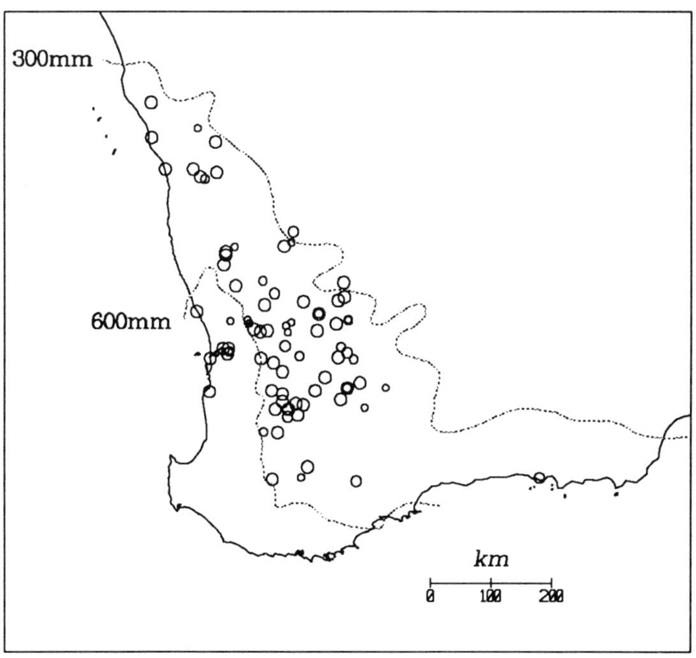

	1900/37	1987	1988	1989	1990
% of observers recording species	10	37	36	40	38
% of recording period species seen	R	22	23	25	29

Laughing turtledove

Streptopelia senegalensis

OTHER NAMES:

Senegal dove, laughing dove

The laughing turtledove is resident in towns, around farms and in woodland adjacent to human settlement throughout the wheatbelt (Fig. 89). It feeds on seed, particularly spilt grain, and fruit. It breeds in spring and early summer on a flimsy stick platform in a tree or shrub. It lays 2 eggs.

Introduced to Perth from overseas, this species has spread along railway lines and roads. Recorded in Northam before 1937 (Jenkins 1931), it has spread rapidly, and will continue to expand its range until it occupies all suitable habitat.

Diamond dove

Geopelia cuneata

The diamond dove is a bird of the arid zone. Occasional increases in abundances see this species occur as an uncommon nomad in woodland in the northern and central wheatbelt (Fig. 90) (Masters and Milhinch 1974). It feeds on seed, breeds in spring, nests on a flimsy stick nest in a tree and lays 2 eggs.

Fig. 90. Distribution 1900–1937

Diamond dove

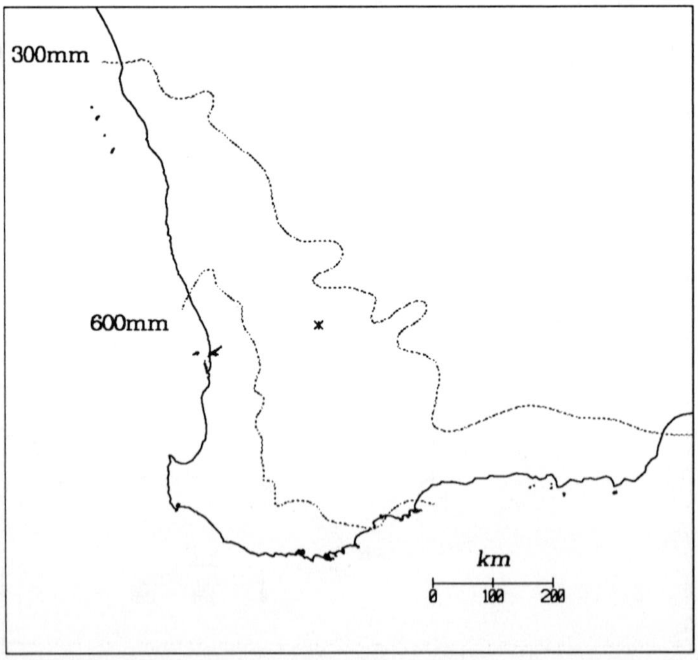

Fig. 90. Distribution 1987–1990

Diamond dove

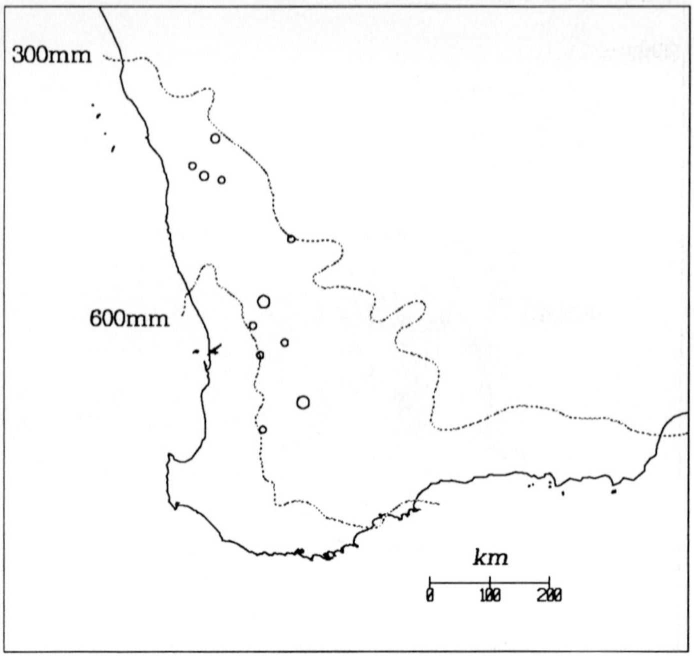

	1900/37	1987	1988	1989	1990
% of observers recording species	10	3.9	4.2	2.6	2.9
% of recording period species seen	—	1.2	1.1	1.8	0.9

Fig. 91. Distribution 1900–1937

Brush bronzewing

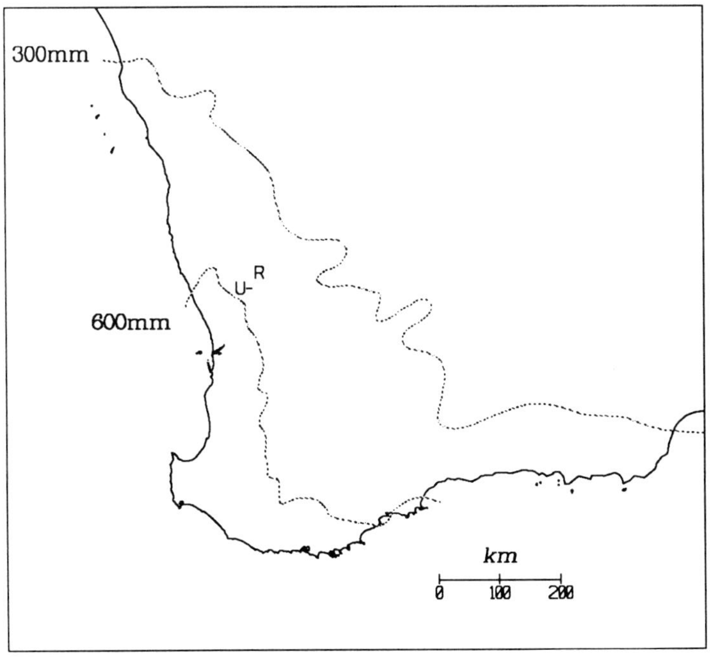

Fig. 91. Distribution 1987–1990

Brush bronzewing

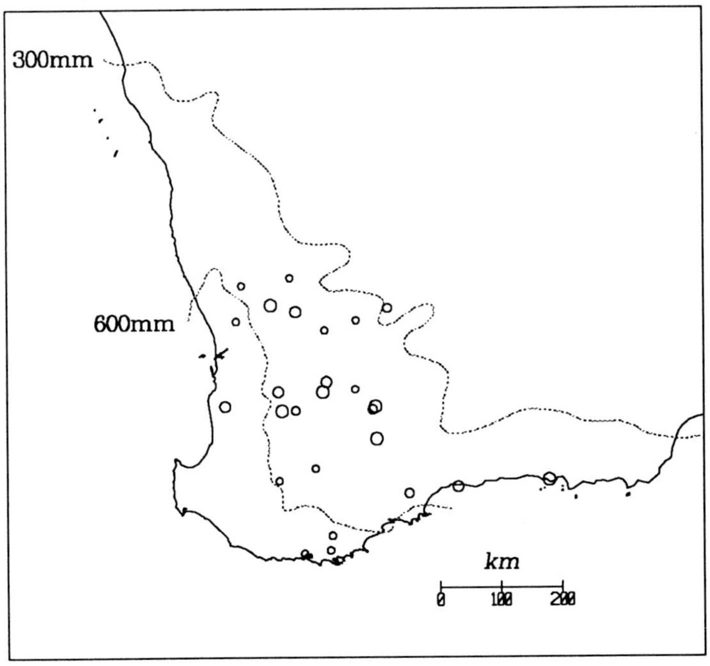

	1900/37	1987	1988	1989	1990
% of observers recording species	20	10	10	9.4	6.7
% of recording period species seen	R-U	3.2	3.7	2.4	2.1

Brush bronzewing

Phaps elegans

The brush bronzewing is an uncommon resident of shrublands and heath of the lower two-thirds of the wheatbelt (Fig. 91). It feeds on seed and breeds in spring and early summer. It builds a stick nest low in a tree or shrub and lays 2 eggs.

Serventy and Whittell (1976) stated that the brush bronzewing had seriously declined in numbers this century, blaming predation by foxes and feral cats on nests because they are constructed on or near the ground. It was formerly a common species around Mogumber, but its disappearance was attributed to the number of domestic cats gone wild (Lawson 1905). J. Masters (in litt.) believes that it is extinct along the Avon River. Clearing for agriculture has also been given as a factor in the decline of the brush bronzewing (Readers' Digest 1976).

Common bronzewing

Phaps chalcoptera

The common bronzewing is a common resident of woodland, mallee and shrublands throughout the wheatbelt (Fig. 92). It feeds on seed and fruit, snails, bugs and mosquitoes. It may breed at any time of year and nests on a stick nest in a shrub or tree. It lays 2 eggs.

The common bronzewing cycles in abundance (Serventy and Whittell 1976). It eats the seed of cereal crops, which in some areas makes up much of its food (Readers' Digest 1976) and so it has been able to cope with the fragmentation of its habitat, although overall it may have decreased in numbers. For example, it was plentiful early this century in the Moora district, but became scarce because of advancing settlement (Orton and Sandland 1913). Similarly, in the Northam district, numbers have decreased since about 1945 (Masters and Milhinch 1974).

Fig. 92. Distribution 1900–1937

Common bronzewing

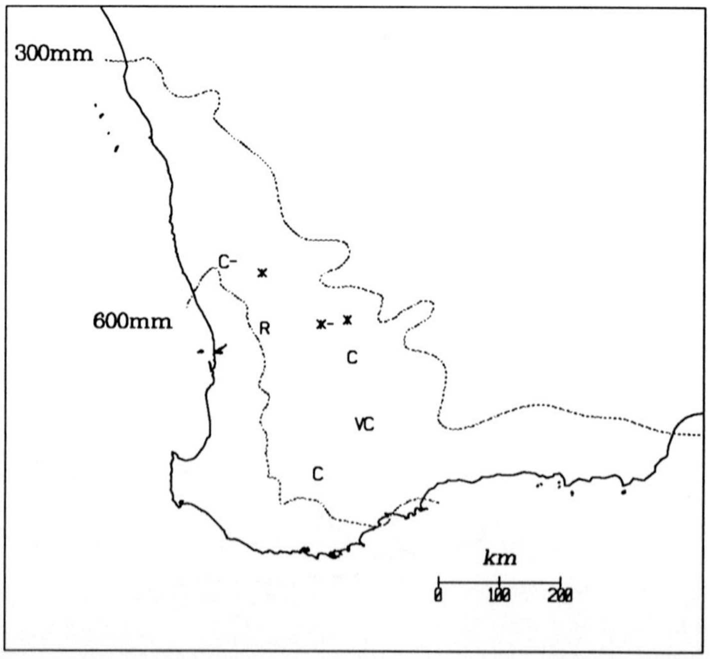

Fig. 92. Distribution 1987–1990

Common bronzewing

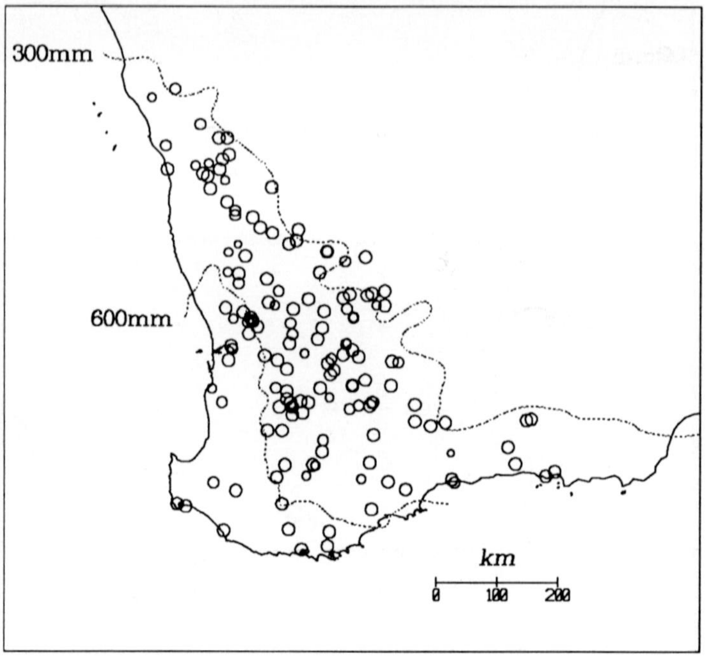

	1900/37	1987	1988	1989	1990
% of observers recording species	80	75	76	79	85
% of recording period species seen	R-VC	45	50	54	55

Crested pigeon

Ocyphaps lophotes

The crested pigeon is a common resident of woodland, mallee, shrubland, heath and remnant native vegetation on farmland throughout the wheatbelt (Fig. 93). It is the most widely distributed and abundant pigeon in the wheatbelt. It feeds on seed including those of cereal crops and weeds, other plant material, snails, dragonflies, grasshoppers, bugs, beetles, flies, caterpillars, ants and spiders. It breeds in spring and early summer on a stick nest in a tree or shrub. It lays 2 eggs.

The crested pigeon was formerly a species of savannah woodlands lining the watercourses of the arid zone. As it eats seed, it requires water every day. In developing the wheatbelt, extensive areas resembling riverine plain have been created, with scattered woodland, introduced grasses and water points in nearly every paddock; habitat ideal for this species. The crested pigeon rapidly colonized the wheatbelt: appearing in the Kellerberrin area in 1946–47 (Leake 1962); the Gabbin area in the latter half of 1951 (Sedgwick 1952); the north-east of the Northam district about 1955, Grass Valley in 1958 and west of the Avon River in 1964 (Masters and Milhinch 1974); the Ongerup area in 1965 or 1966 (K. Newbey, in litt.); the Pingelly-Brookton area between 1967 and 1972 (D. Shepherd, in litt.); and around Yornup in the mid-1980s (N. White, in litt.). It is increasing its range in the extreme south (R. Hammond, in litt.).

Fig. 93. Distribution 1987–1990

Crested pigeon

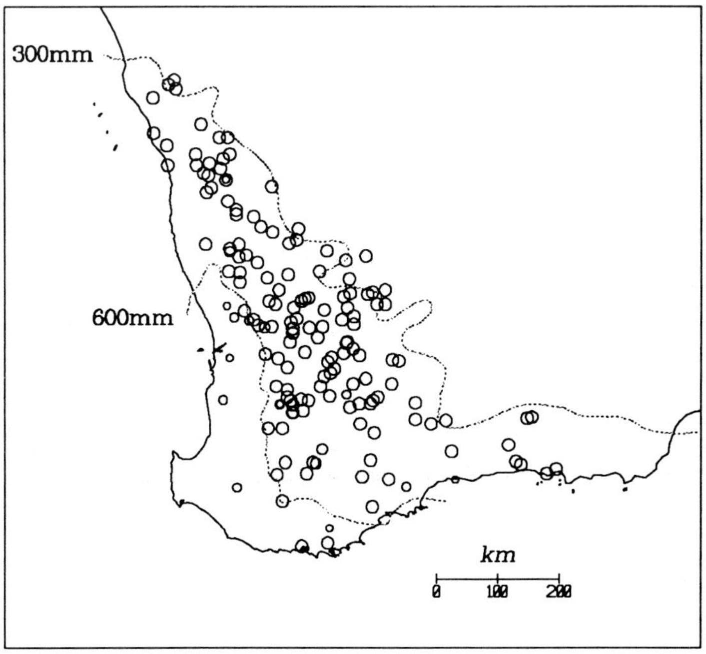

	1900/37	1987	1988	1989	1990
% of observers recording species	0	92	85	87	86
% of recording period species seen	NR	78	73	74	73

Red-tailed black cockatoo

Calyptorhynchus magnificus

There are two taxa of black cockatoos with red tails in southwestern Australia. *Calyptorhynchus magnificus naso*, the forest red-tailed black cockatoo occurs in the jarrah *Eucalyptus marginata,* marri *E. calophylla* and karri *E. diversicolor* forests of the southwest corner, in the area receiving more than 600 mm of annual average rain (Fig. 94). This bird feeds on eucalypt seed with marri seed being an important source of food (Saunders *et al.* 1985). It nests in a hollow in a tree and lays 1 egg. Masters and Milhinch (1974) noted that the forest red-tailed black cockatoo had contracted in the northeastern part of its range over the past 50 years. For example, J. Masters (in litt.) recorded that this bird had disappeared from areas where he formerly encountered flocks of greater than 40. Serventy and Whittell (1976) stated that the red-tailed black cockatoo had declined seriously in numbers in the southern part of the state since European settlement began. Forestry practice, particularly clear-felling and 80-year cutting rotations have the potential to restrict the availability of large tree hollows which provide nest sites for this cockatoo.

C. m. samueli, the inland red-tailed black cockatoo occurs in the northern and north-eastern wheatbelt (Fig. 94). It is a common resident in areas with heavy infestations of doublegee *Emex australis*, an agricultural weed. Originally a bird of the riverine plains of the Murchison and Gascoyne regions, this subspecies has invaded the adjacent wheatbelt as it was cleared (Saunders *et al.* 1985). It is currently expanding its range. Any successful control of doublegee will see the loss of this bird from those areas it has invaded. The inland red-tailed black cockatoo nests in a tree hollow, particularly those found in broken-off straight, dead trees (Saunders *et al.* 1982). It lays 1 egg and is the only cockatoo known to have two distinct breeding seasons; autumn and spring (Saunders 1977).

Fig. 94. Distribution 1900–1937

Red-tailed black cockatoo

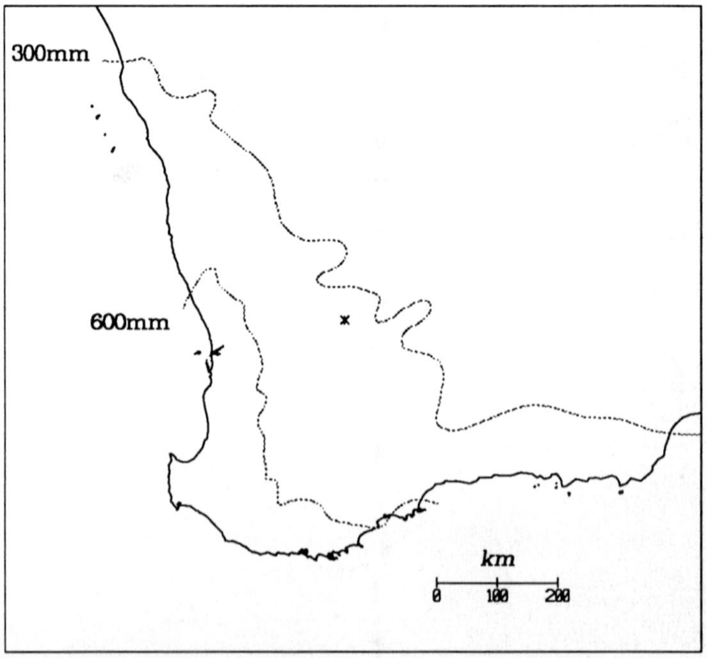

Fig. 94. Distribution 1987–1990

Red-tailed black cockatoo

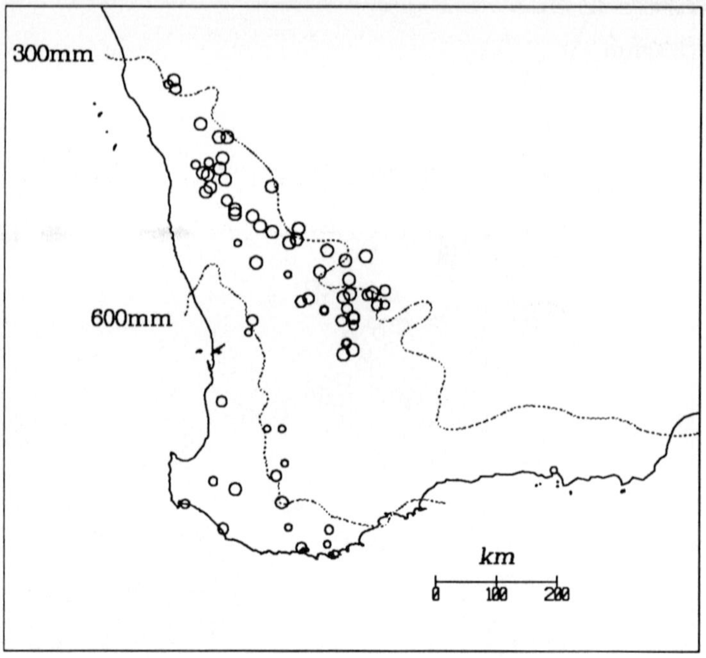

	1900/37	1987	1988	1989	1990
% of observers recording species	10	33	35	32	30
% of recording period species seen	—	21	19	19	18

PLATE 2a. Galah chewing bark from its nest tree north of Kellerberrin in the central wheatbelt (photo D. A. Saunders).

PLATES 2b and c. Bark removed from a galah nest tree in the northern wheatbelt. As shown in Plate 2c, the tree has been almost girdled (photo D. A. Saunders).

Plate 2c

Carnaby's cockatoo

Calyptorhynchus funereus latirostris

There are two taxa of black cockatoos with white tails in southwestern Australia (Saunders 1979). Carnaby's cockatoo (or short-billed form of the white-tailed black cockatoo) occurs in woodland and shrubland through much of the wheatbelt. It is also found in the forests of the southwest. It feeds on seed of the Proteaceae (hakeas, grevilleas, banksias), eucalypts (jarrah, marri) and some weeds. It also feeds on the larvae of wood-boring insects (Saunders 1980). It breeds in spring in a hollow in a eucalypt and lays 2 eggs.

Baudin's cockatoo *(C. baudinii)* (or long-billed form of the white-tailed black cockatoo) occurs in the forests of the southwest, in the area receiving more than 650 mm of annual average rainfall (Saunders 1979; Saunders *et al.* 1985). It feeds on the seed of eucalypts, particularly marri, and of the Proteaceae. It also feeds on insect larvae of wood-boring insects. It breeds in spring in a hollow in a eucalypt and lays 2 eggs.

It is difficult to distinguish between these two cockatoos, unless familiar with the different calls or their feeding habits. When feeding on marri nuts, Carnaby's cockatoo, with its shorter bill, always damages the rim of the nut, whereas Baudin's cockatoo rarely does so. Because of the difficulty in identification, the two forms are combined in Figure 95.

It is possible that Baudin's cockatoo has not changed in distribution and abundance over the past 50 years. However, like the forest red-tailed black cockatoo, forestry operations have the potential to affect this species by reducing the availability of nest sites (Saunders *et al.* 1985). On the other hand, Carnaby's cockatoo has disappeared from over a third of its breeding range over the last 20 years (Saunders 1980, 1982, 1986, 1990). For example, before 1940 large numbers of this black cockatoo visited the sandplains in the Tammin area in winter and nested in wandoos (J. Masters, in litt.). Carnaby's cockatoo has been extirpated from the Tammin area since then.

Fig. 95. Distribution 1900–1937

Carnaby's cockatoo

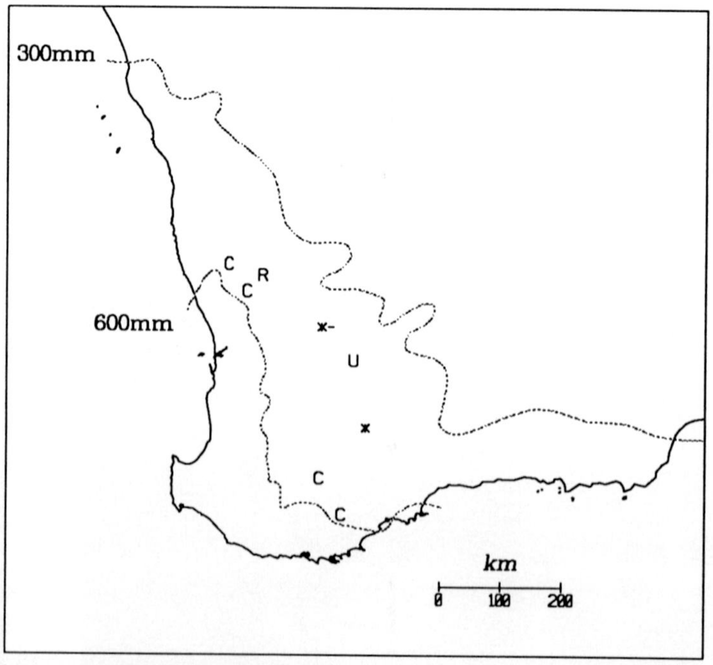

Fig. 95. Distribution 1987–1990

Carnaby's cockatoo

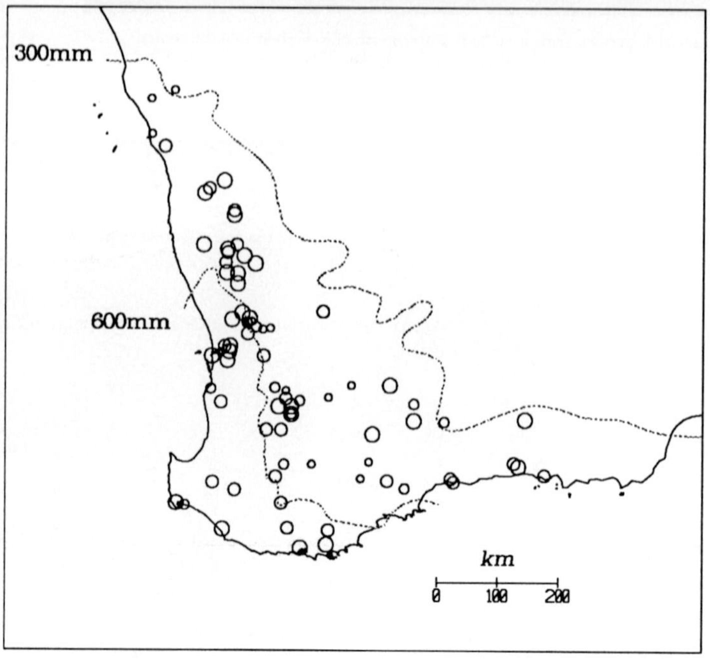

	1900/37	1987	1988	1989	1990
% of observers recording species	80	30	36	41	45
% of recording period species seen	R-C	11	13	13	17

Carnaby's cockatoo is almost entirely dependent on native vegetation for food in the wheatbelt. Extensive clearing and fragmentation of native vegetation have left this species unable to find sufficient food to breed in much of its former range. It is resident in the higher rainfall districts where much native vegetation remains, however, it is a breeding migrant in the drier parts of its range (see Fig. below). Its pattern of appearing in late winter and spring, remaining to breed and moving to the higher rainfall districts in summer has earned it the name of "rain bird".

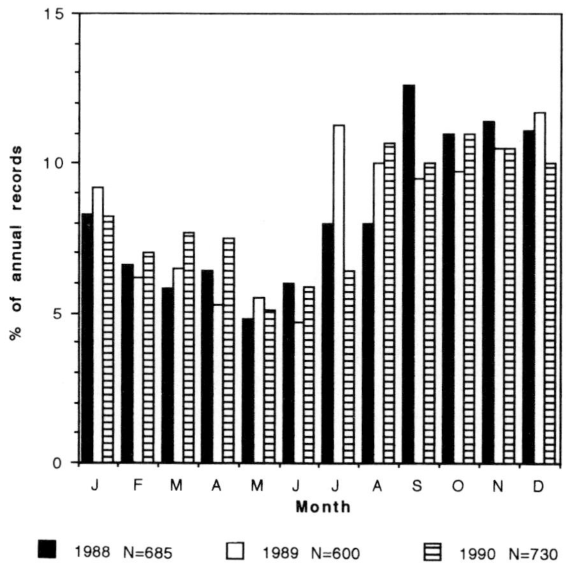

Galah

Cacatua roseicapilla

Widespread and common in woodland, mallee and farmland throughout most of the wheatbelt (Fig. 96), the galah is one of the most successful species of native bird. It feeds on seed, particularly those of cereal crops and agricultural weeds, fruit, grubs and caterpillars. It breeds in spring in a hollow in a eucalypt and lays 2–4 eggs (Rowley 1990).

Like the crested pigeon and inland red-tailed black cockatoo, the galah was formerly a bird of the riverine plains of the arid zone. With development of the wheatbelt, humans created the equivalent of riverine plain by clearing native vegetation, planting cereal crops and providing water. The galah rapidly colonized this habitat. For example, Orton and Sandland (1913) noted that small flocks were seen in the Moora district occasionally, but they did not remain for more than a few hours. The galah invaded the north-eastern parts of the Northam district about 1946 and was first noted west of the Avon River in 1952 (Masters and Milhinch 1974). It was around Lake Hinds in the Wongan Hills district (de Rebeira and de Rebeira 1977) and the Kellerberrin area (Leake 1962) in the early 1920s, and in small parties around Mogumber in 1950 (Loaring and Serventy 1952), although not in the area in the early 1900s (Lawson 1905). The galah became established near Gingin and Lower Chittering in the 1950s (Loaring and Serventy 1952), colonized the Dalwallinu district around 1928 (G. Hyde, in litt.) and the Manmanning district in the late 1930s (B. Smith, in litt.). It is common in all of these areas and is spreading south and southwest. It will continue to do so until it occupies all suitable habitat.

The galah competes with other hole nesters for nest sites. It also chews the bark off the trunk of its nest tree to form a type of shield. In extreme cases this removes a girdle of bark resulting in the death of the tree (Plates 2a, b and c). In some areas in the wheatbelt, the galah is a significant cause of tree death.

Fig. 96. Distribution 1900–1937

Galah

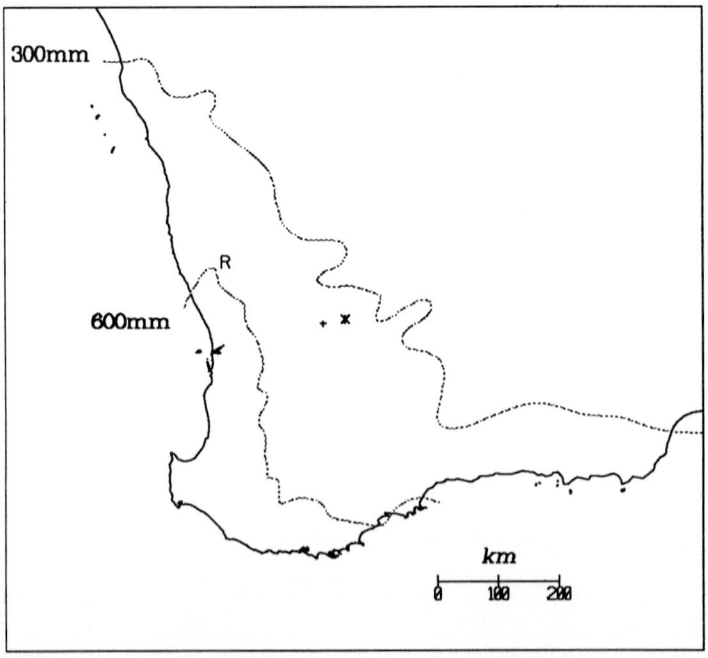

Fig. 96. Distribution 1987–1990

Galah

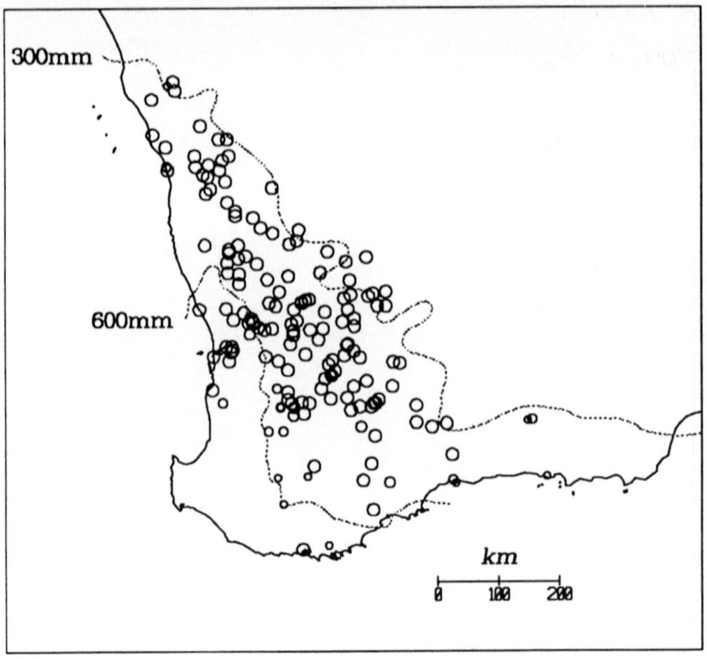

	1900/37	1987	1988	1989	1990
% of observers recording species	30	96	87	88	88
% of recording period species seen	R	83	73	74	73

Little corella

Cacatua sanguinea

The little corella is a bird of the arid zone river systems which has invaded the adjacent northern wheatbelt (Saunders *et al.* 1985) (Fig. 97). It is an uncommon resident in woodland and on farmland. It feeds on seed of cereal crops, weeds and also on insect larvae. It breeds in spring in a hollow in a eucalypt and lays 1–4 eggs.

The little corella is extending its range south and overlaps the range of the long-billed corella.

Fig. 97. Distribution 1987–1990

Little corella

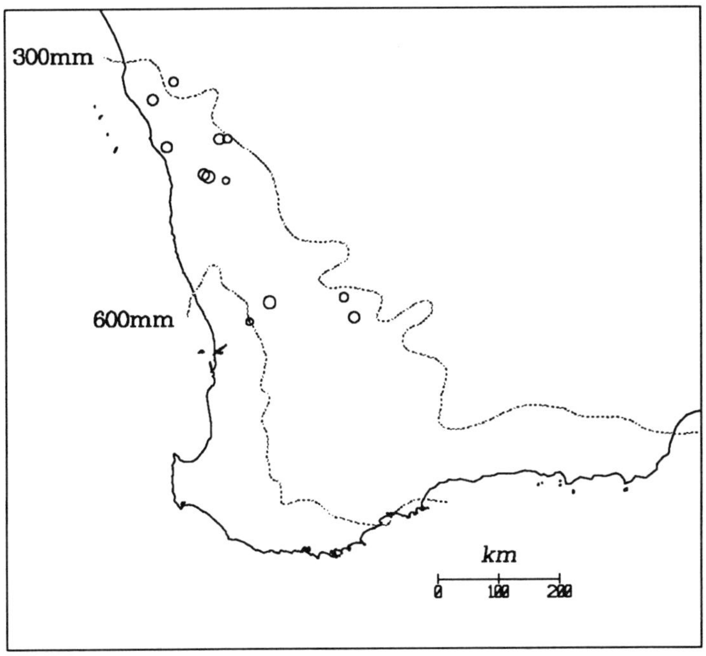

	1900/37	1987	1988	1989	1990
% of observers recording species	0	3.1	4.9	6.8	9.6
% of recording period species seen	NR	0.8	2.3	1.9	1.4

Long-billed corella

Cacatua tenuirostris

The long-billed corella is a resident, which may be locally common in woodland and farmland throughout the northern half of the wheatbelt (Smith 1991) (Fig. 98), where it was estimated there were 5 000 to 10 000 birds (Smith and Moore 1992). This population is extending its range. For example, along the Avon River between York and Toodyay there has been a rapid increase in this species over the past 10 years (J. Masters, in litt.). There is also a small, isolated population around Lake Muir on the edge of the forest country (Smith and Moore 1992). This population is listed as vulnerable (Garnett 1992) and the Western Australian Wildlife Conservation Act classifies it "Rare or likely to become extinct". During this survey there were no observers in that area. Accordingly, no record of that population appears on Figure 98.

The long-billed corella feeds on seed of cereal crops and weeds, digs in the ground for tubers of plants, and eats insect larvae. It breeds in spring in a eucalypt hollow and lays 1–4 eggs.

After breeding, the long-billed corella joins summer foraging flocks. These flocks may become large and may be regarded as a local nuisance (e.g., at Dongara and Dalwallinu) because of damage to trees and because of the noise associated with such large congregations. These flocks break up as the birds disperse to their breeding areas in late summer (Smith and Moore 1992).

Fig. 98. Distribution 1900–1937

Long-billed corella

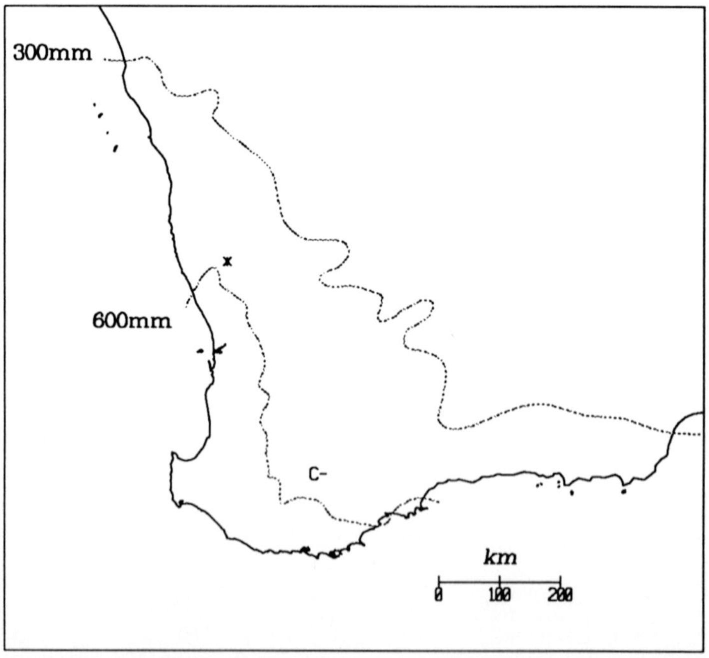

Fig. 98. Distribution 1987–1990

Long-billed corella

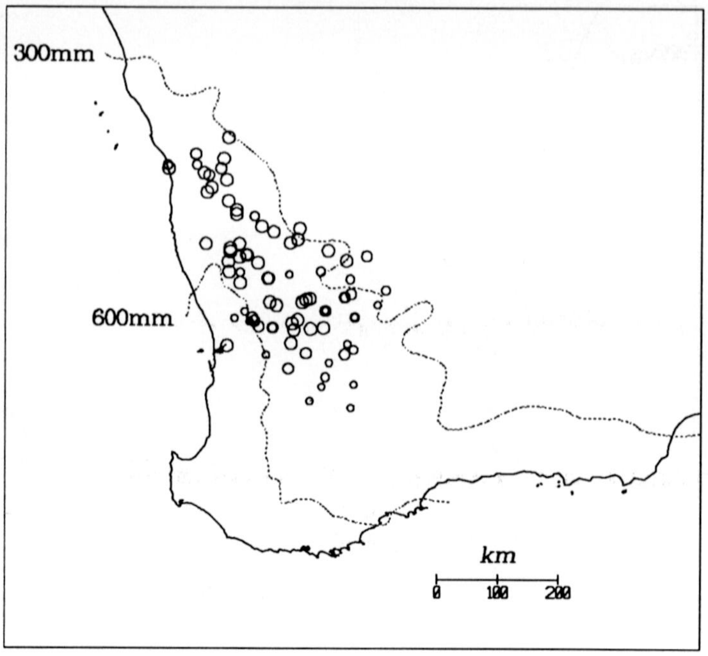

	1900/37	1987	1988	1989	1990
% of observers recording species	20	38	44	36	36
% of recording period species seen	C	22	19	16	21

Pink cockatoo

Cacatua leadbeateri

OTHER NAMES:

Major Mitchell, Major Mitchell's cockatoo.

The pink cockatoo is an uncommon resident of woodlands in the northern and northeastern wheatbelt (Fig. 99). It feeds on fruit of native vegetation, seed of cereal crops and weeds, and also on insect larvae. It apparently requires access to a variety of native foods in order to thrive. It breeds in spring in a eucalypt hollow and lays 1–4 eggs (Rowley and Chapman 1991).

The range of the pink cockatoo is declining. It formerly bred in the Wongan Hills district, but the species was never numerous (de Rebeira and de Rebeira 1977). It does not occur there now. Extensive clearing and fragmentation of native vegetation throughout its range in the wheatbelt, together with robbing of nests by aviculturalists (Rowley and Chapman 1991) have been major factors in its decline.

Fig. 99. Distribution 1987–1990

Pink cockatoo

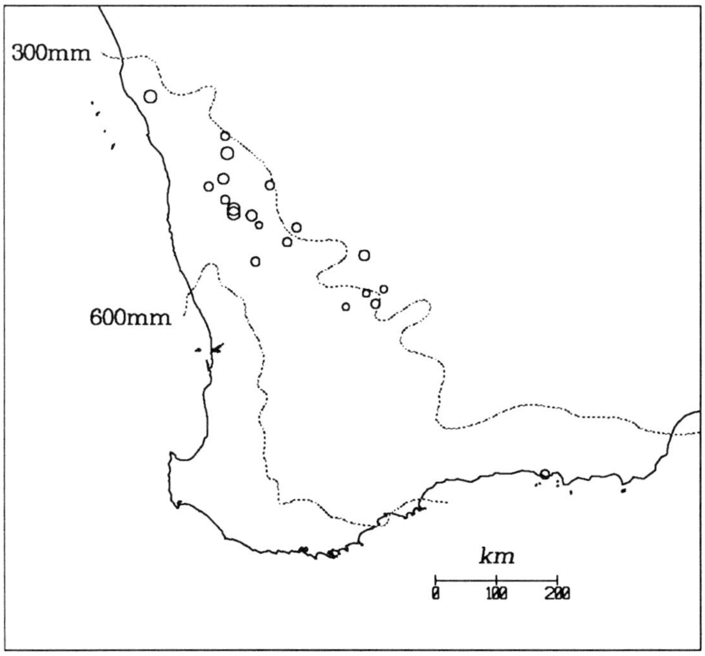

	1900/37	1987	1988	1989	1990
% of observers recording species	0	12	6.3	6.0	4.8
% of recording period species seen	NR	3.0	2.4	1.9	2.5

Sulphur-crested cockatoo

Cacatua galerita

The sulphur-crested cockatoo is not found in the wheatbelt, but does occur in two small populations; one near Guildford on the outskirts of Perth, and the other near Pinjarra south of Perth (Saunders *et al.* 1985) (Fig. 100). Both of these populations probably started with birds escaping from aviaries, or from deliberate introductions. The Agriculture Protection Board regards this species as having the potential to be a major pest of agriculture and has been controlling both populations.

The sulphur-crested cockatoo feeds on seed of cereal crops and agricultural weeds, roots and tubers, and on insect larvae. It breeds in spring in a eucalypt hollow and lays 2 eggs.

Fig. 100. Distribution 1987–1990

Sulphur-crested cockatoo

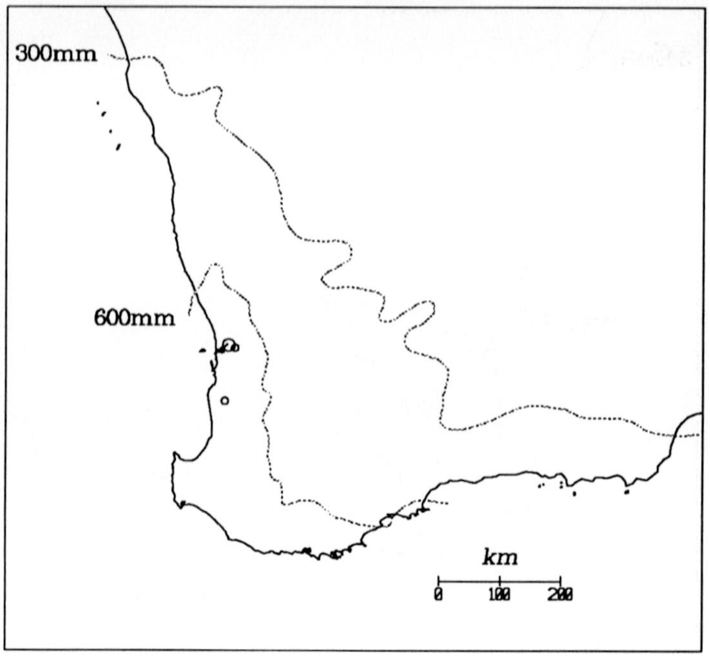

	1900/37	1987	1988	1989	1990
% of observers recording species	0	0.8	1.4	1.7	1.9
% of recording period species seen	NR	0.07	0.7	0.6	0.1

Purple-crowned lorikeet

Glossopsitta porphyrocephala

The purple-crowned lorikeet is resident in woodland and mallee in parts of the wheatbelt (Fig. 101) where there are still extensive areas of suitable habitat. It has declined to become an uncommon nomad, or has disappeared from the central and parts of the southern wheatbelt where there has been extensive loss of woodlands. It feeds on pollen, nectar, fruit and insects such as thrips. Its movements are dictated by the flowering of woodland species (Masters and Milhinch 1974). The purple-crowned lorikeet breeds in spring in a tree hollow and lays 3–4 eggs.

The purple-crowned lorikeet was numerous in woodland in the Wongan Hills at the turn of the century (Milligan 1904), but it had disappeared by the 1970s (de Rebeira and de Rebeira 1977). Leake (1962) wrote that in the Kellerberrin area "Practically gone are the vast flocks of purple-crowned lorikeets . . . that used to feed on the nectar of the gimlet, salmon, white and York gums, when they are in bloom." Similarly, in the Northam district during the 1930s where it was seen in thousands, "it is now seldom seen and then only a few" (Masters and Milhinch 1974). There have been no sightings of the purple-crowned lorikeet in the Hyden district where it used to be plentiful 10 years previously (V. Brown, in litt.). It was plentiful north of Dowerin in the 1930s, but has disappeared since then (G. Evans, in litt.).

Fig. 101. Distribution 1900–1937

Purple-crowned lorikeet

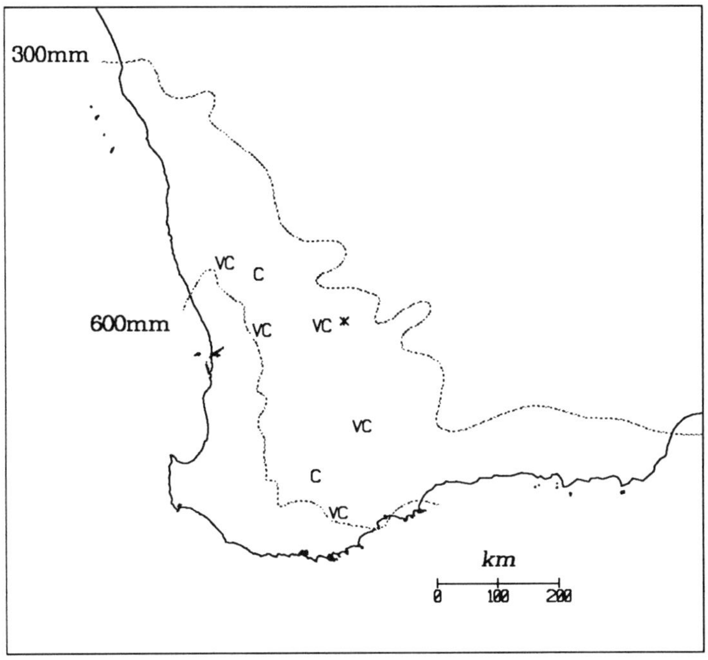

Fig. 101. Distribution 1987–1990

Purple-crowned lorikeet

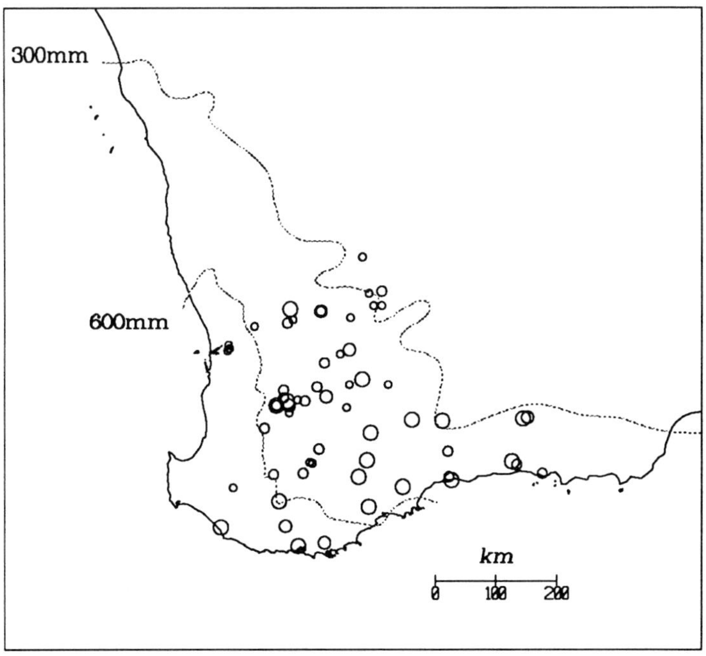

	1900/37	1987	1988	1989	1990
% of observers recording species	80	13	26	26	21
% of recording period species seen	C-VC	2.5	7.0	9.6	7.4

Regent parrot

Polytelis anthopeplus

OTHER NAME:

Smoker parrot

The regent parrot occurs in woodland, mallee and shrubland in the wheatbelt (Fig. 102). It is not found in the forests of the southwest. It is an uncommon resident in the higher rainfall areas and in more recently cleared areas, where there is more remnant vegetation. It is a nomad in the extensively cleared districts.

During this survey, the regent parrot was more commonly recorded over a wider area in the period May to December than in the period January to March (see Fig. on facing page). It may be a seasonal migrant in some of its range. It feeds on seed of grasses and cereal crops, and fruits. It breeds in spring in a large tree hollow and lays 4–6 eggs.

Serventy and Whittell (1976) noted that the regent parrot had increased in abundance and range since Europeans settled the wheatbelt. It was declared vermin in a number of agricultural shires in the 1930s and 1940s. Since then it has undergone a steady decline. In the 1950s the regent parrot was plentiful in the Kellerberrin district, particularly in *Eucalyptus* and *Acacia* trees (Ford and Stone 1957); it no longer occurs in the area except as a rare nomad (Saunders and Curry 1990). It was fairly common in the Northam district up to about 1950 and has been scarce since about 1955 (Masters and Milhinch 1974). It was a common species in the Wongan Hills district formerly, but declined in numbers and had left the area by the 1940s (de Rebeira and de Rebeira 1977). Up to 1940 in the Tammin area, "very large numbers of regent parrots were always about the countryside" (J. Masters, in litt.); they no longer occur there. A number of observers remarked during this survey that the regent parrot had disappeared from their area. For example: it was plentiful around Goomalling in the 1930s and 1940s but does not occur there now (F. Boase); it was plentiful to the north of Dowerin in the 1940s, but has disappeared (G. Evans); it has been extinct north-east of Brookton for over 30 years (E. Fleay); it was almost as common as Port Lincoln ringnecks around Pingelly, then it

Fig. 102. Distribution 1900–1937

Regent parrot

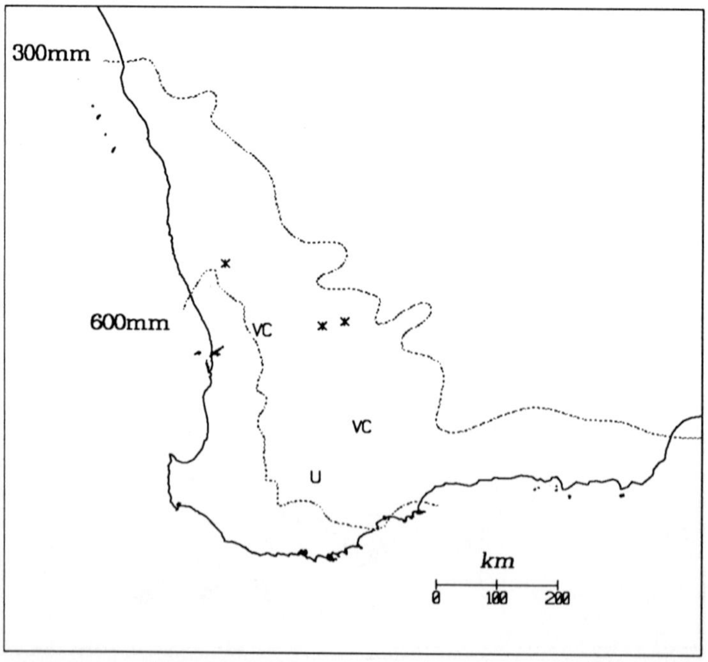

Fig. 102. Distribution 1987–1990

Regent parrot

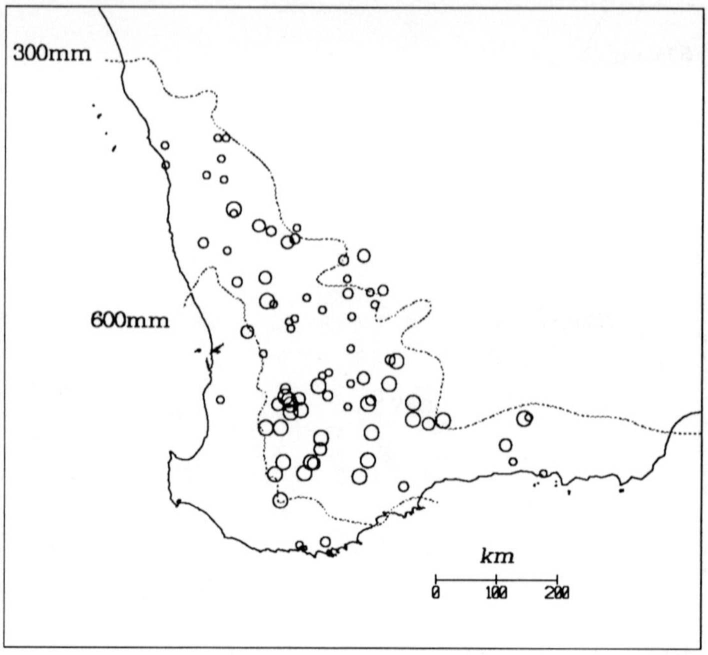

	1900/37	1987	1988	1989	1990
% of observers recording species	60	30	37	34	36
% of recording period species seen	U-VC	8.0	11	14	13

Fig. 102. Distribution 1987–1990 — January–March

Regent parrot

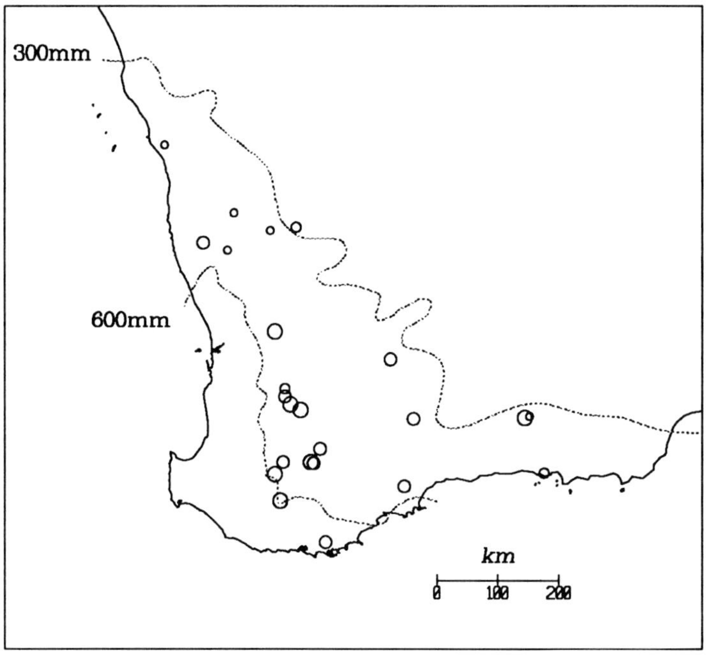

Fig. 102. Distribution 1987–1990 — August–December

Regent parrot

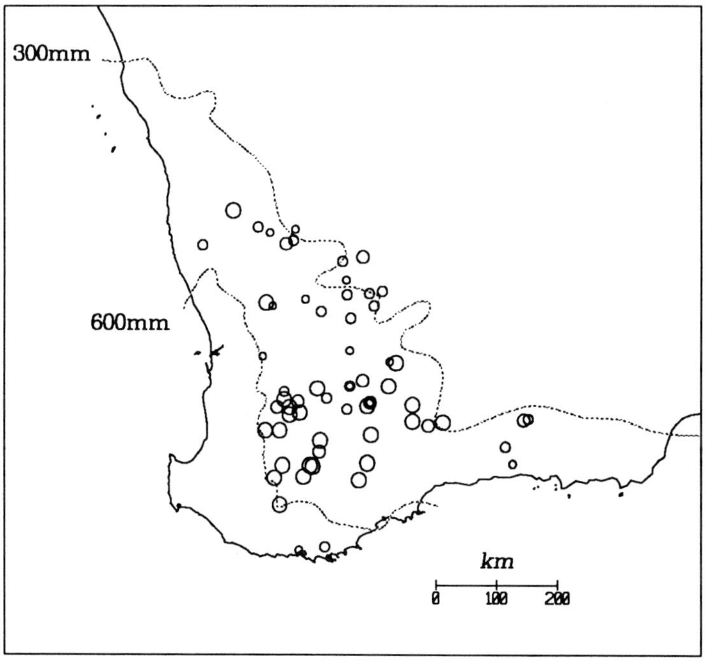

disappeared (D. Shepherd); it disappeared from the Manmanning area in the 1960s (B. Smith); from the Merredin area (B. Willis); it was very prevalent in the Dalwallinu district during the 1940s but is seldom seen now (G. Hyde); and Morawa Shire removed it from their vermin list in the early 1970s, after it had disappeared (J. White).

The regent parrot breeds in tree hollows which are larger than those used by the similar sized Port Lincoln parrot. In fact the hollows favoured by regent parrots are similar in size to those chosen by the galah and it is possible that interaction with galahs for nest sites may have also been a factor in the decline of the regent parrrot.

Over the last few years, this parrot is being seen more often and may be increasing in numbers.

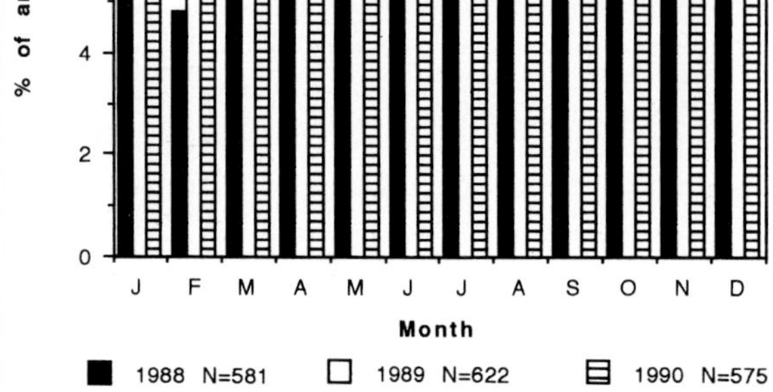

Cockatiel

Nymphicus hollandicus

OTHER NAME:

Weero

The cockatiel is a regular spring and summer breeding migrant in woodland and farmland of the northern and northeastern wheatbelt (Fig. 103). It is an uncommon nomad in adjacent areas. It feeds on seed, including agricultural grains, and on berries. It breeds in spring in a tree hollow and lays 4–7 eggs. During this survey, the cockatiel was observed throughout the year in 1989, whereas it demonstrated a marked seasonal pattern during 1988 and 1990 (see Fig. below).

There is no evidence to suggest that the cockatiel has markedly changed in distribution over the last 50 years.

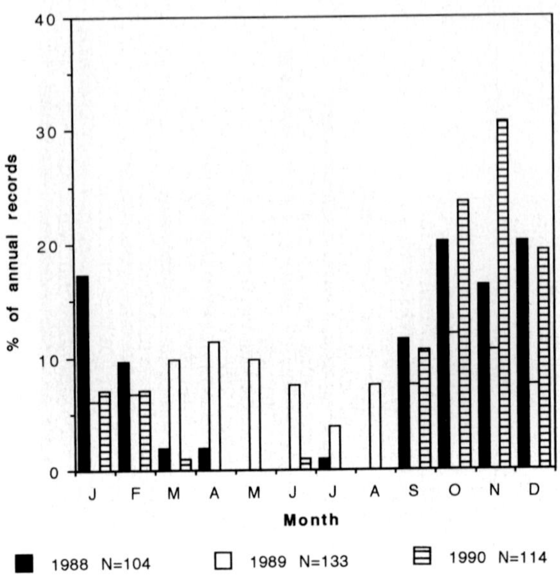

Fig. 103. Distribution 1900–1937

Cockatiel

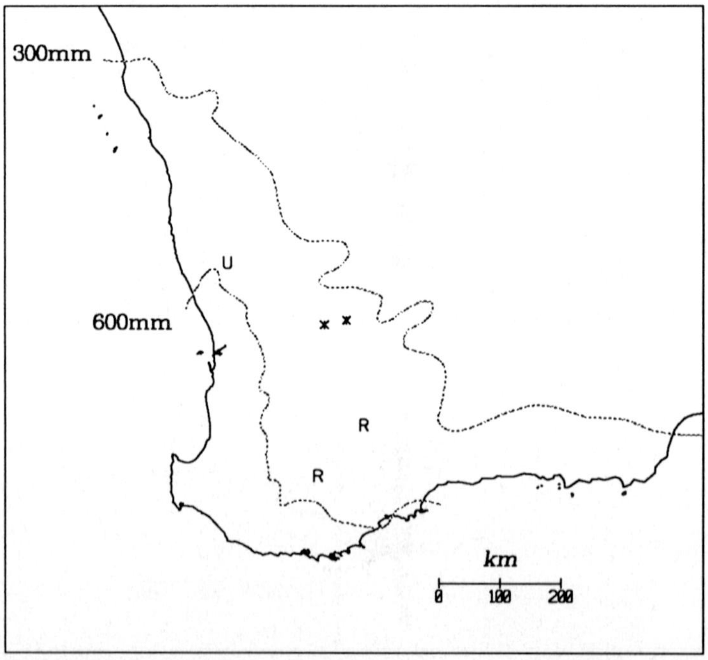

Fig. 103. Distribution 1987–1990

Cockatiel

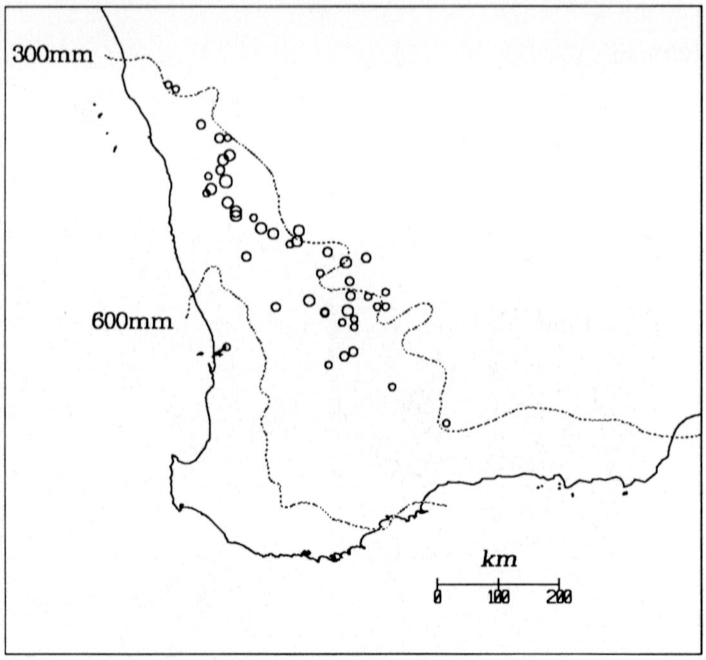

	1900/37	1987	1988	1989	1990
% of observers recording species	50	19	15	16	16
% of recording period species seen	R-U	3.1	2.2	2.1	2.6

Shrubland habitat (photos by D. A. Saunders).

Budgerigar

Melopsittacus undulatus

The budgerigar is a bird of the arid zone and an uncommon, seasonal nomad in woodland, shrubland and farmland of the wheatbelt (Fig. 104). It feeds on seed, breeds in spring in a tree hollow and lays 4–6 eggs.

Regarded as highly nomadic, the distribution on the edge of its range varies widely (Fig. 104). During the spring and early summer of 1990 the budgerigar invaded much of the wheatbelt, being seen by twice as many observers as during previous years. The recording rate for 1990 was three times those of previous years; in fact 51 per cent of the recordings of this species were made between the beginning of September and the end of December 1990. In addition, the bird was recorded more widely and generally further south during the last 4 months of 1990.

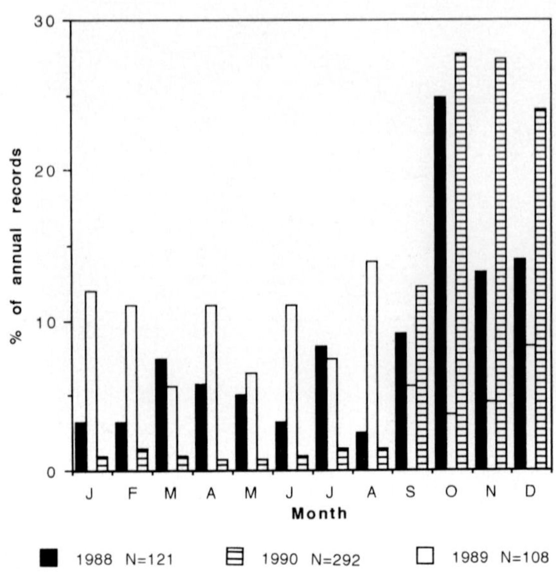

Fig. 104. Distribution 1900–1937

Budgerigar

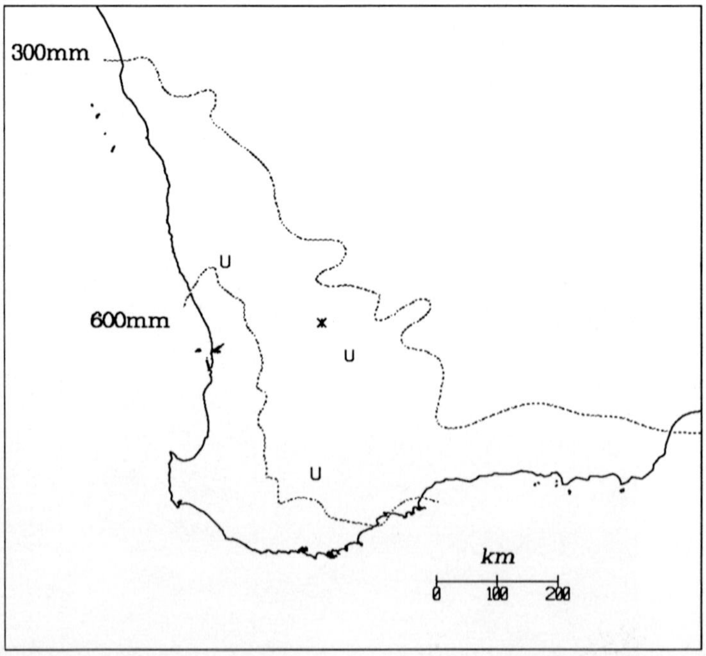

Fig. 104. Distribution 1987–1988

Budgerigar

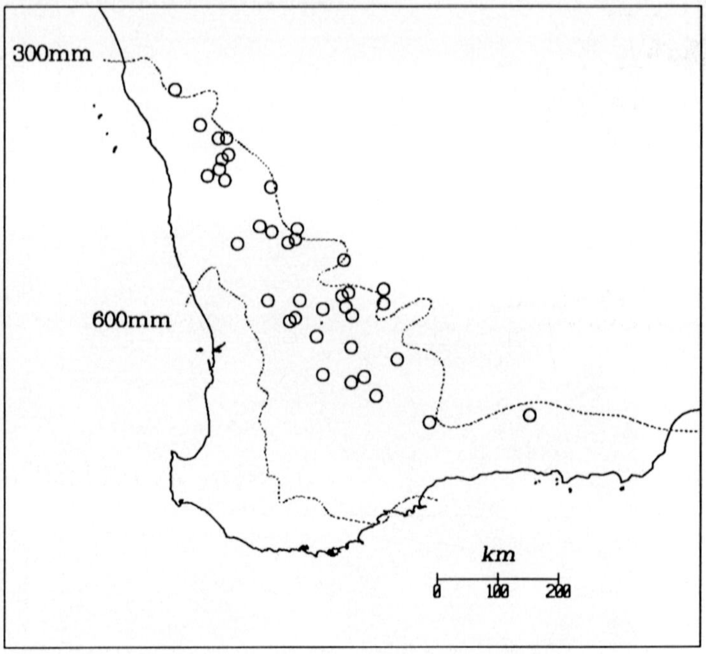

	1900/37	1987	1988	1989	1990
% of observers recording species	40	20	19	20	42
% of recording period species seen	U	3.1	2.6	2.2	6.7

Fig. 104. Distribution 1989

Budgerigar

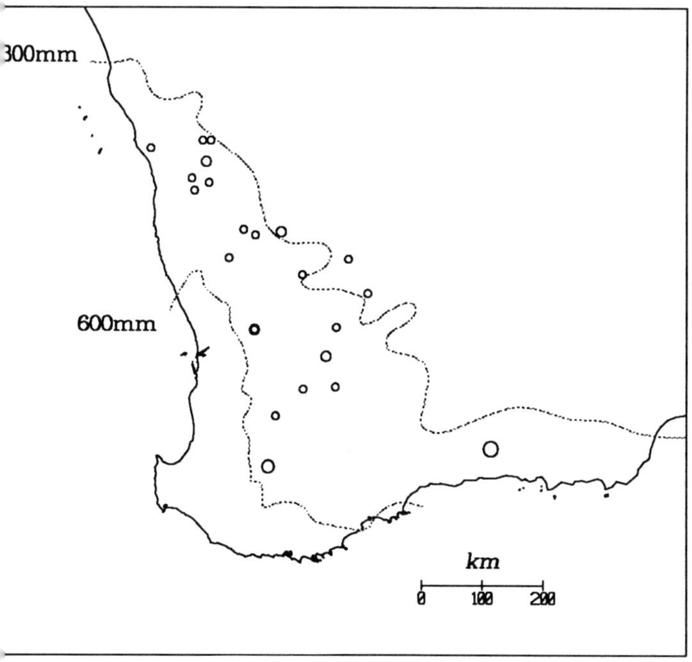

Fig. 104. Distribution 1987–1990 — January–May

Budgerigar

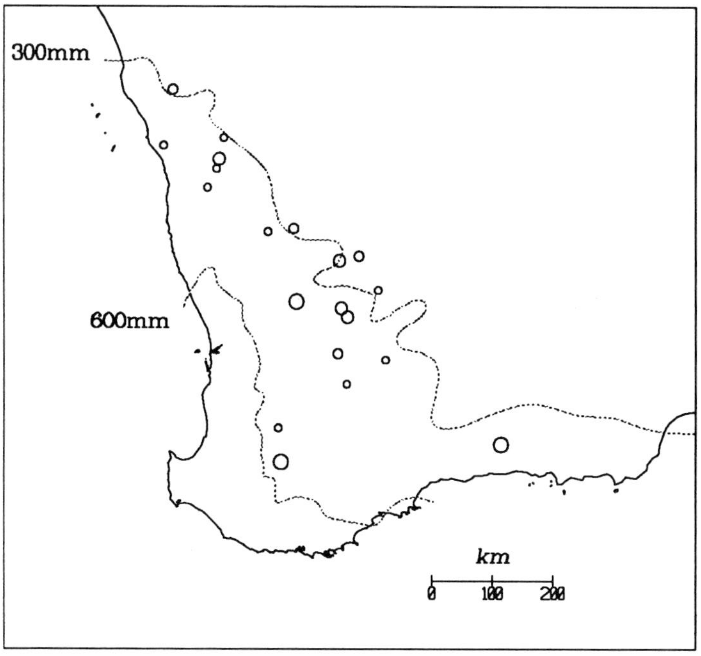

Fig. 104. Distribution 1990

Budgerigar

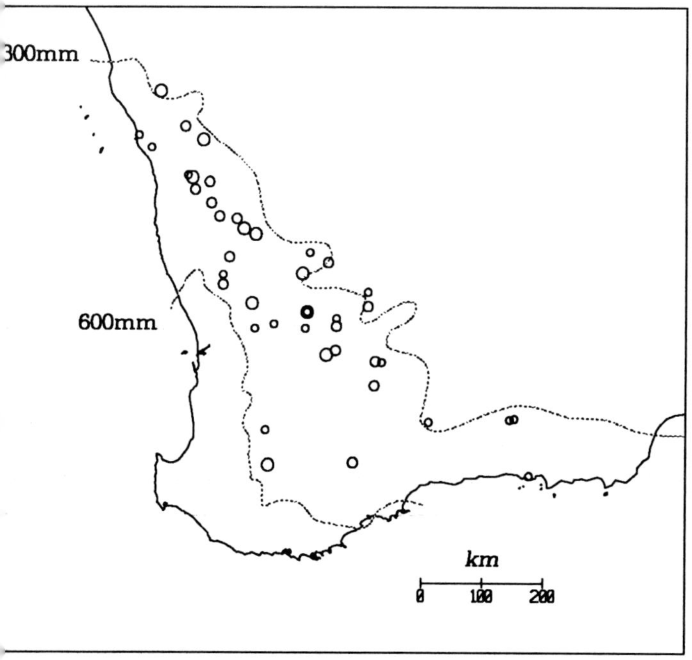

Fig. 104. Distribution 1987–1990 — July–November

Budgerigar

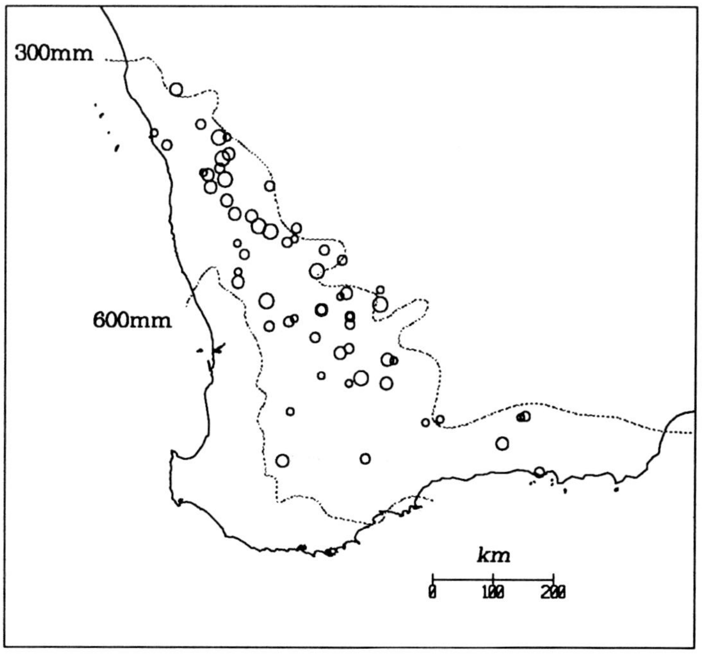

Red-capped parrot

Purpureicephalus spurius

The red-capped parrot is an uncommon to common resident in woodland of the higher rainfall districts to the west and south of the wheatbelt (Fig. 105). It is an uncommon nomad on the drier edges of this range. The red-capped parrot feeds on seed, fruit and is sometimes a pest in orchards. It also feeds on blossom, bugs and lerps. It breeds in spring in a tree hollow and lays 4–7 eggs.

The range of the red-capped parrot has contracted at the drier margins as a result of the loss of suitable habitat. For example, Leake (1962) noted that the red-capped parrot occurred in the Kellerberrin district, but was declining; it is a rare nomad there.

Fig. 105. Distribution 1900–1937

Red-capped parrot

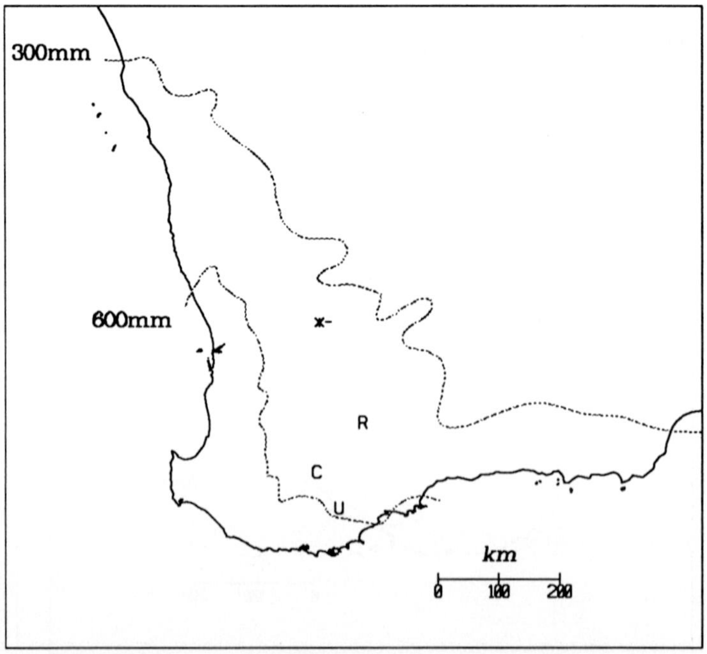

Fig. 105. Distribution 1987–1990

Red-capped parrot

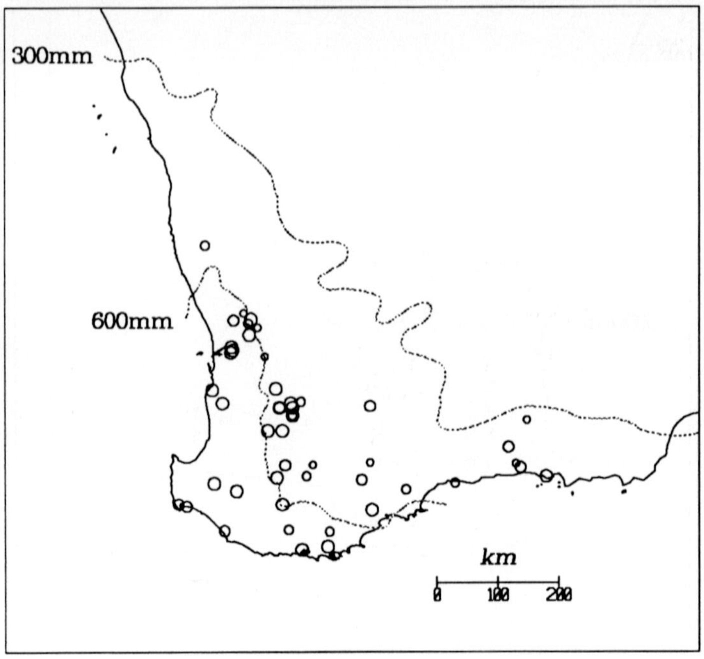

	1900/37	1987	1988	1989	1990
% of observers recording species	40	14	25	26	30
% of recording period species seen	R-C	7.6	15	16	16

Fig. 106. Distribution 1900–1937

Western rosella

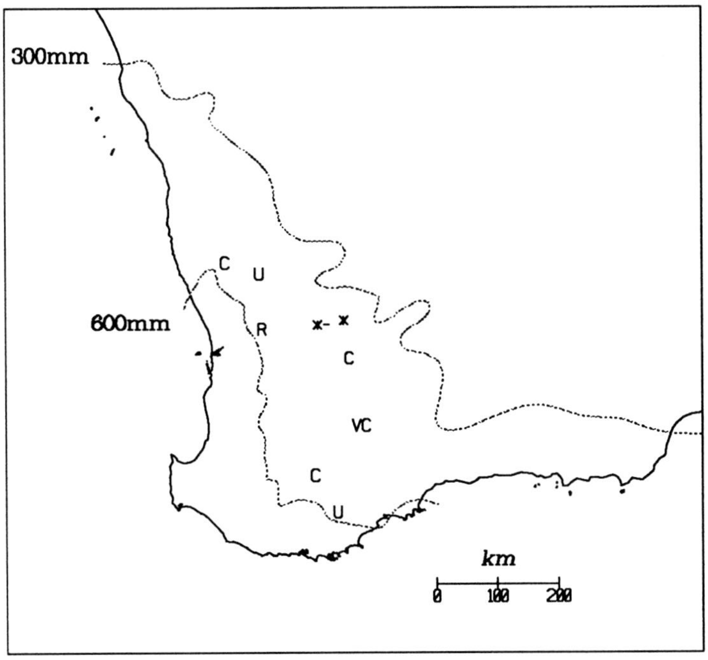

Fig. 106. Distribution 1987–1990

Western rosella

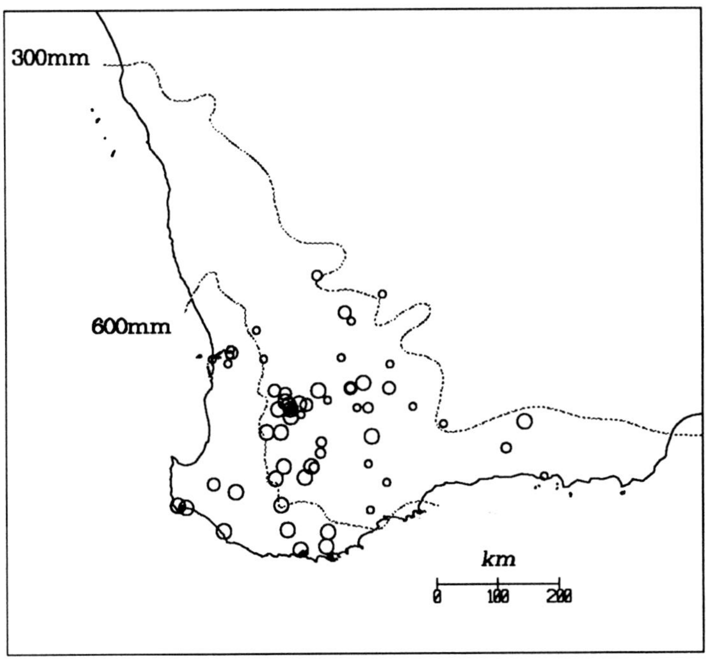

	1900/37	1987	1988	1989	1990
% of observers recording species	90	20	27	28	29
% of recording period species seen	R-VC	6.2	13	16	18

Western rosella

Platycercus icterotis

The western rosella is a common resident in woodland of the higher rainfall districts in the southern wheatbelt (Fig. 106). It is an uncommon nomad in the central wheatbelt, where it may occur during any part of the year. It feeds on seed and fruit and is regarded as a pest in orchards. It breeds in spring in a tree hollow and lays 3–7 eggs.

The range of the western rosella has contracted in the central wheatbelt because of loss of woodland. Leake (1962) noted the western rosella was declining around Kellerberrin; it is a rare nomad there. Milligan (1904) collected it in the Wongan Hills, yet it was not recorded there in the 1970s (de Rebeira and de Rebeira 1977). The western rosella has not been common in the Northam district since 1930 (Masters and Milhinch 1974).

Port Lincoln parrot

Barnardius zonarius

OTHER NAMES:

Port Lincoln ringneck, twenty-eight parrot, ring-necked parrot

The Port Lincoln parrot is a common resident in woodland, mallee, shrubland and farmland throughout the wheatbelt (Fig. 107). During this survey it was recorded by nearly all observers, over 90 per cent of the time, and is the most abundant parrot, and possibly the most common bird in the wheatbelt. It feeds on seed (including those of cereal crops), fruit, wood-boring grubs, lerps, caterpillars and spitfire larvae, and is widely regarded as a pest in orchards and gardens. Seed of introduced species may account for 82 per cent of its diet in some seasons, particularly winter (Long 1984). It breeds in spring in a tree hollow and lays 4–7 eggs.

There are several races of the Port Lincoln parrot, however, there is no basis for seperating them for this atlas. Serventy and Whittell (1976) noted that most of the wheatbelt is a transition zone where the southwest *semitorquatus* and the arid zone *zonarius* grade into each other.

Nine of the published lists before 1937 classified this species as common or very common, however, there is some evidence to suggest that it may have increased in the early part of this century. For example, in the Stirling Range Milligan (1903) did not record the Port Lincoln parrot and Whitlock (1911) reported that it was "distinctly rare" there. By the early 1950s it had increased markedly and Sedgwick (1964) recorded that it was the most common parrot in the Stirling Range. Other anecdotal evidence from older residents of the wheatbelt suggests that it may have increased markedly in abundance in many districts.

Fig. 107. Distribution 1900–1937

Port Lincoln parrot

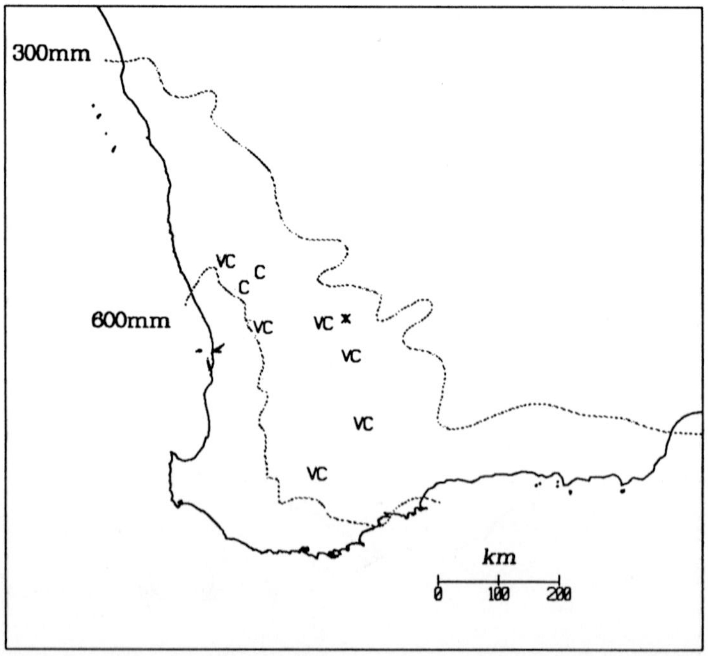

Fig. 107. Distribution 1987–1990

Port Lincoln parrot

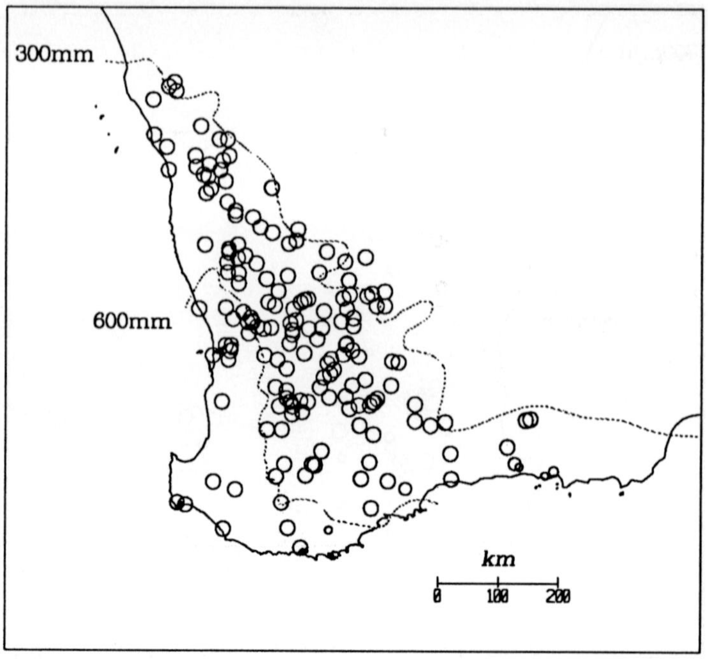

	1900/37	1987	1988	1989	1990
% of observers recording species	90	98	96	97	96
% of recording period species seen	C-VC	91	91	90	89

Fig. 108. Distribution 1900–1937

Mulga parrot

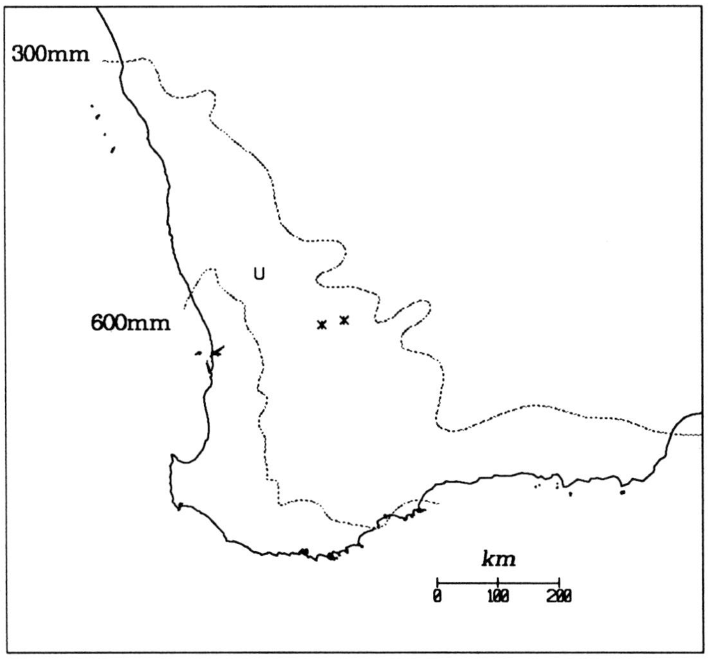

Fig. 108. Distribution 1987–1990

Mulga parrot

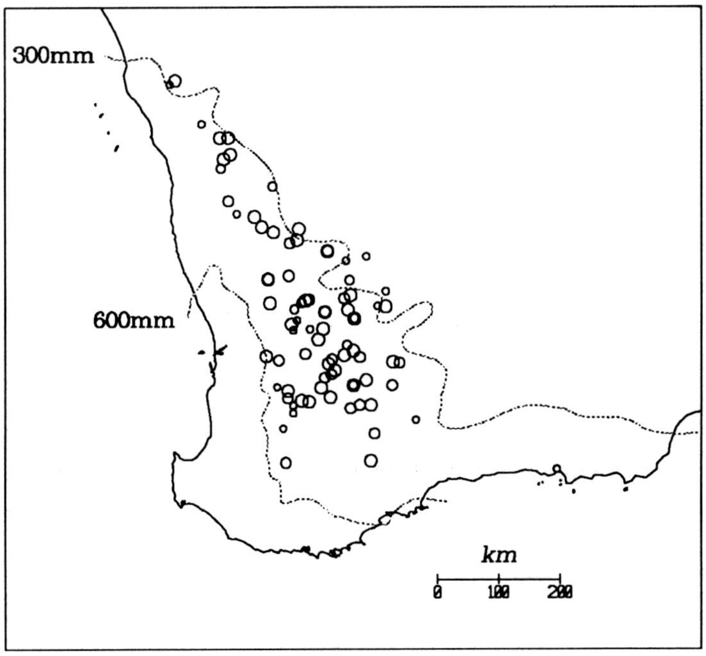

	1900/37	1987	1988	1989	1990
% of observers recording species	30	42	42	43	38
% of recording period species seen	U	21	19	20	20

Mulga parrot

Psephotus varius

The mulga parrot is an uncommon resident in woodland, shrubland and farmland of the central and northern wheatbelt (Fig. 108). It feeds on seed (including those of cereal crops), fruit and grubs. It breeds in spring in a tree hollow and lays 4–6 eggs.

Of the wheatbelt bird lists published before 1937, this species was only mentioned at Yandanooka (Milligan 1905), Wongan Hills (Milligan 1904), Kellerberrin (invaded the area in 1916; Leake 1962) and Nangeenan (Sedgwick 1937). Serventy and Whittell (1962) noted it occurred north of a line from Moora to Grass Valley, Quairading, Kweda and Dundas. This species has expanded its range over the last 30 years as it occurs well south of this line.

Elegant parrot

Neophema elegans

OTHER NAMES:

Grass parrot, elegant grass parrot

The elegant parrot is an uncommon resident in woodland, shrubland and farmland throughout the wheatbelt (Fig. 109). It feeds on seed, particularly of grasses, and also eats fruit. The elegant parrot breeds in spring in a tree hollow and lays 4–5 eggs.

According to Serventy and Whittell (1976) the elegant parrot was only found in the southwest corner, north to Moora, east to Merredin and near Esperance. It did not occur in the forest of the southwest nor on the Swan Coastal Plain. Since 1937 it has extended its range into the northern wheatbelt. It arrived in Kellerberrin in 1942 (Leake 1962) and increased in abundance (Ford and Stone 1957). It was rare in the Northam district before 1940, but increased to be classified as uncommon (Masters and Milhinch 1974). It has invaded the southwest forest area where clearing for agriculture has allowed it to penetrate (Readers' Digest 1976) and has expanded north into the pastoral zone (M. Brooker, pers. comm.).

Fig. 109. Distribution 1900–1937

Elegant parrot

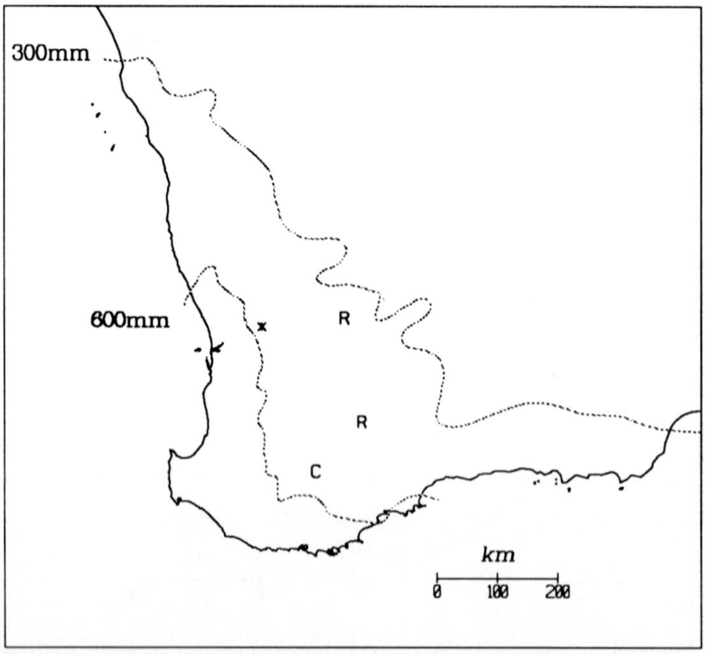

Fig. 109. Distribution 1987–1990

Elegant parrot

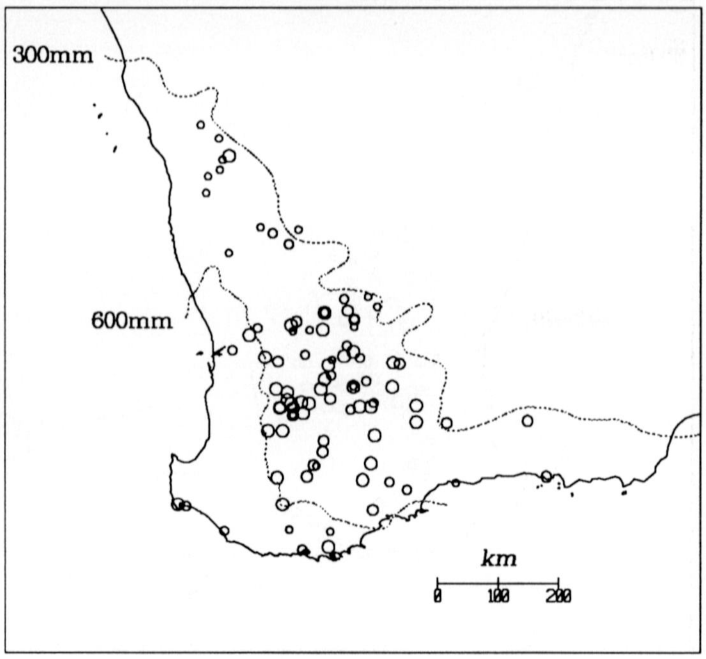

	1900/37	1987	1988	1989	1990
% of observers recording species	40	42	46	43	42
% of recording period species seen	R-C	17	18	18	20

Bourke's parrot

Neophema bourkii

Bourke's parrot is a bird of the arid zone and a rare vagrant in the northeastern margins of the wheatbelt (Fig. 110) adjacent to its range in the pastoral zone. It feeds on seed, particularly of grasses.

Fig. 110. Distribution 1987–1990

Bourke's parrot

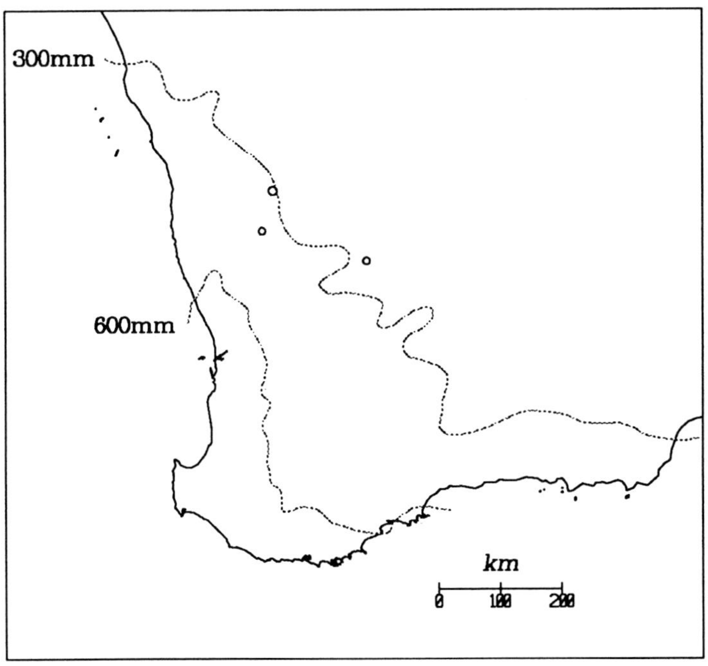

	1900/37	1987	1988	1989	1990
% of observers recording species	0	0.8	0	0.9	1.0
% of recording period species seen	NR	0.06	0	0.02	0.02

Pallid cuckoo

Cuculus pallidus

The pallid cuckoo is an uncommon seasonal migrant in woodland, mallee, shrubland and farmland throughout the wheatbelt (Fig. 111). It feeds on mistletoe berries, millipedes, centipedes, cockroaches, grasshoppers, cicadas, bugs, beetles, caterpillars, ants and spiders. It breeds in spring and early summer and lays its eggs in a cup-shaped nest of another species, for that species to raise. The major hosts of the pallid cuckoo in the wheatbelt are the red wattlebird, singing honeyeater and yellow-throated miner (Brooker and Brooker 1989). The latter two species are common throughout the wheatbelt, which accounts for the abundance of the pallid cuckoo.

The pallid cuckoo varies in its time of arrival each year and in the number appearing; in some years they may not be seen in some districts. For example, historically it arrived in the Broome Hill district between the end of April and July, and departed between October and January (Carter 1923). It was plentiful between July and March in the Lake Grace area (Carnaby 1933). In the Northam district, it arrived occasionally in March, but usually later, departed in early December through to January and varied greatly in numbers and time of arrival (Masters and Milhinch 1974). In the Kellerberrin area it arrived in June or July and left by December (Leake 1962). This seasonal pattern is unchanged as shown below by arrival and departure times for five locations during this survey.

	1988	1989	1990
Tardun	Jun. Week 3 Sep. Week 4	Apr. Week 2 Nov. Week 1	May Week 2 Nov. Week 1
Arrino	Jun. Week 3 Oct. Week 2	Apr. Week 3 Nov Week 4	Mar. Week 4 *
Avon River between Northam and Toodyay	May Week 1 Jan. Week 2	May Week 3 Dec. Week 4	Apr. Week 3 Dec. Week 1
Guildford	Aug. Week 3 Dec. Week 4	Jul. Week 1 Dec. Week 1	Jun. Week 3 *
Pingrup	Jul. Week 2 Nov. Week 4	Apr. Week 4 Nov. Week 2	Mar. Week 4 Nov. Week 2

*No data on departure time as the survey ended on 31 December, 1990.

Fig. 111. Distribution 1900–1937

Pallid cuckoo

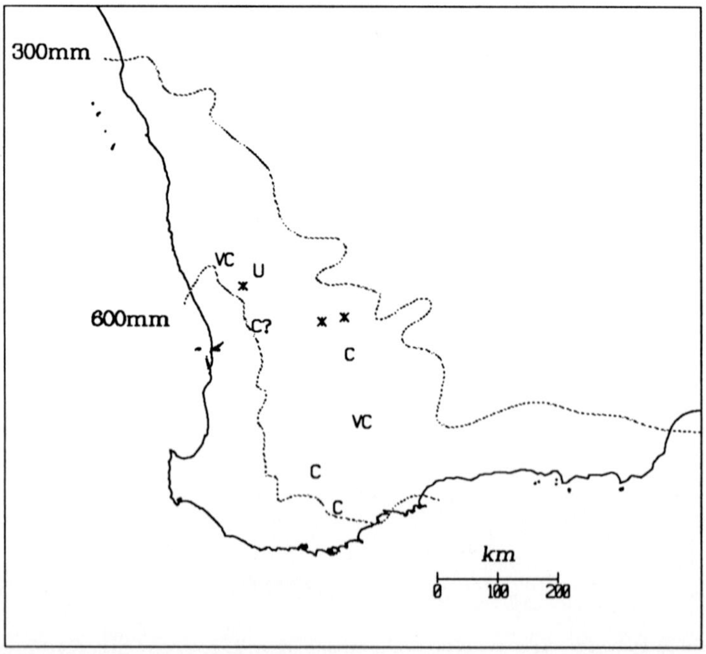

Fig. 111. Distribution 1987–1990

Pallid cuckoo

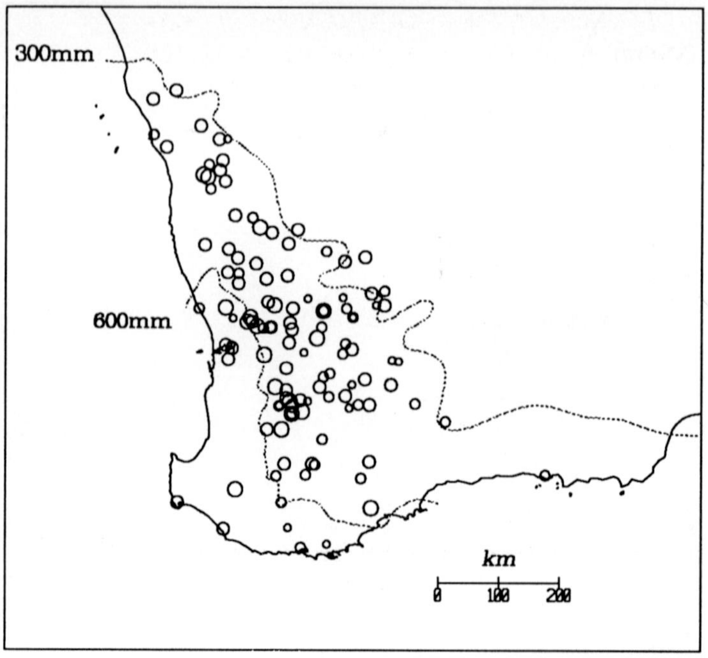

	1900/37	1987	1988	1989	1990
% of observers recording species	100	62	52	58	67
% of recording period species seen	U-VC	20	12	11	20

Fig. 111. Distribution 1987–1990 — January–February

Pallid cuckoo

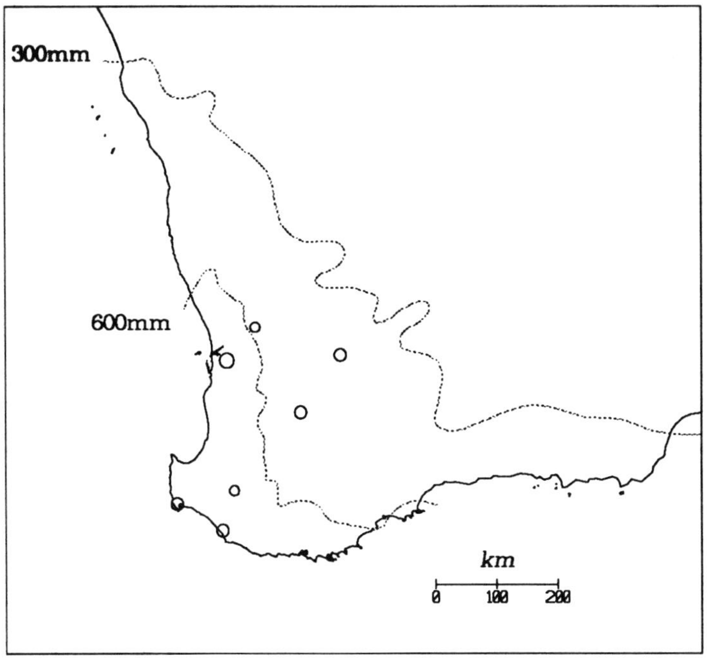

Fig. 111. Distribution 1987–1990 — March–April

Pallid cuckoo

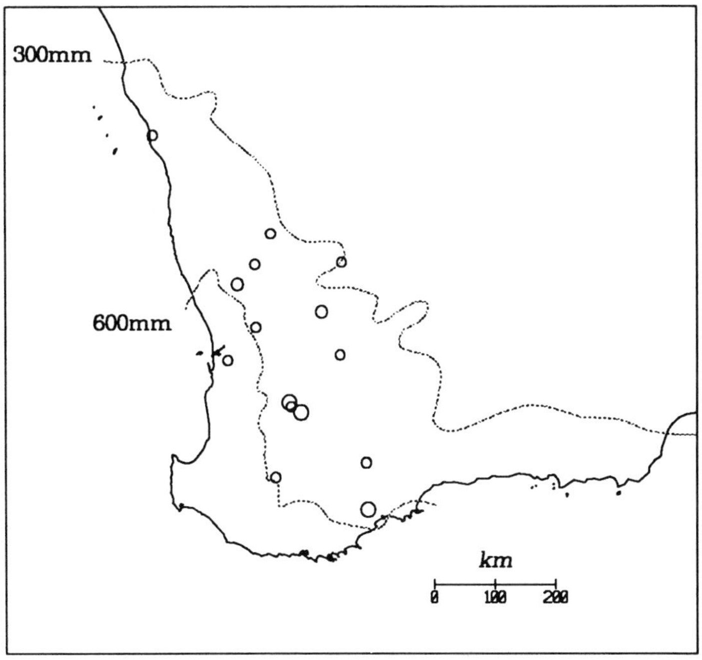

The difference in abundance between years was demonstrated during this survey as the pallid cuckoo was recorded by more observers in 1990 than in the previous two years. In addition, recording rates were nearly twice those of 1988 and 1989.

There is no evidence to suggest that the status of the pallid cuckoo has changed over the last 50 years.

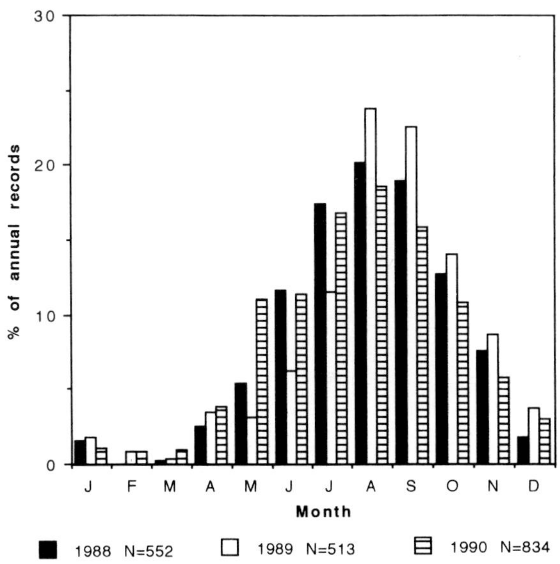

Fig. 111. Distribution 1987–1990 — May–December

Pallid cuckoo

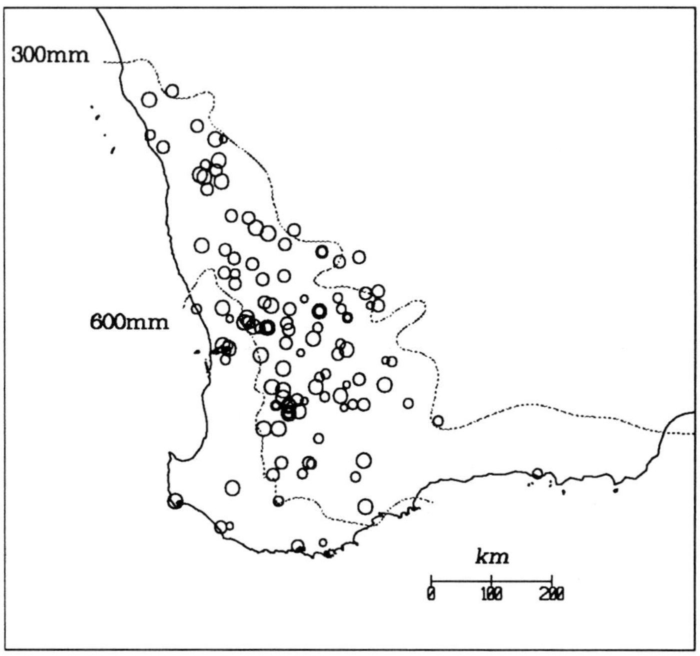

Fan-tailed cuckoo

Cuculus pyrrhophanus

The fan-tailed cuckoo is an uncommon seasonal migrant in woodland, mallee and farmland in the wheatbelt (Fig. 112). It is resident in the higher rainfall zone of the lower southwest. It feeds on centipedes, cockroaches, grasshoppers, bugs, beetles, caterpillars, moths, sawfly larvae, ants and spiders.

The fan-tailed cuckoo breeds in spring and parasitizes species which build dome-shaped nests. The white-browed scrub wren and broad-tailed thornbill are the preferred hosts, however, the redthroat, field wren and yellow-rumped thornbill are also parasitized (Brooker and Brooker 1989). All of the host species have declined in the wheatbelt, so it is not surprising that this cuckoo is rare or uncommon in locations where it was regarded as common, e.g., Wongan Hills (Milligan 1904), near Bruce Rock (Crossman 1909), Broome Hill (Carter 1923) and Lake Grace (Carnaby 1933) (Fig. 112).

Historically, the seasonal movement pattern of the fan-tailed cuckoo was noted for the Broome Hill district (Carter 1923) and the Northam district (late March to mid-October) (Masters and Milhinch 1974), but it was resident around Lake Grace (Carnaby 1933). At present, the fan-tailed cuckoo does not appear every year at the outer limits of its range. The timing of its arrival varies between years and districts and this is illustrated for four locations in the wheatbelt during this survey (Table below and Fig. on facing page).

	1988	1989	1990
Tardun	Not seen	Not seen	Seen Aug. Wk 2
Arrino	Jun. Wk 4 Aug. Wk 4	Seen Aug. Wk 1	Not seen
Avon River between Northam and Toodyay	Apr. Wk 3 Sep. Wk 2	Apr. Wk 1 Sep. Wk 3	May Wk 2 Jun. Wk 4
Pingrup	Jun. Wk 4 Nov. Wk 3	Jul. Wk 4 Sep. Wk 4	Jul. Wk 1 Oct. Wk 1

Fig. 112. Distribution 1900–1937

Fan-tailed cuckoo

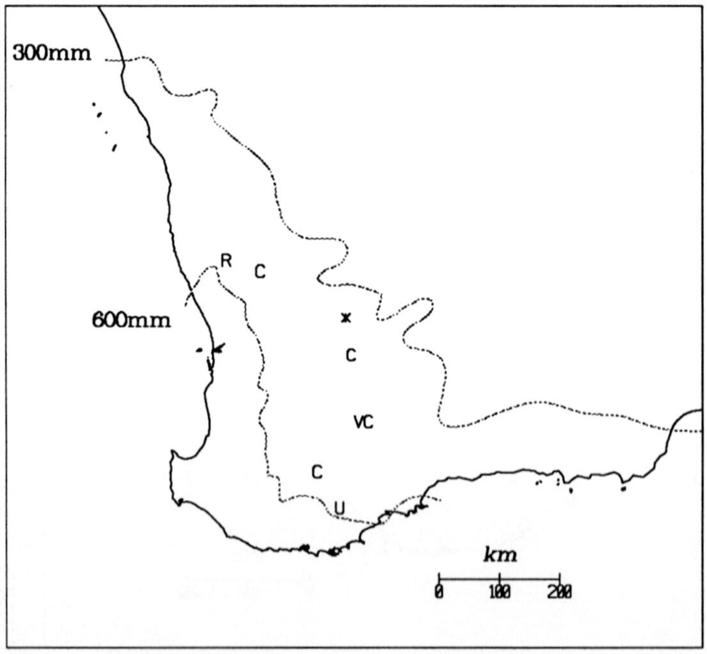

Fig. 112. Distribution 1987–1990

Fan-tailed cuckoo

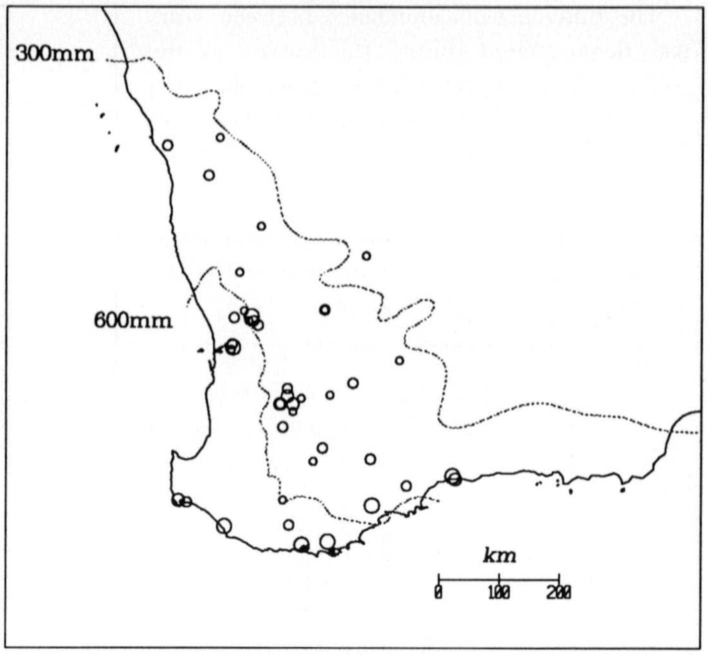

	1900/37	1987	1988	1989	1990
% of observers recording species	70	13	19	22	20
% of recording period species seen	R-VC	3.6	5.1	4.9	4.3

Fig. 112. Distribution 1987–1990 — April–November

Fan-tailed cuckoo

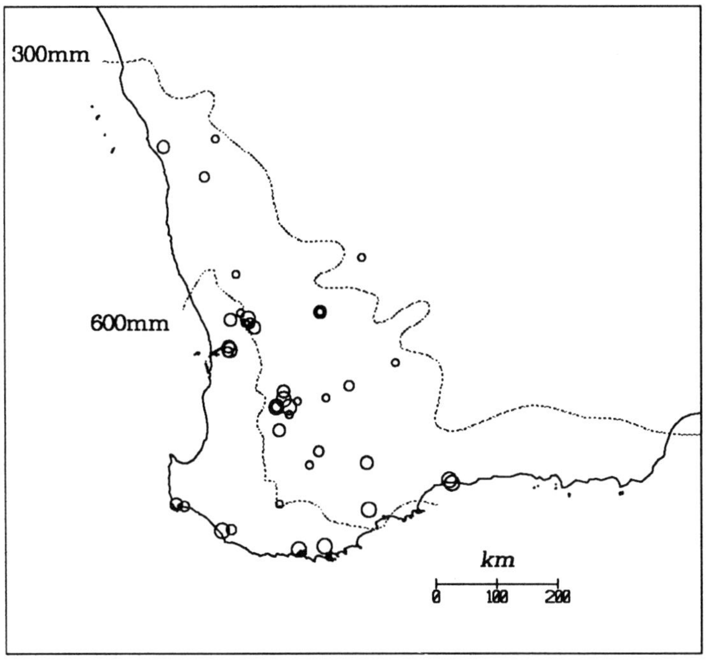

Fig. 112. Distribution 1987–1990 — December–March

Fan-tailed cuckoo

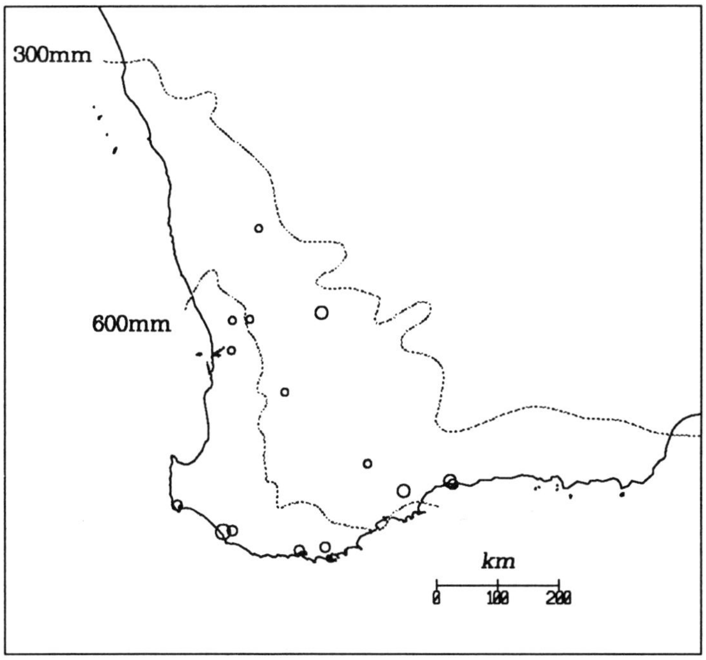

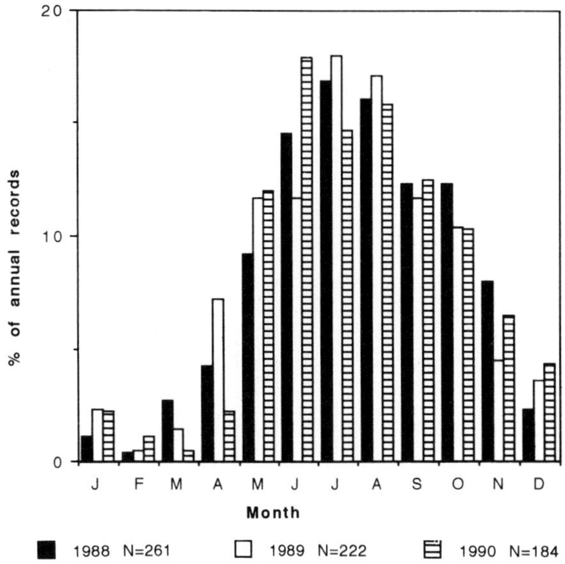

Black-eared cuckoo

Chrysococcyx osculans

The black-eared cuckoo is an arid zone bird that is a rare, seasonal nomad in woodland, mallee and shrubland of the northern and central wheatbelt (Fig. 113). It feeds on bugs, beetles, sandflies, caterpillars and moths. The black-eared cuckoo breeds in spring and early summer, and lays its eggs in the nests of the redthroat (Brooker and Brooker 1989) and broad-tailed thornbill (J. Masters, in litt.). Both species have declined markedly and this cuckoo may have declined as a consequence.

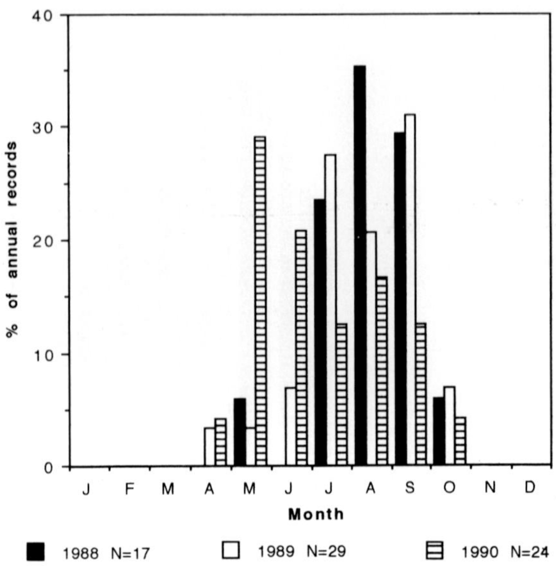

Fig. 113. Distribution 1900–1937

Black-eared cuckoo

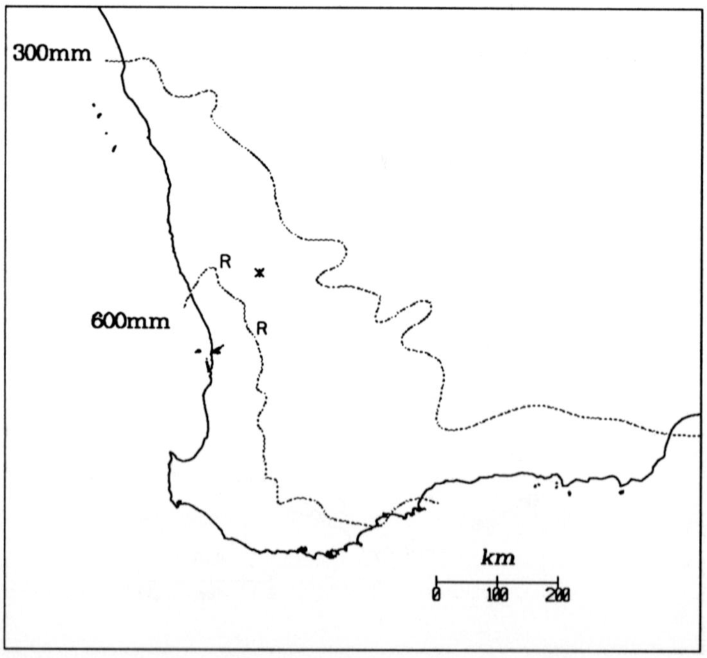

Fig. 113. Distribution 1987–1990

Black-eared cuckoo

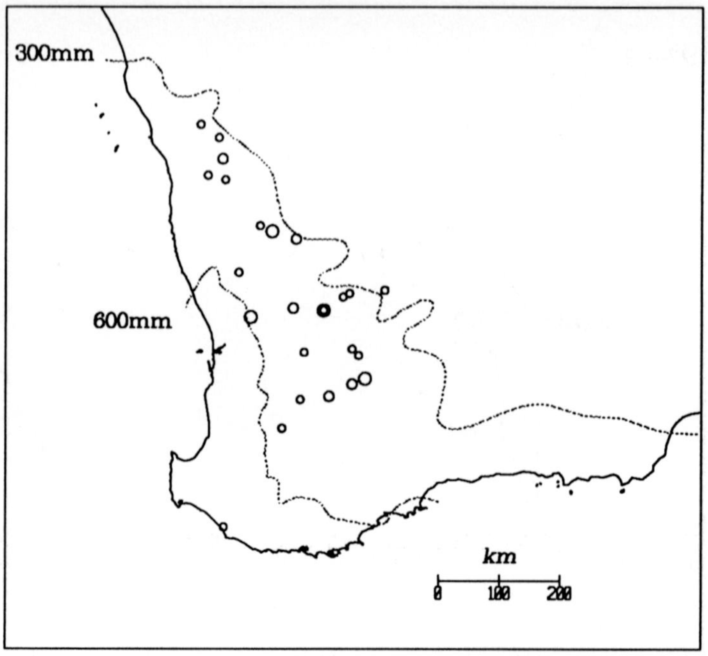

	1900/37	1987	1988	1989	1990
% of observers recording species	30	10	4.9	6.8	8.7
% of recording period species seen	R	1.5	0.3	0.6	0.6

Fig. 113. Distribution 1987–1990 — January–April

Black-eared cuckoo

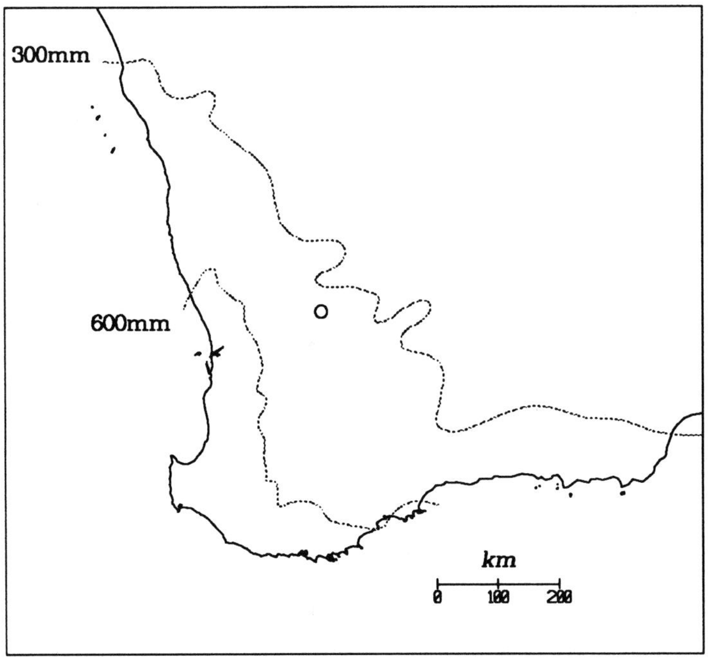

Fig. 113. Distribution 1987–1990 — June–December

Black-eared cuckoo

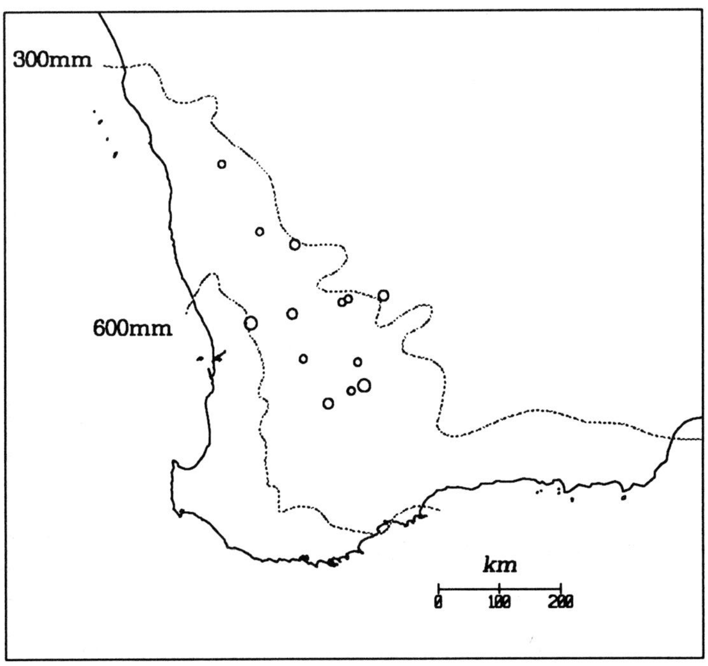

Horsfield's bronze cuckoo

Chrysococcyx basalis

Horsfield's bronze cuckoo is ubiquitous in Australia, and an uncommon seasonal, breeding migrant in woodland, mallee, shrubland and heath throughout the wheatbelt (Fig. 114). It feeds on fruit, berries, seed, termites, earwigs, grasshoppers, crickets, bugs, beetles, fly larvae, caterpillars, moths, wasps, ants and spiders. This cuckoo breeds in spring and lays its eggs in dome-shaped nests of wrens, particularly the splendid wren, and thornbills, including the yellow-rumped thornbill (Brooker and Brooker 1989). It occasionally lays in open nests of species like the red-capped and hooded robins. All of these species have declined and, as a result, this cuckoo has probably decreased in abundance. Historically, the seasonal pattern of occurrence was noted in the Kellerberrin/ Kwolyin districts (early June to end of January; earliest record in late April) (Ford and Stone 1957) and around Northam (earliest record 17 March and latest 1 January) (Masters and Milhinch 1974). This pattern is unchanged as shown by arrival and departure dates for five wheatbelt locations during this survey (Table below and Fig. on facing page).

	1988	1989	1990
Tardun	Jul. Wk 4 Nov. Wk 2	Jun. Wk 4 Nov. Wk 4	Jun. Wk 3 Oct. Wk 2
Arrino	Seen Jul. Wk 3	Seen Aug. Wk 3	Sep. Wk 3 Dec. Wk 2
Avon River between Northam and Toodyay	Jul. Wk 4 Nov. Wk 4	Jul. Wk 4 Dec. Wk 2	May Wk 1 Nov. Wk 3
Bruce Rock	Seen Sep. Wk 3	Not seen	Not seen
Pingrup	Jun. Wk 4 Jan. Wk 4	Aug. Wk 2 Jan. Wk 2	Jul. Wk 4 *

*No data available as survey ended 31 December, 1990.

Fig. 114. Distribution 1900–1937

Horsfield's bronze cuckoo

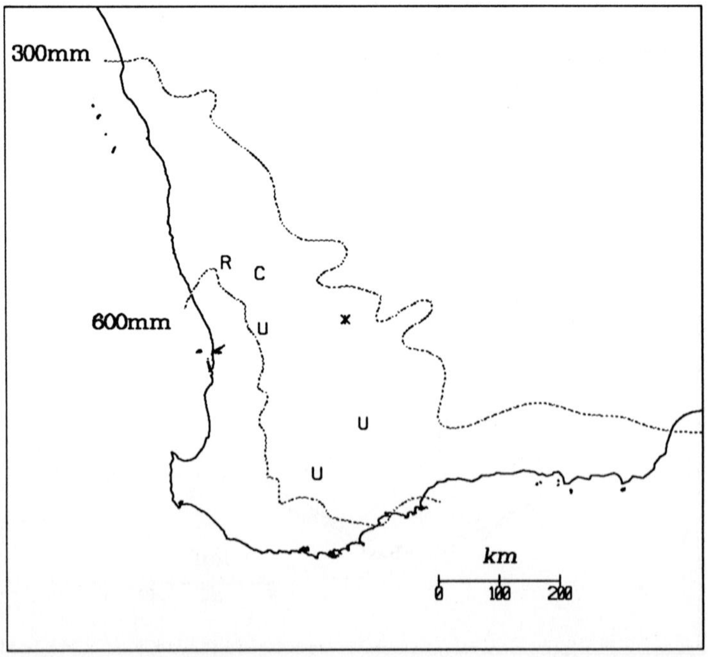

Fig. 114. Distribution 1987–1990

Horsfield's bronze cuckoo

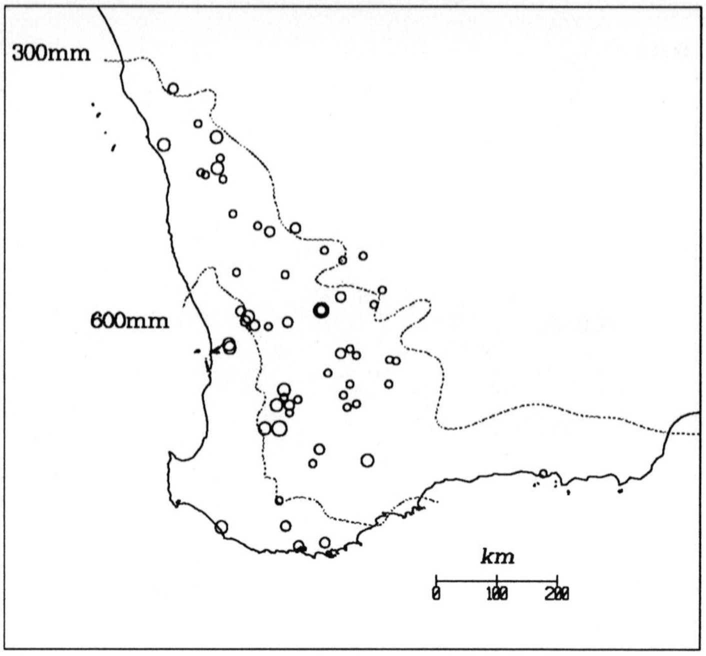

	1900/37	1987	1988	1989	1990
% of observers recording species	60	26	25	20	32
% of recording period species seen	R-C	5.2	4.1	3.7	4.7

Fig. 114. Distribution 1987–1990 — January–April

Horsfield's bronze cuckoo

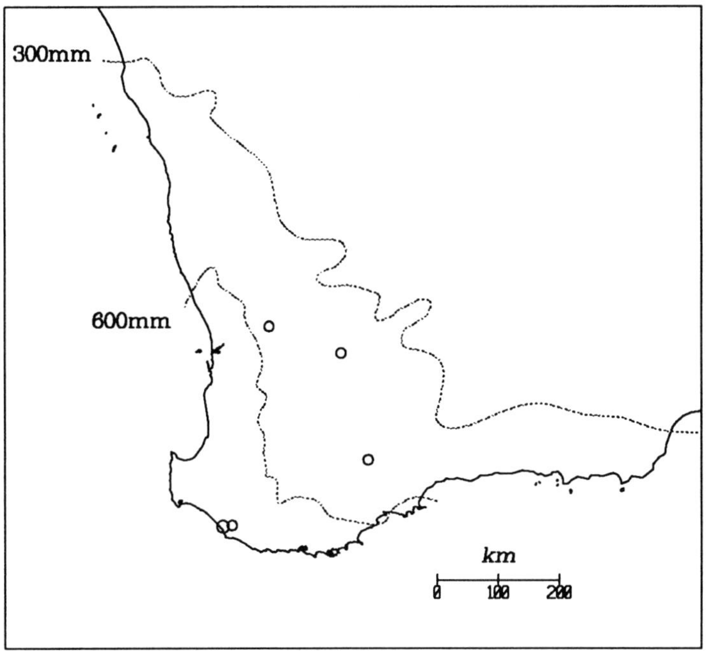

Fig. 114. Distribution 1987–1990 — May–December

Horsfield's bronze cuckoo

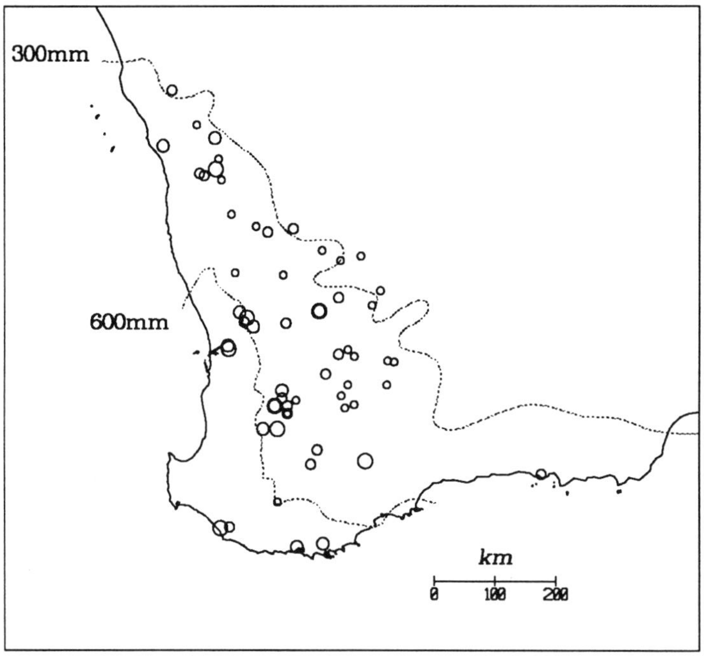

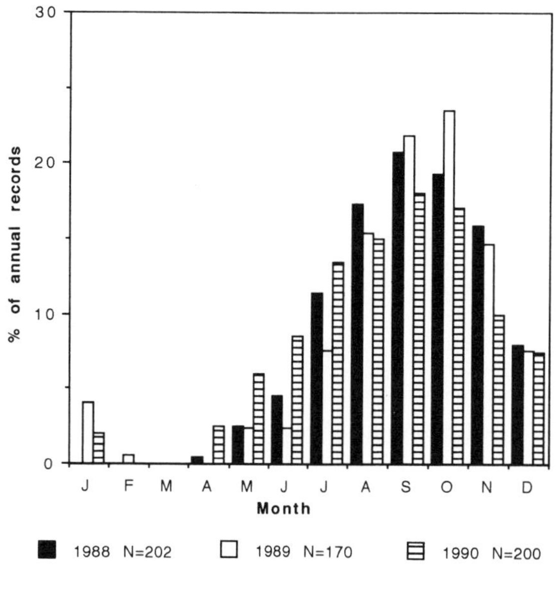

Golden bronze cuckoo

Chrysococcyx lucidus

OTHER NAME:

Shining bronze cuckoo

The golden bronze cuckoo winters in Indonesia (Brooker and Brooker 1989) and is an uncommon, seasonal breeding migrant in woodlands of the wheatbelt (Fig. 115). It feeds on bugs, beetles, caterpillars and spiders. It breeds in spring and early summer and lays its eggs in the enclosed nests of thornbills, particularly yellow-rumped and broad-tailed thornbills (Brooker and Brooker 1989).

Historically, the golden bronze cuckoo was the most common of the bronze cuckoos in the central valley of the Northam district. It occurred there from late April to December, with most arriving in June and July (Masters and Milhinch 1974). This seasonal pattern is shown for two wheatbelt locations during this survey in the Table below and in Figure below.

	1988	1989	1990
Avon River between Northam and Toodyay	Jul. Wk 1 Nov. Wk 3	Jul. Wk 3 Dec. Wk 3	May Wk 3 Dec. Wk 3
Pingrup	Jun. Wk 3 Jan. Wk 2	Aug. Wk 1 Nov. Wk 3	Jul. Wk 4 *

*No data available on departure times as the survey ended on 31 December, 1990.

Regarded by observers as common to very common over a wide area of the wheatbelt, this species has declined in abundance over the last 50 years.

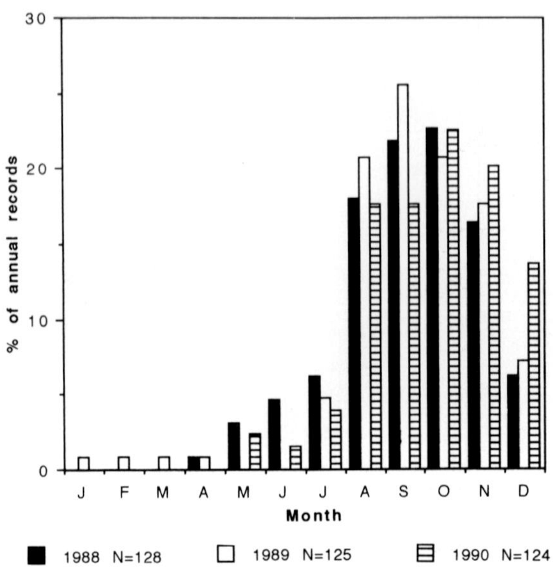

Fig. 115. Distribution 1900–1937

Golden bronze cuckoo

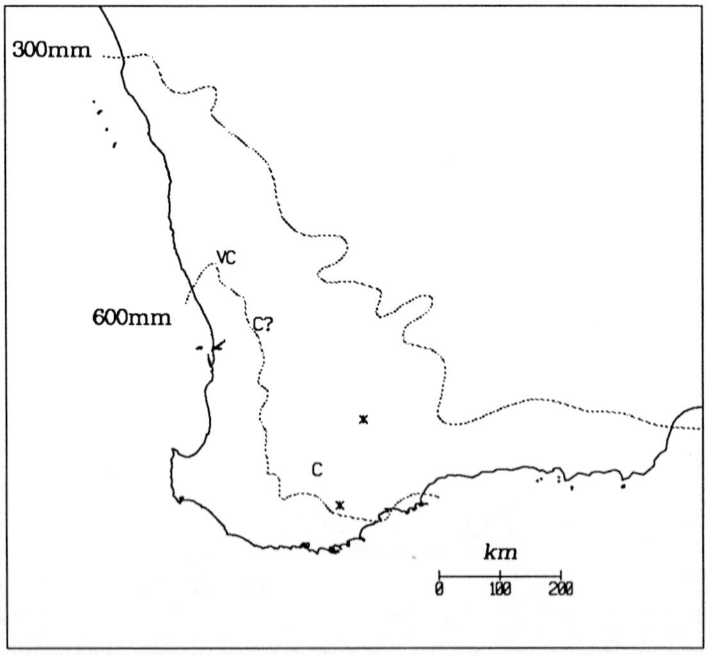

Fig. 115. Distribution 1987–1990

Golden bronze cuckoo

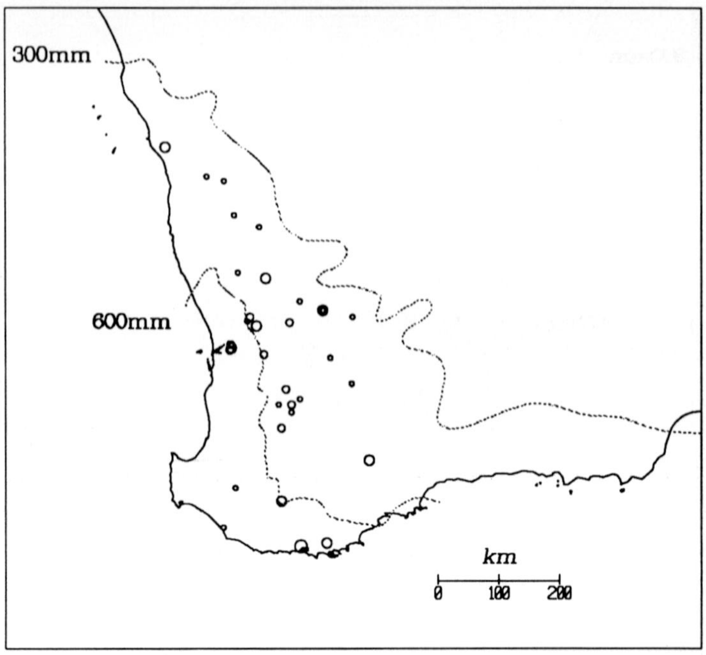

	1900/37	1987	1988	1989	1990
% of observers recording species	50	12	15	16	15
% of recording period species seen	C-VC	2.9	2.7	2.8	2.9

Fig. 115. Distribution 1987–1990 — February–April

Golden bronze cuckoo

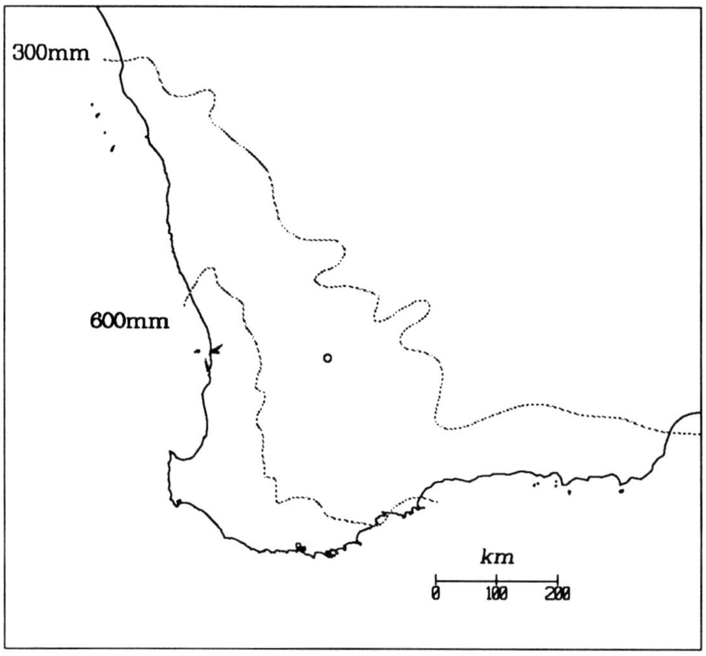

Fig. 115. Distribution 1987–1990 — May–December

Golden bronze cuckoo

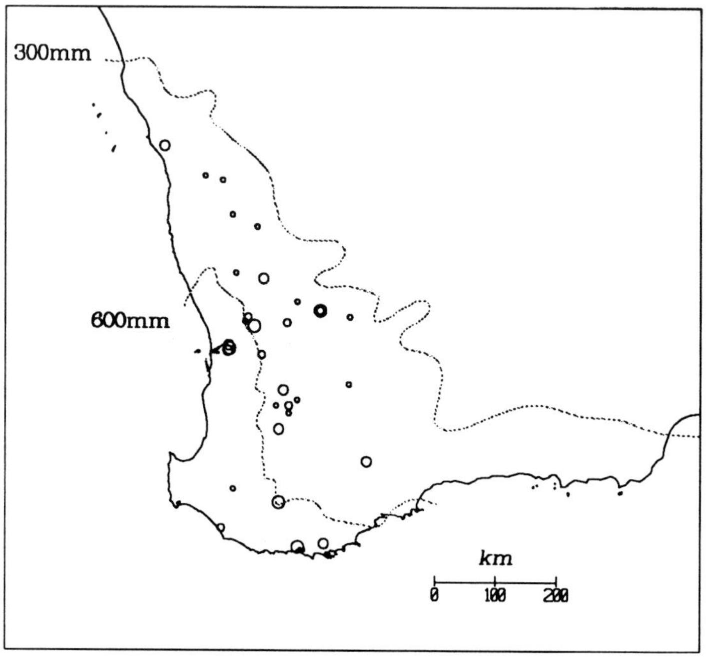

Boobook owl

Ninox novaeseelandiae

The boobook owl is an uncommon resident in woodland, mallee and shrubland of the wheatbelt (Fig. 116). It feeds on centipedes, cockroaches, mantids, grasshoppers, crickets, cicadas, beetles, moths, freshwater crayfish, spiders, frogs, geckoes, birds and small mammals. It breeds in spring in a tree hollow and lays 2–3 eggs.

The boobook owl is nocturnal and is heard more often than seen. This species may have declined in abundance over the last 50 years as a result of destruction of its habitat, as pointed out by Leake (1962) around Kellerberrin. Masters and Milhinch (1974) noted that the boobook owl was not heard as often around Northam as it was in the 1930s.

Fig. 116. Distribution 1900–1937

Boobook owl

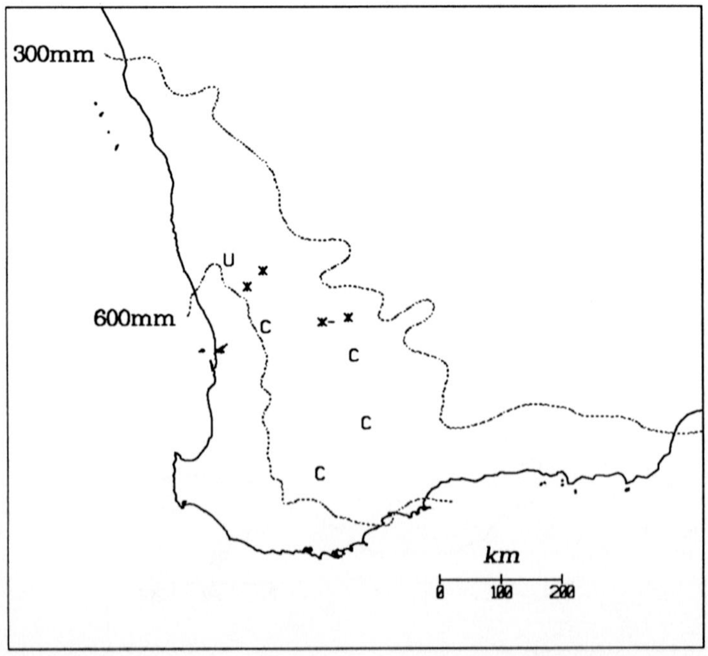

Fig. 116. Distribution 1987–1990

Boobook owl

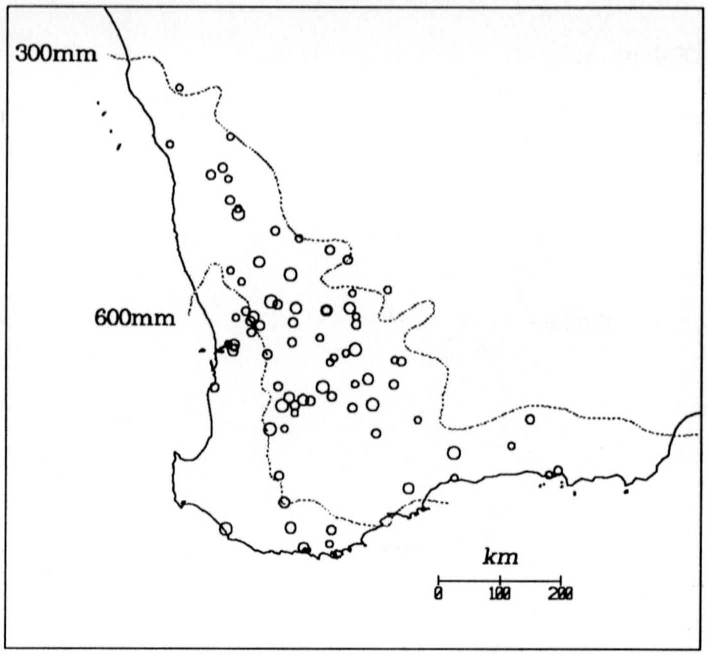

	1900/37	1987	1988	1989	1990
% of observers recording species	90	29	37	35	39
% of recording period species seen	U-C	6.3	6.7	7.2	7.2

Barking owl

Ninox connivens

The barking owl is a rare resident in woodland and mallee in the wheatbelt (Fig. 117). It feeds on grasshoppers, beetles, lizards, birds and small mammals. It breeds in late winter and spring in a tree hollow and lays 2–3 eggs.

Serventy and Whittell (1962) noted that this species is definitely rarer than the boobook owl and was only occasionally reported. There is no evidence to suggest that this species has changed in abundance over the last 50 years.

Fig. 117. Distribution 1987–1990

Barking owl

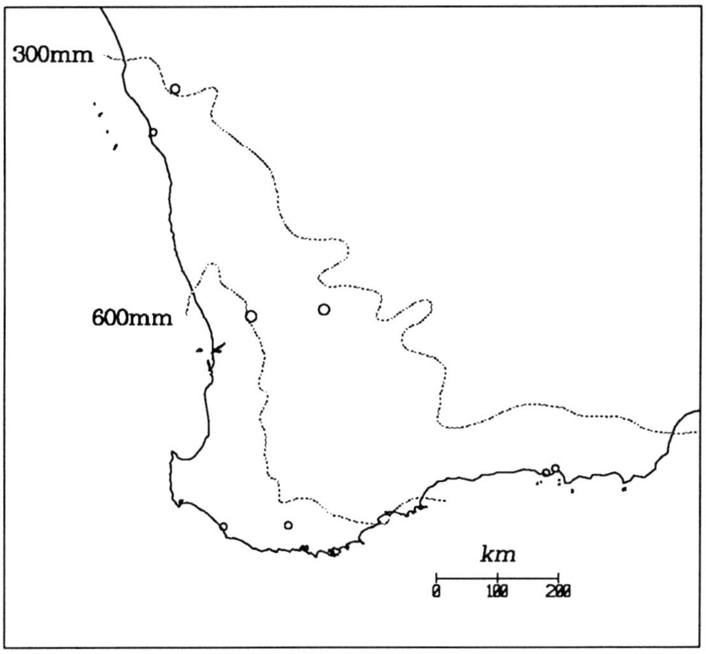

	1900/37	1987	1988	1989	1990
% of observers recording species	0	2.4	3.5	4.3	4.8
% of recording period species seen	NR	0.4	0.3	0.5	0.3

Barn owl

Tyto alba

The barn owl is an uncommon resident of woodland and farmland throughout the wheatbelt (Fig. 118). It feeds on cockroaches, grasshoppers, crickets, beetles, caterpillars, moths, bees, wasps, spiders, frogs, lizards, birds and mammals, including bats. The barn owl breeds in winter and spring in a tree hollow. It lays 3–4 eggs.

There is no evidence to suggest that the barn owl has changed in status over the last 50 years.

Fig. 118. Distribution 1900–1937

Barn owl

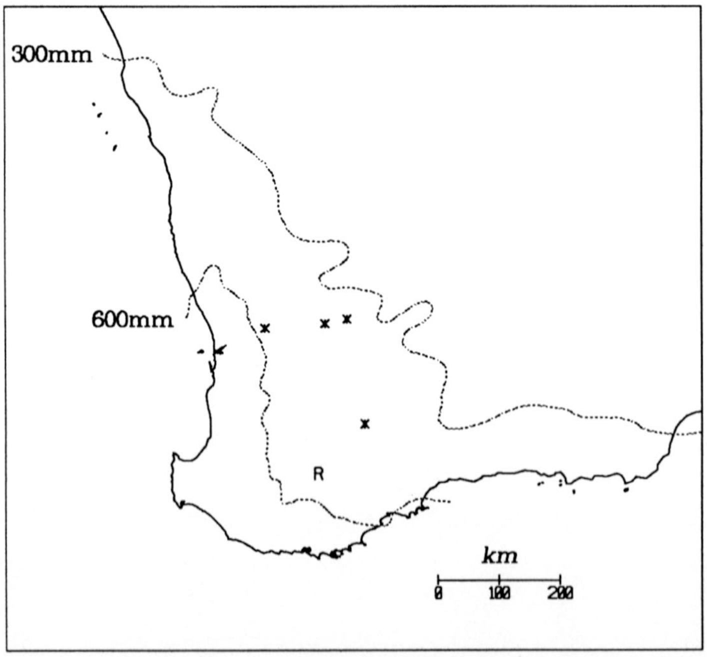

Fig. 118. Distribution 1987–1990

Barn owl

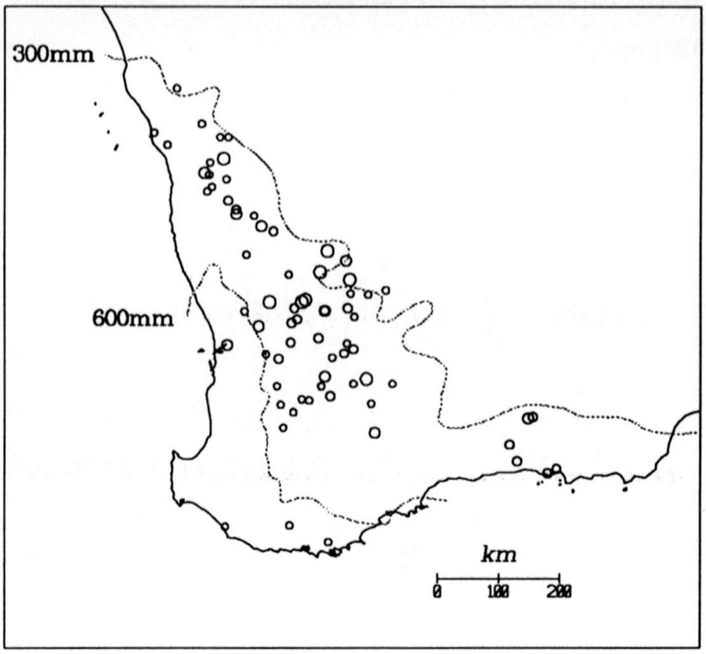

	1900/37	1987	1988	1989	1990
% of observers recording species	50	20	27	32	33
% of recording period species seen	R	3.6	5.5	5.5	5.7

Masked owl

Tyto novaehollandiae

The masked owl is a rare resident of woodland in the wheatbelt (Fig. 119). It feeds on beetles, moths and mammals. It breeds in spring in a tree hollow and lays 2–3 eggs.

According to Serventy and Whittell (1962), the masked owl was much less common than the barn owl. Like most nocturnal species, this species is probably under-sampled and there are insufficient data to assess if there have been any changes in status.

Fig. 119. Distribution 1987–1990

Masked owl

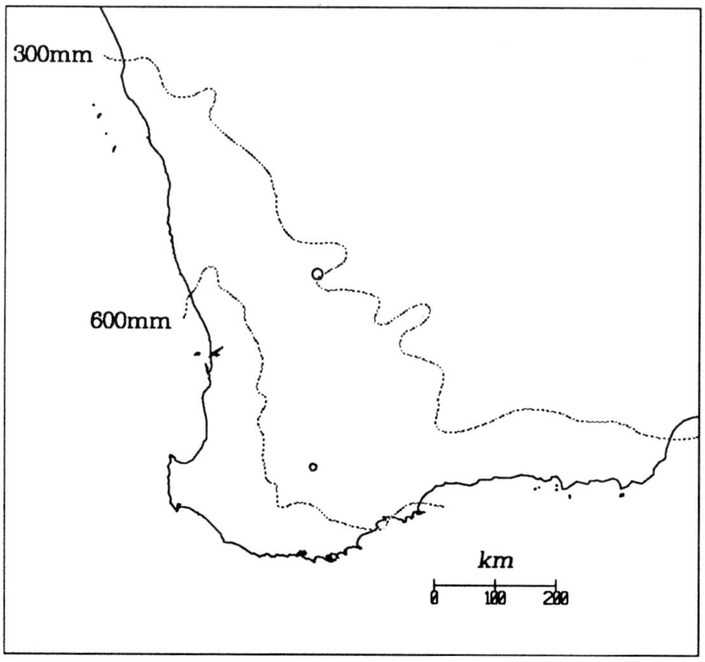

	1900/37	1987	1988	1989	1990
% of observers recording species	0	0	1.4	0.9	1.0
% of recording period species seen	NR	0	0.2	0.2	0.4

Tawny frogmouth

Podargus strigoides

The tawny frogmouth is an uncommon resident of woodland, mallee and shrubland throughout the wheatbelt (Fig. 120). It feeds on seed, millipedes, centipedes, dragonflies, cockroaches, mantids, earwigs, grasshoppers, crickets, phasmids, bugs, cicadas, beetles, moths, caterpillars, wasps, sawfly larvae, ants, scorpions, spiders, frogs, lizards, birds and small mammals. The tawny frogmouth breeds in spring on a stick nest in a tree and lays 2–3 eggs.

The tawny frogmouth is most often seen along roads at night and they are occasionally killed by cars. Five of the annotated bird lists of the wheatbelt published before 1937 listed this species as common, however, it is no longer common. It has probably decreased in abundance over the last 50 years as a result of clearing for agriculture (Leake 1962; Masters and Milhinch 1974).

Fig. 120. Distribution 1900–1937

Tawny frogmouth

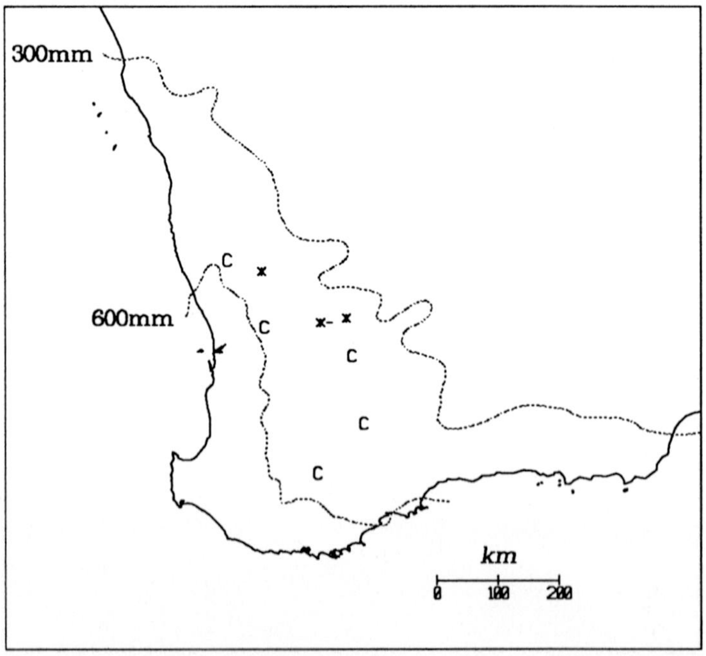

Fig. 120. Distribution 1987–1990

Tawny frogmouth

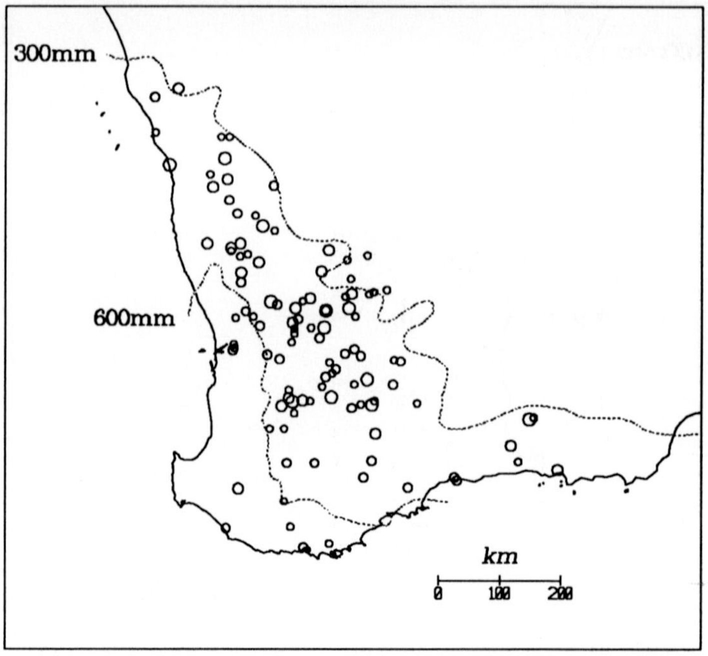

	1900/37	1987	1988	1989	1990
% of observers recording species	80	44	45	52	52
% of recording period species seen	C	9.6	9.0	11	8.5

Fig. 121. Distribution 1900–1937

Owlet-nightjar

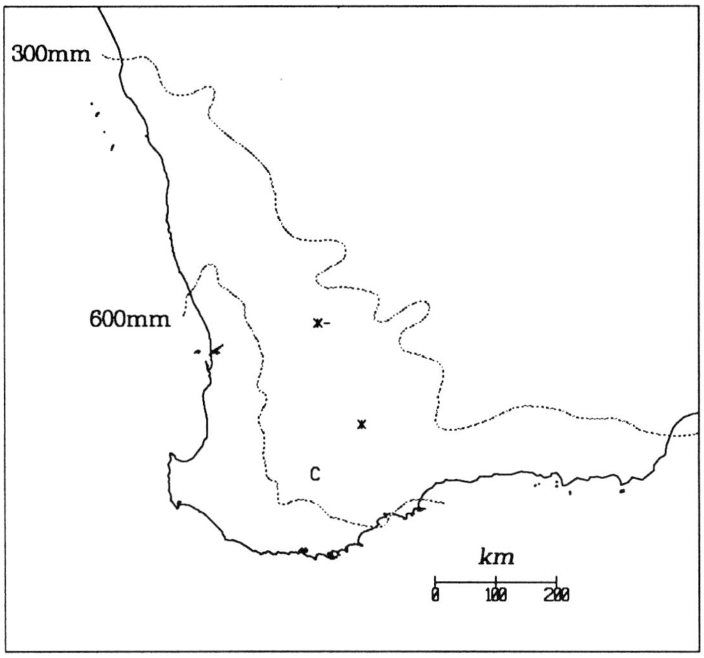

Fig. 121. Distribution 1987–1990

Owlet-nightjar

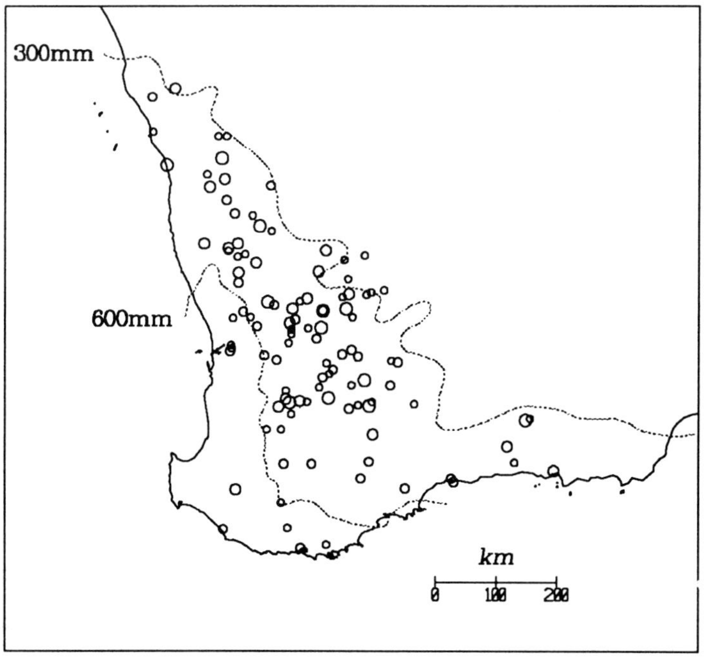

	1900/37	1987	1988	1989	1990
% of observers recording species	30	8.7	19	15	15
% of recording period species seen	C	1.5	1.8	2.0	1.2

Owlet-nightjar

Aegotheles cristatus

OTHER NAME:

Australian owlet-nightjar

The owlet-nightjar is ubiquitous in Australia, and an uncommon resident in woodland and mallee throughout the wheatbelt (Fig. 121). It feeds on millipedes, cockroaches, earwigs, grasshoppers, bugs, beetles, maggots, caterpillars, ants and spiders. The owlet-nightjar breeds in spring in a tree hollow and lays 3–4 eggs.

Like other nocturnal birds, the owlet-nightjar is probably under-sampled as it is heard more often than seen. Anyone unfamiliar with its call may not record the species, unless they see it at night along a road, or scare it out of a tree hollow by day. Leake (1962) commented that it had declined as a result of clearing for agriculture.

Spotted nightjar

Caprimulgus guttatus

The spotted nightjar is rare in woodland and mallee in the wheatbelt (Fig. 122). It feeds on mantids, grasshoppers, crickets, bugs, lacewings, beetles, moths and flying ants. It breeds in winter and spring, nests on the ground and lays one egg.

This survey did not reveal any pattern of occurrence. Being a nocturnal species, the spotted nightjar may have been under-sampled. There are no data to indicate that this species has changed in status.

Fig. 122. Distribution 1900–1937

Spotted nightjar

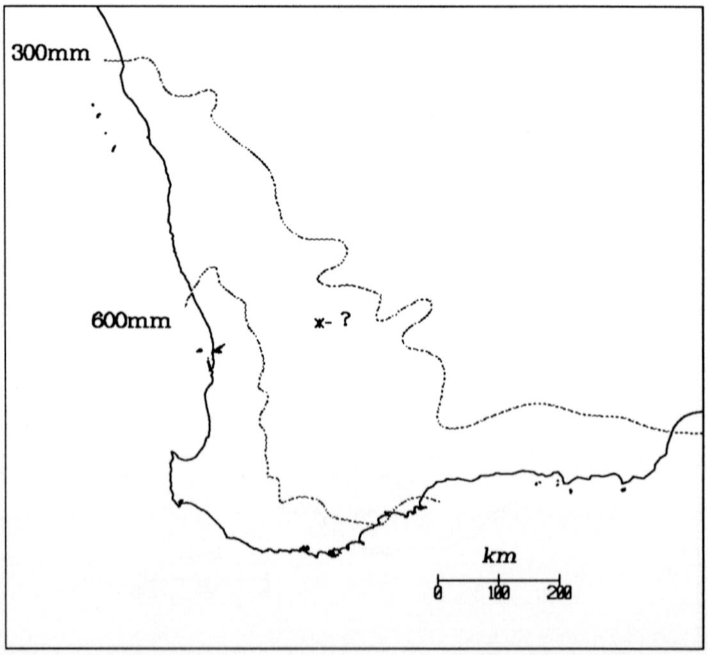

Fig. 122. Distribution 1987–1990

Spotted nightjar

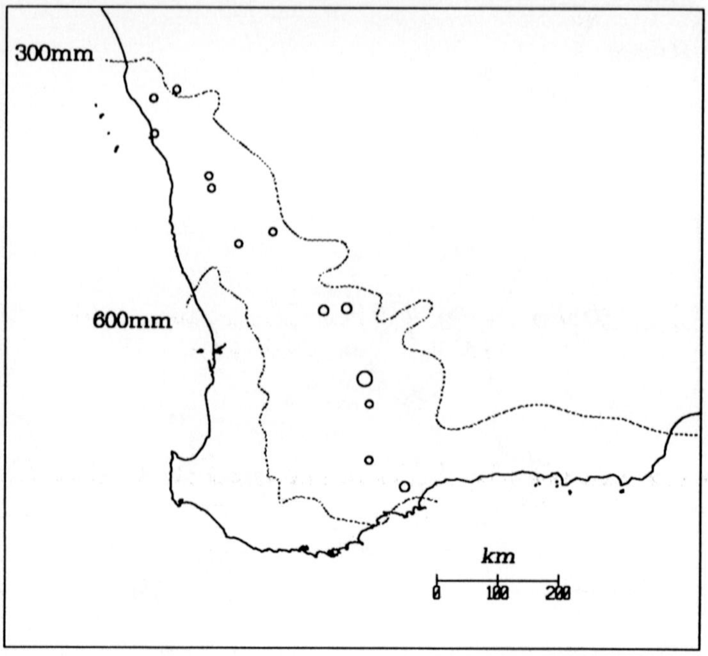

	1900/37	1987	1988	1989	1990
% of observers recording species	20	3.9	4.9	6.0	4.8
% of recording period species seen	—	0.3	0.8	1.1	1.2

Fig. 123. Distribution 1900–1937

Fork-tailed swift

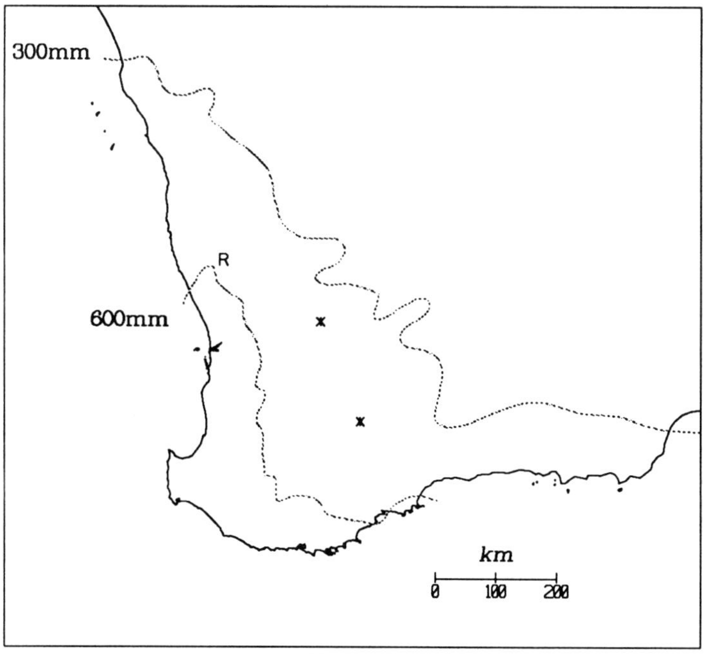

Fig. 123. Distribution 1987–1990

Fork-tailed swift

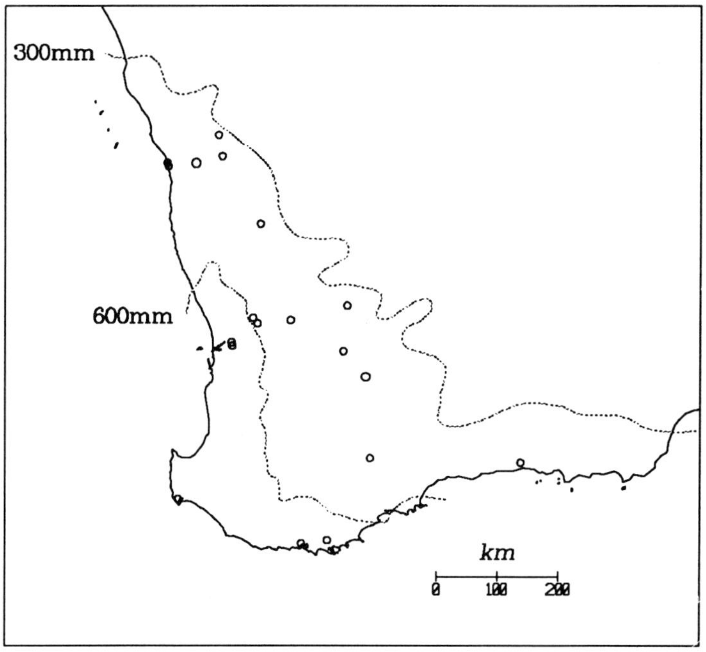

	1900/37	1987	1988	1989	1990
% of observers recording species	30	3.1	4.9	4.3	7.6
% of recording period species seen	R	0.2	0.2	0.1	0.3

Fork-tailed swift

Apus pacificus

The fork-tailed swift is a rare aerial vagrant which appears in summer and autumn (Fig. 123), particularly during stormy weather caused by northern low-pressure systems (Serventy and Whittell 1962). It is a transequatorial migrant which breeds in the Northern Hemisphere and has never been seen alighting in Australia (Readers' Digest 1976). It feeds on flying insects such as termites, bugs, beetles, midges, moths, bees and flying ants.

There are no data to suggest any changes in status.

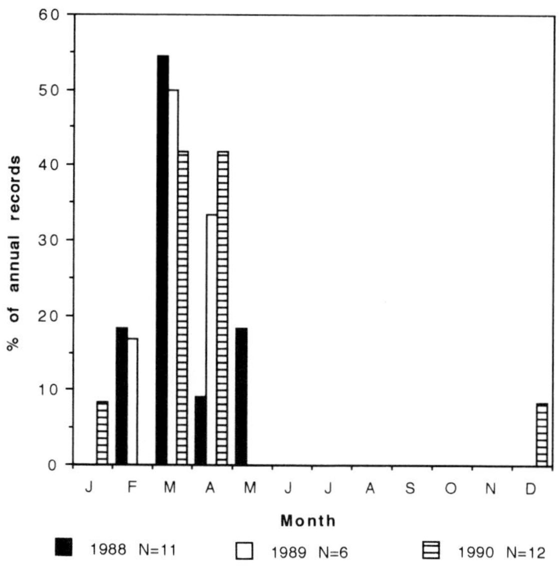

Laughing kookaburra

Dacelo gigas

The laughing kookaburra is a common resident in woodland in the higher rainfall districts of the wheatbelt (Fig. 124) and a rare vagrant in the drier margins adjacent to this range. It eats snails, slugs, earthworms, millipedes, centipedes, damselflies, mantids, grasshoppers, crickets, cicadas, beetles, caterpillars, ants, yabbies, spiders, fish, lizards, snakes, birds and small mammals. It breeds in spring and early summer in a tree hollow that has the floor level with the base of the entrance, and lays 2–4 eggs.

The laughing kookaburra was introduced into Western Australia at South Perth after 1897 (Serventy and Whittell 1962) and had invaded all suitable habitat by the 1940s. For example, it started breeding along the Avon River by 1926 and its population has been steady since about 1940 (Masters and Milhinch 1974).

Fig. 124. Distribution 1900–1937

Laughing kookaburra

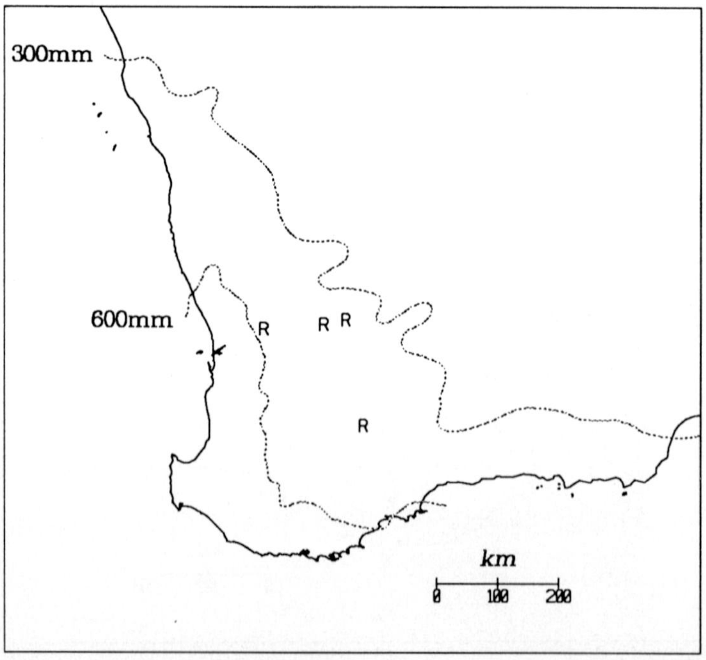

Fig. 124. Distribution 1987–1990

Laughing kookaburra

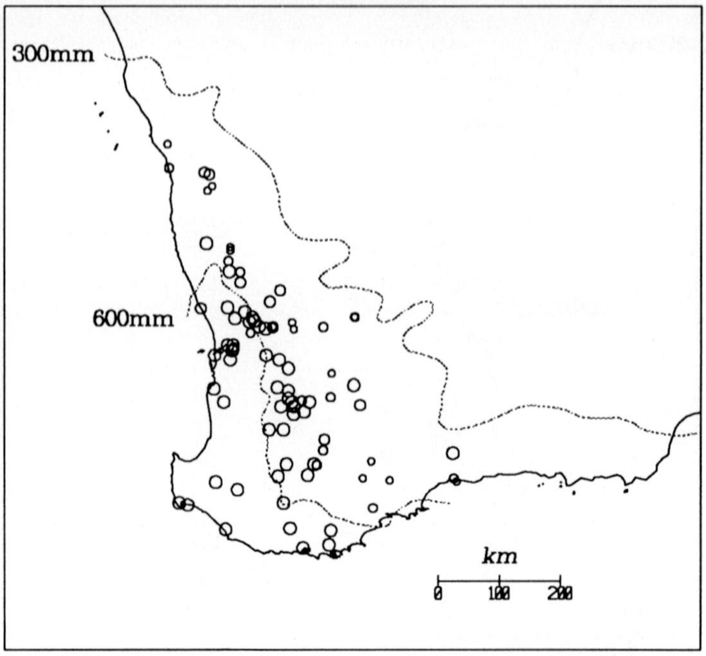

	1900/37	1987	1988	1989	1990
% of observers recording species	40	36	42	39	44
% of recording period species seen	R	18	28	30	31

Heath habitat (photos by D. A. Saunders).

Red-backed kingfisher

Halcyon pyrrhopygia

The red-backed kingfisher is a rare, seasonal breeding migrant in woodland and mallee of the wheatbelt (Fig. 125). It feeds on centipedes, mantids, grasshoppers, crickets, backswimmers, water boatmen, beetles, flies, caterpillars, crustacea, spiders, frogs, lizards and small mammals. It breeds in spring, nests in a burrow and lays 3–5 eggs.

There is no evidence to indicate that the status of the red-backed kingfisher has changed.

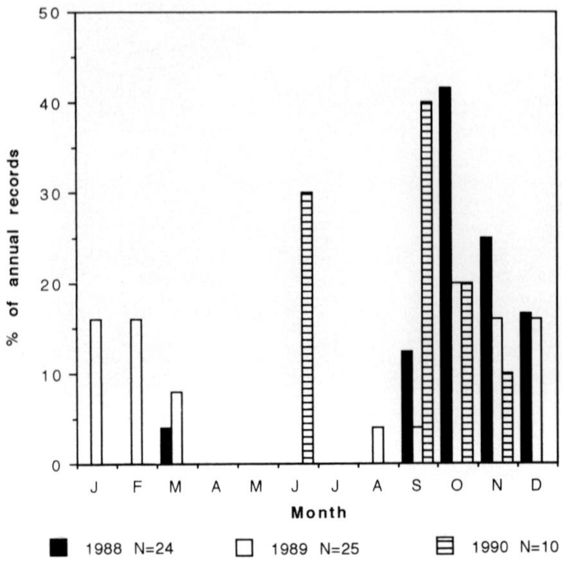

Fig. 125. Distribution 1900–1937

Red-backed kingfisher

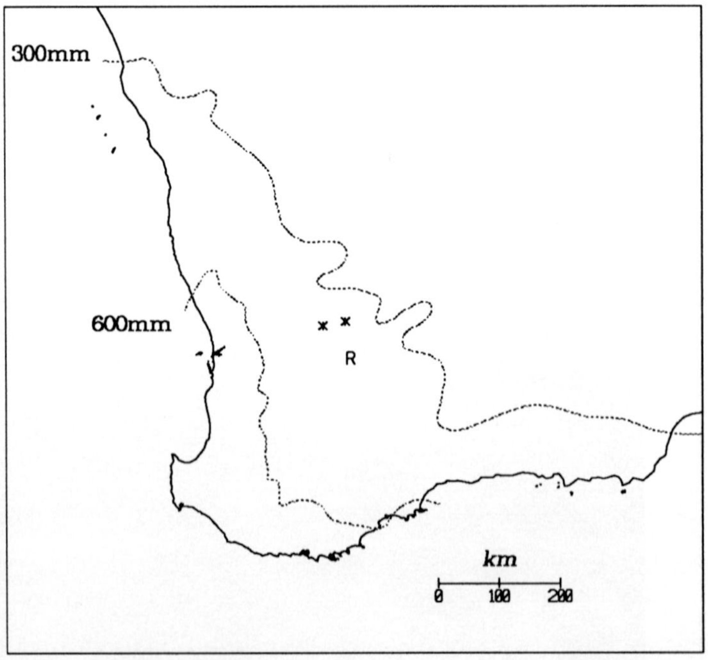

Fig. 125. Distribution 1987–1990

Red-backed kingfisher

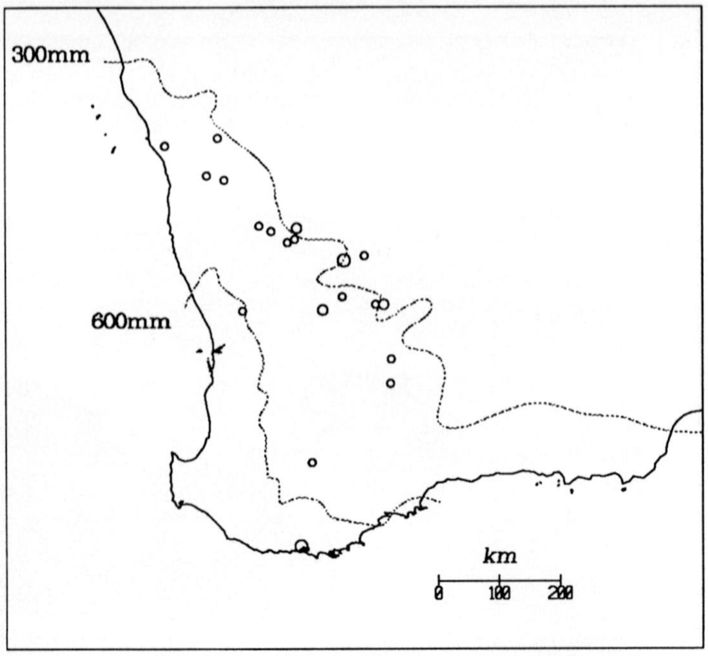

	1900/37	1987	1988	1989	1990
% of observers recording species	30	7.1	5.6	2.3	3.8
% of recording period species seen	R	0.6	0.5	0.5	0.2

Fig. 125. Distribution 1987–1990 — January–June

Red-backed kingfisher

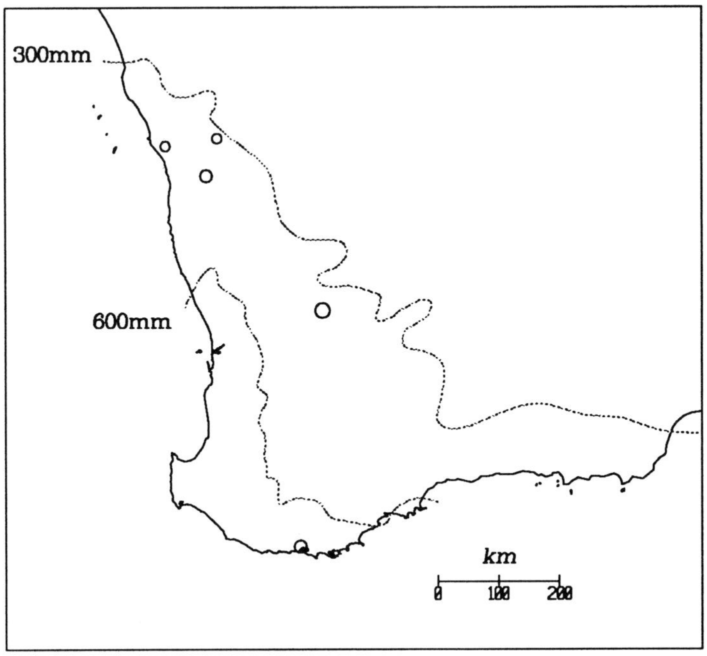

Fig. 125. Distribution 1987–1990 — July–December

Red-backed kingfisher

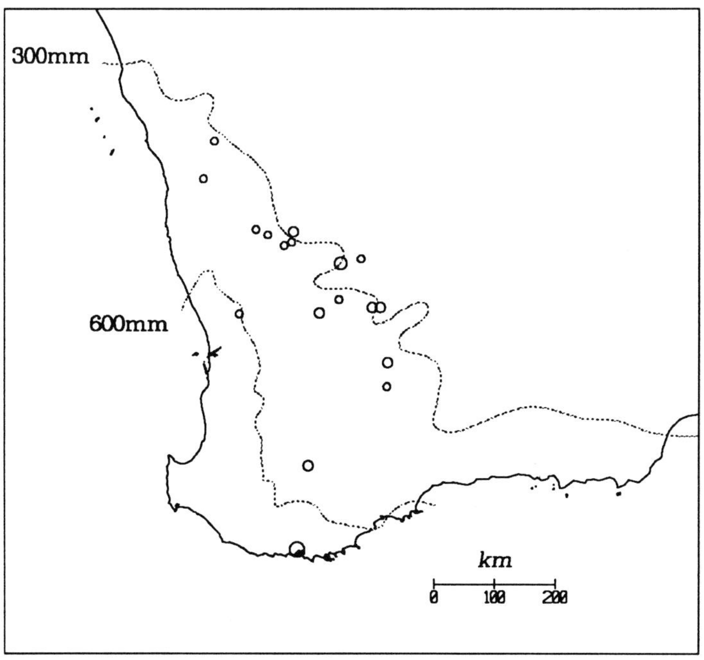

Sacred kingfisher

Halcyon sancta

The sacred kingfisher is an uncommon, breeding migrant in woodland and shrubland throughout the wheatbelt (Fig. 126). It migrates from Timor and some Indonesian islands (Serventy and Whittell 1962). It feeds on centipedes, mantids, earwigs, crickets, cicadas, beetles, caterpillars, native bees, ants, horseshoe shrimps, slaters, freshwater crayfish, spiders, fish, tadpoles, lizards, snakes and small mammals. It breeds in spring in a burrow in the ground or a hollow in a tree and lays 3–6 eggs.

Historically, the sacred kingfisher arrived in the Broome Hill district in late September or early October and remained until January or February (Carter 1923). In the Northam district it arrived about the end of September and departed in January (Jenkins 1931; Masters and Milhinch 1974). This seasonal pattern is also shown by the arrival and departure dates from four locations during this survey (Table below and Fig. on facing page). It is possible that some birds remain in the southwest all year.

	1988	1989	1990
Arrino	Oct. Wk 2 Dec. Wk 4	Oct. Wk 2 Jan. Wk 3	Sep. Wk 2 *
Avon River between Northam and Toodyay	Oct. Wk2 Jan. Wk 3	Oct. Wk 1 Jan. Wk 3	Sep. Wk 3 *
Guildford	Sep. Wk 4 Feb. Wk 1	Sep. Wk 3 Jan. Wk 2	Sep. Wk 2 *
Pingrup	Sep. Wk 3 Jan. Wk 4	Did not appear	Seen Mar.–Jun.

*No data available on departure times as the survey ended on 31 December, 1990.

Regarded as common to very common before 1937, the sacred kingfisher has declined and is now uncommon throughout much of the wheatbelt. For example, it was common in the Wongan Hills (Milligan 1904), however, it was only recorded once during four years of surveys in the same area 70 years later (de Rebeira and de Rebeira 1977).

Fig. 126. Distribution 1900–1937

Sacred kingfisher

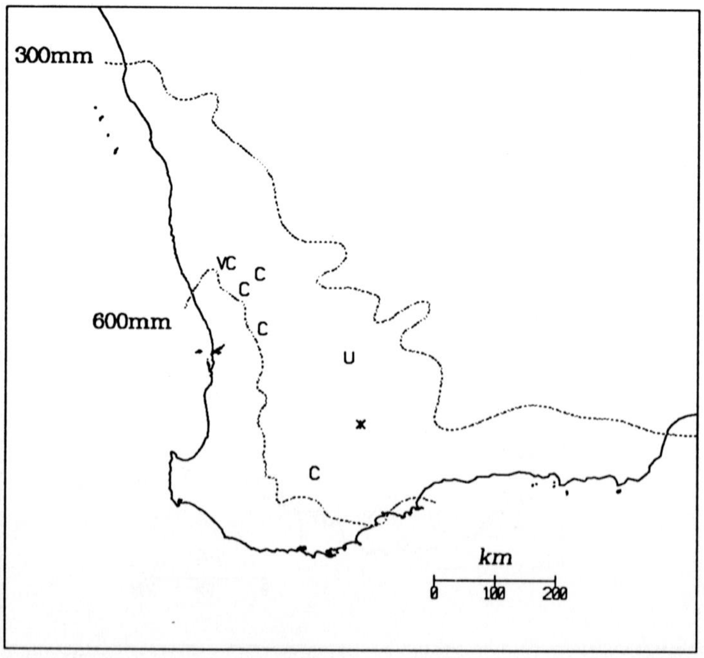

Fig. 126. Distribution 1987–1990

Sacred kingfisher

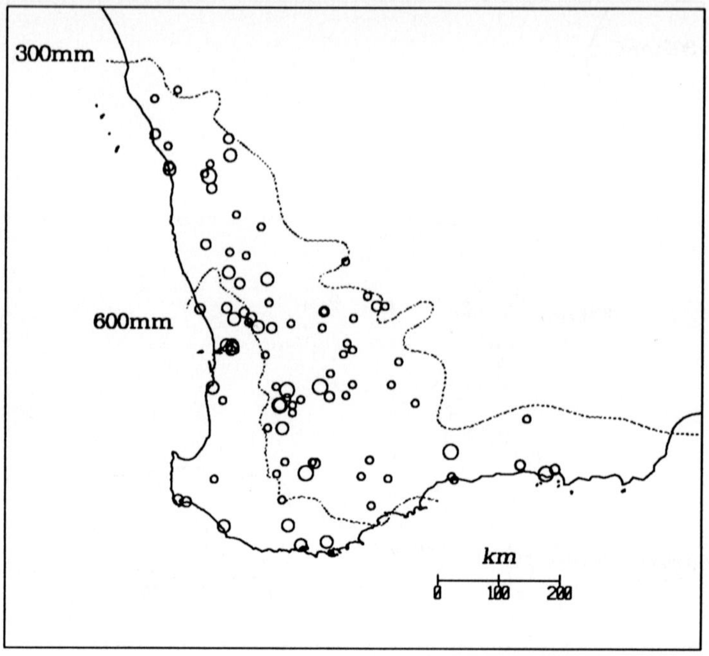

	1900/37	1987	1988	1989	1990
% of observers recording species	70	33	34	30	38
% of recording period species seen	C-VC	4.8	5.0	5.3	6.5

Fig. 126. Distribution 1987–1990 — February–August

Sacred kingfisher

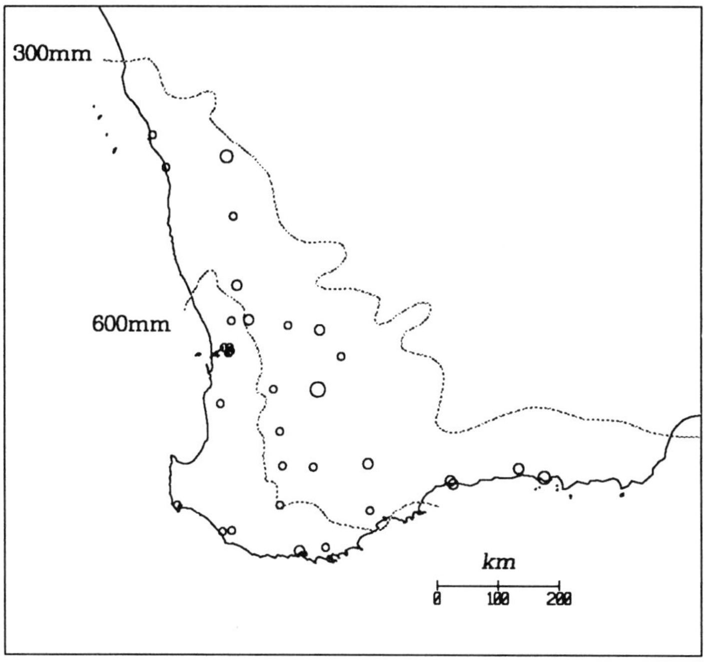

Fig. 126. Distribution 1987–1990 — September–January

Sacred kingfisher

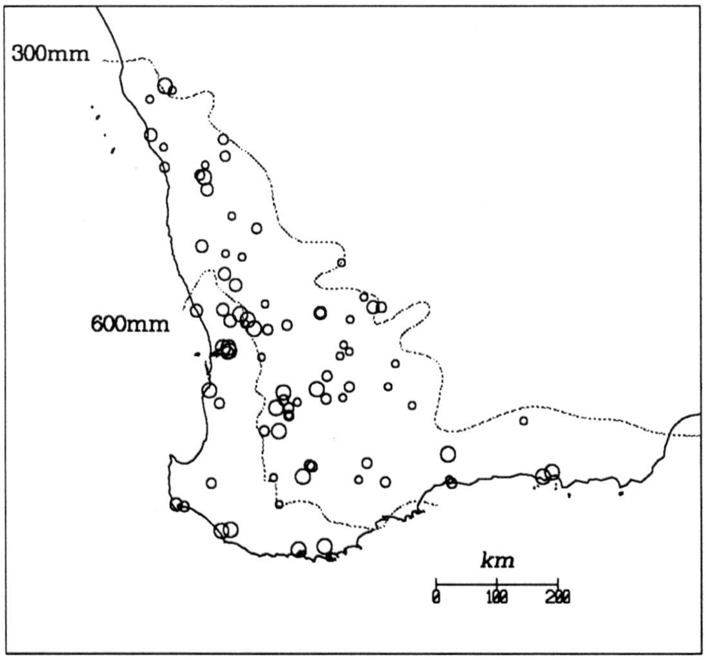

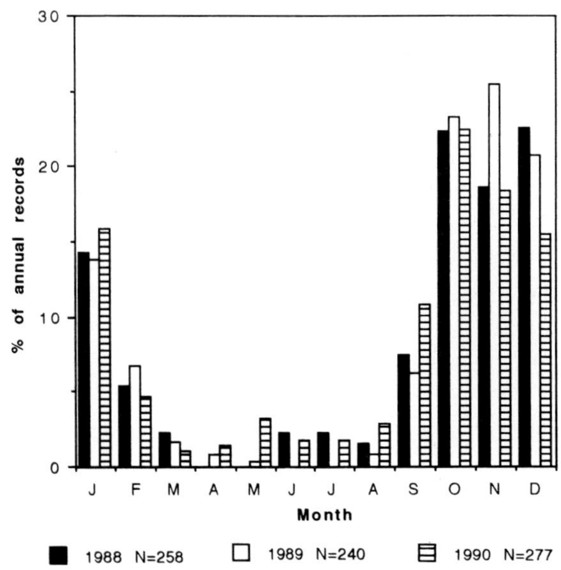

Rainbow bee-eater

Merops ornatus

The rainbow bee-eater is an uncommon, seasonal breeding migrant from the north which is found in woodland, mallee, shrubland, heath and on farmland throughout the wheatbelt (Fig. 127). It feeds on dragonflies, flying termites, bugs, cicadas, beetles, scorpionflies, flies, moths, wasps, bees, ants, slaters, spiders and tadpoles. Bees and related insects make up the bulk of its food. The rainbow bee-eater breeds in spring in a burrow in the ground and lay 4–5 eggs.

Historically, the seasonal pattern of occurance was noted for the Northam district (end of September to February or March) (Jenkins 1931; Masters and Milhinch 1974), the Kellerberrin area (arrived in October or the beginning of November and left by end of March) (Leake 1962), the Broome Hill district (early October to end of February) (Carter 1923), and the Lake Grace area (August to March) (Carnaby 1933). This pattern was also shown during this survey by arrival and departure dates for five locations (Table below and Fig. on facing page).

	1988	1989	1990
Tardun	Oct. Wk 3 Mar. Wk 1	Oct. Wk 2 Mar. Wk 2	Oct. Wk 2 *
Arrino	Oct. Wk 2 Mar. Wk 1	Oct. Wk 2 Feb. Wk 4	Oct. Wk 2 *
Avon River between Northam and Toodyay	Sep. Wk 4 Mar. Wk 2	Sep. Wk 4† Mar. Wk 2	Sep. Wk 4 *
Bruce Rock	Oct. Wk 2 Mar. Wk 4	Oct. Wk 1 Feb. Wk 4	Oct. Wk 2 *
Pingrup	Oct. Wk 2 Nov. Wk 4	Did not appear	Did not appear

*No departure dates as survey finished 31 December, 1990. †Observer noted low numbers of rainbow bee-eater in 1989.

There are no data to indicate any change in the status of the rainbow bee-eater.

Fig. 127. Distribution 1900–1937

Rainbow bee-eater

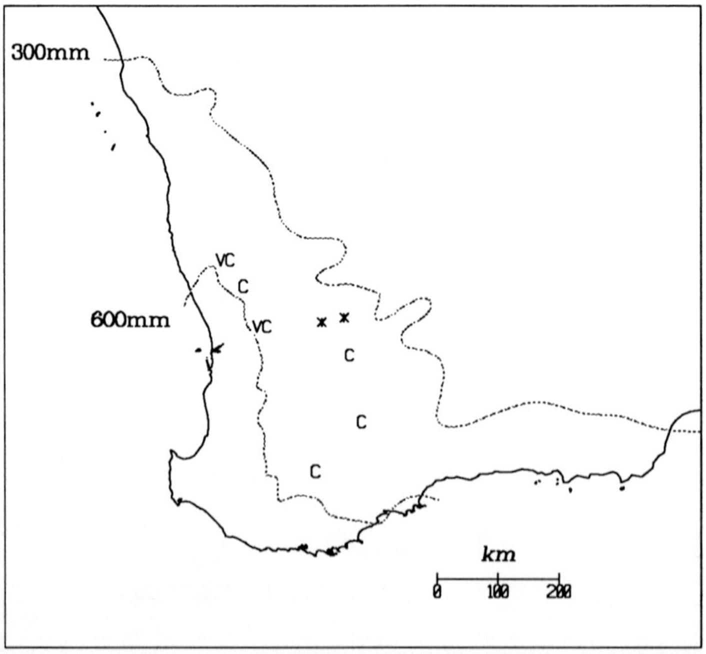

Fig. 127. Distribution 1987–1990

Rainbow bee-eater

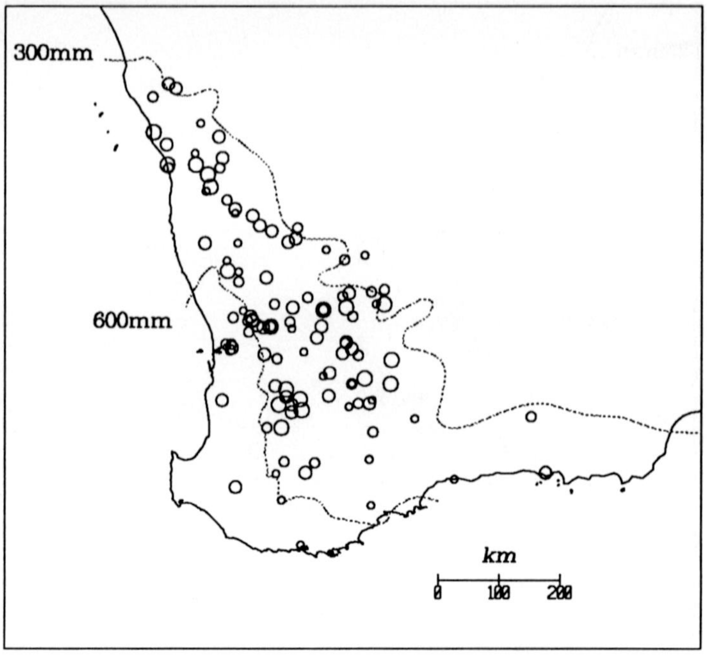

	1900/37	1987	1988	1989	1990
% of observers recording species	80	55	59	56	54
% of recording period species seen	C-VC	18	14	15	12

Fig. 127. Distribution 1987–1990 — April–August

Rainbow bee-eater

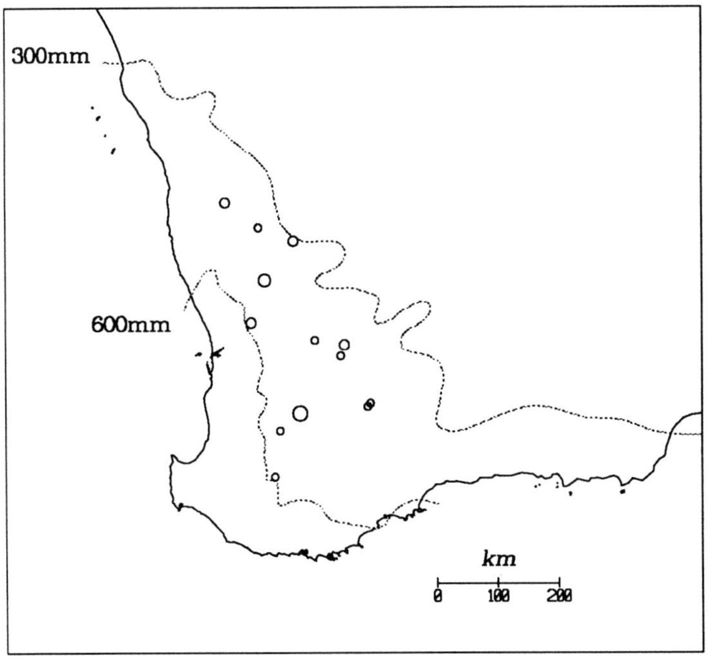

Fig. 127. Distribution 1987–1990 — September–March

Rainbow bee-eater

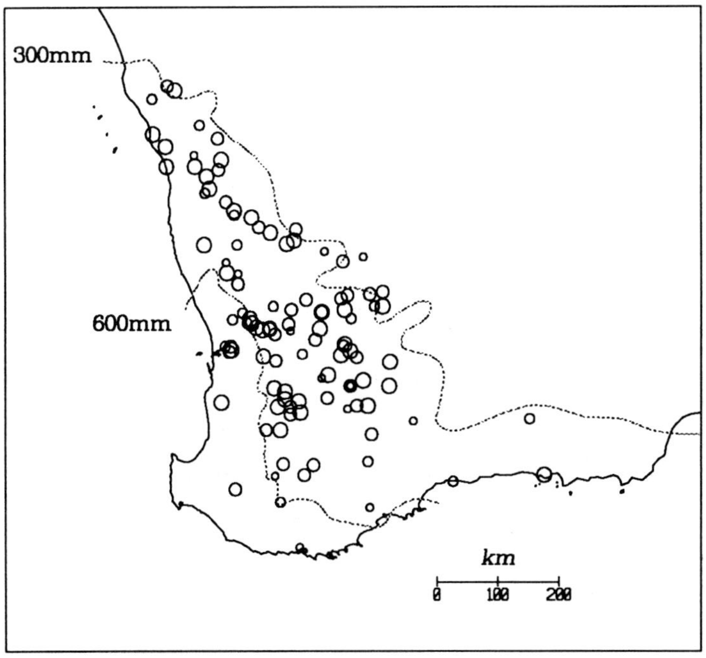

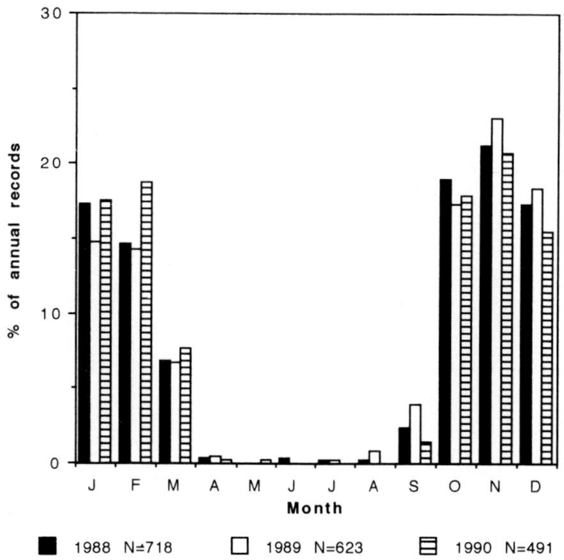

Typical farmland habitat in the wheatbelt (photos by D. A. Saunders).

PASSERINES

White-backed swallow

Cheramoeca leucosternum

The white-backed swallow is an uncommon resident of heath and farmland throughout the wheatbelt (Fig. 128). It feeds on bugs, beetles, flies, moths, bees, wasps, ants and spiders. It breeds in spring in a burrow in the ground and lays 4–6 eggs.

Serventy and Whittell (1962) noted that this species was not usually found south of Mogumber, nor west of the Great Southern railway, however, it was spreading southward. This range extension has continued with the bird penetrating the forest to Augusta, probably via areas cleared for agriculture.

Fig. 128. Distribution 1900–1937

White-backed swallow

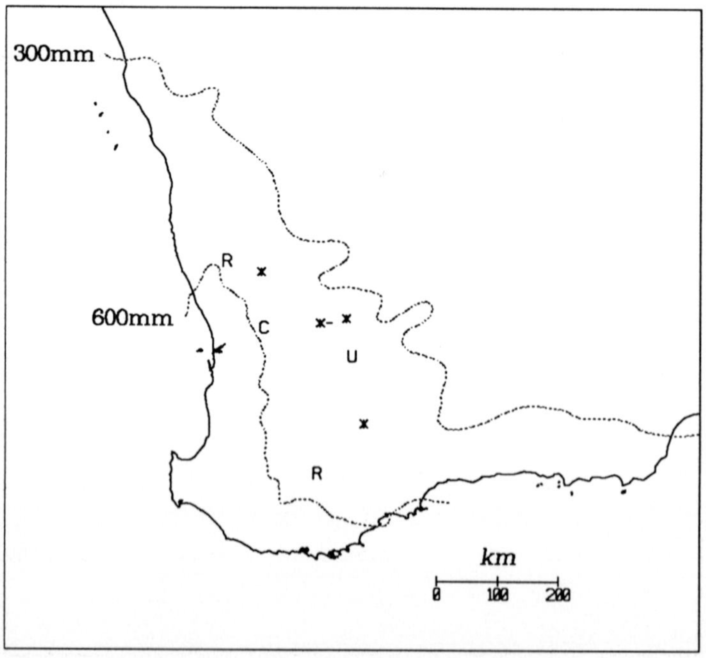

Fig. 128. Distribution 1987–1990

White-backed swallow

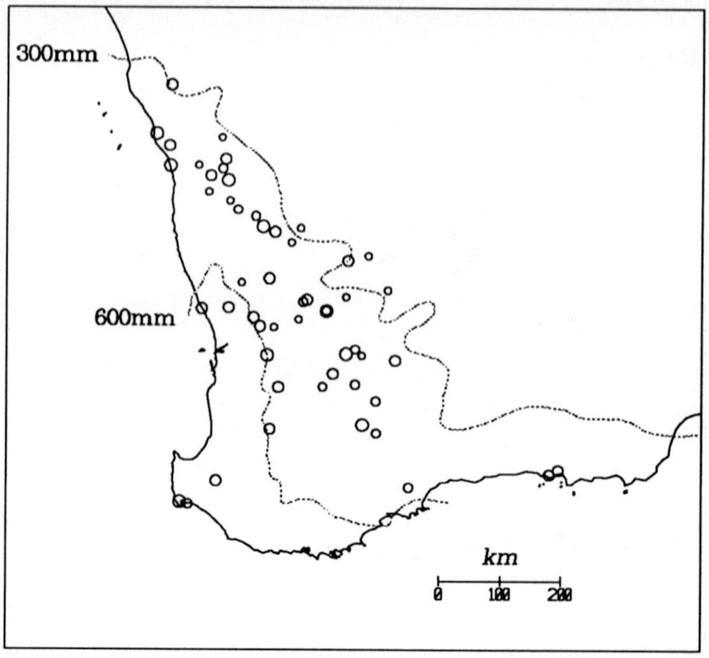

	1900/37	1987	1988	1989	1990
% of observers recording species	80	17	27	25	24
% of recording period species seen	R-C	5.0	6.5	6.8	7.5

Fig. 129. Distribution 1900–1937

Welcome swallow

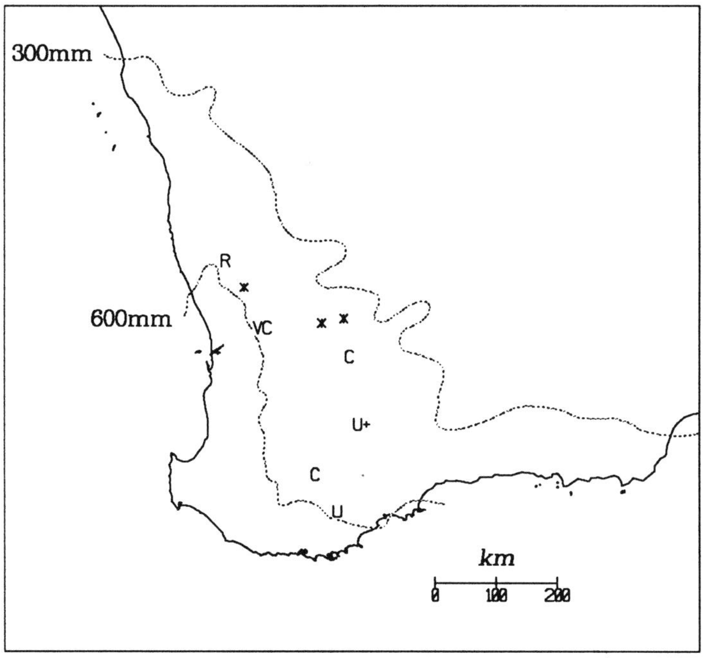

Fig. 129. Distribution 1987–1990

Welcome swallow

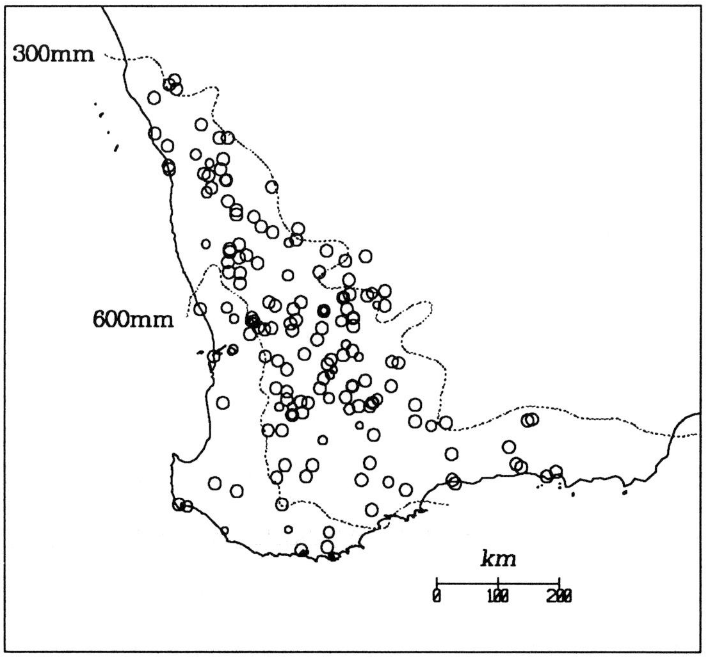

	1900/37	1987	1988	1989	1990
% of observers recording species	90	83	85	88	86
% of recording period species seen	R-VC	59	60	69	69

Welcome swallow

Hirundo neoxena

The welcome swallow is a common resident in towns and on farms, particularly around buildings, and an uncommon resident in woodland throughout the wheatbelt (Fig. 129). It feeds on stone flies, water boatmen, beetles, flies, moths and ants. It breeds in spring in an open nest of mud and vegetation, often attached to buildings and other human-provided structures, and lays 4–5 eggs.

Serventy and Whittell (1962) noted that the welcome swallow seemed to have profited greatly from settlement, mainly because of an abundance of suitable nest sites. They pointed out that one rarely saw a welcome swallow nest in other than a human-provided situation; this applies today. For example, the welcome swallow colonized a farm shed within a year of its construction on a freshly-cleared farm in the eastern wheatbelt (Sedgwick 1974).

Tree martin

Cecropis nigricans

The tree martin is ubiquitous in Australia. Throughout the wheatbelt (Fig. 130) it is a common resident in woodland, mallee, towns and farmland. It feeds on bugs, water boatmen, beetles, flies, wasps and ants. It breeds in late winter and spring in a tree hollow and lays 3–5 eggs.

Recording rates from this survey show that there is some seasonality in occurrence of the tree martin, with recording rates being highest in late spring and early summer (see Fig. below). Serventy and Whittell (1962) stated that there was evidence of a long drawn out passage movement with parties travelling north in autumn and winter, and south in spring and summer. Mass roosting flocks of over 2 000 occur along the Avon River during summer and autumn. This may involve birds from more distant forests and farms and give the appearance of migration (J. Masters, in litt.). Similarly, the tree martin visits Rottnest Island in summer in flocks that may exceed 15 000 (Saunders and de Rebeira 1993).

It is possible that the tree martin has declined in abundance as a result of loss of woodland.

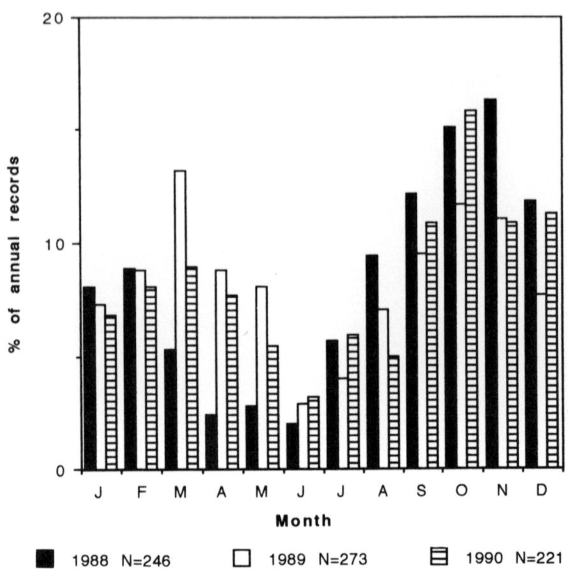

Fig. 130. Distribution 1900–1937

Tree martin

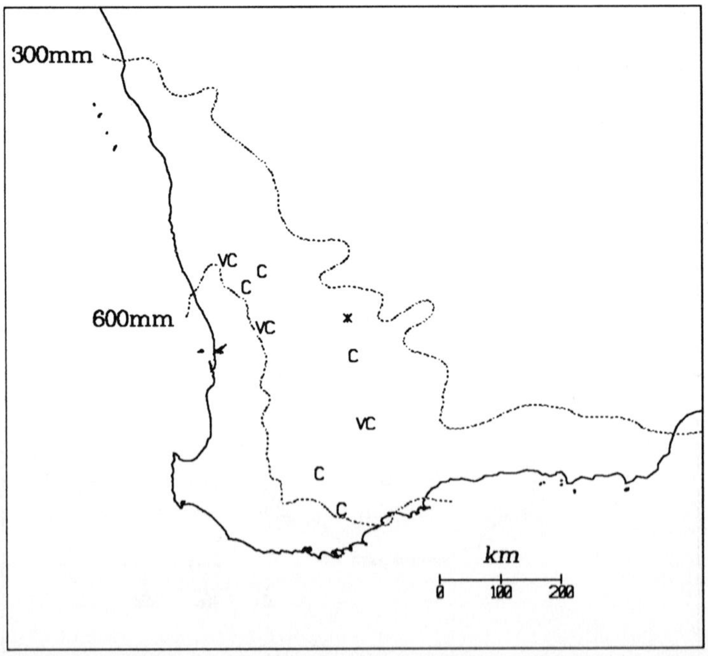

Fig. 130.' Distribution 1987–1990

Tree martin

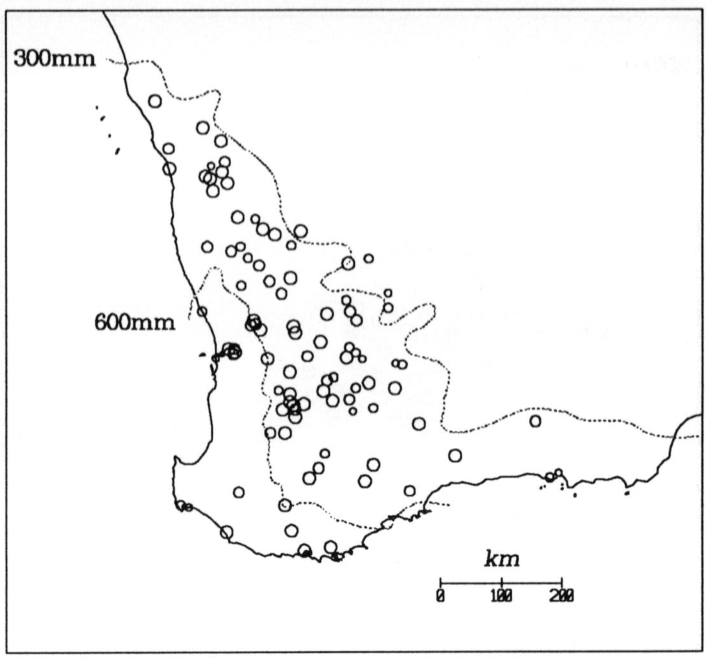

	1900/37	1987	1988	1989	1990
% of observers recording species	90	37	51	56	50
% of recording period species seen	C-VC	18	24	29	27

Fig. 131. Distribution 1900–1937

Fairy martin

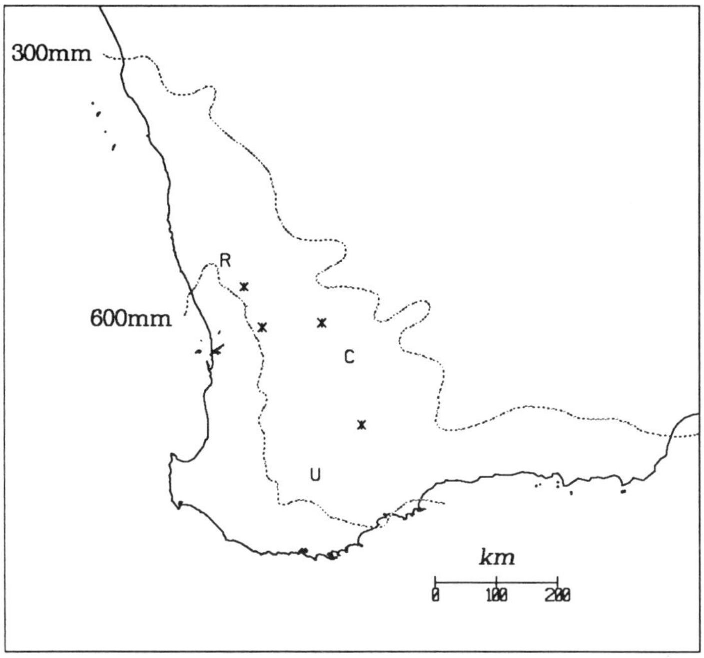

Fig. 131. Distribution 1987–1990

Fairy martin

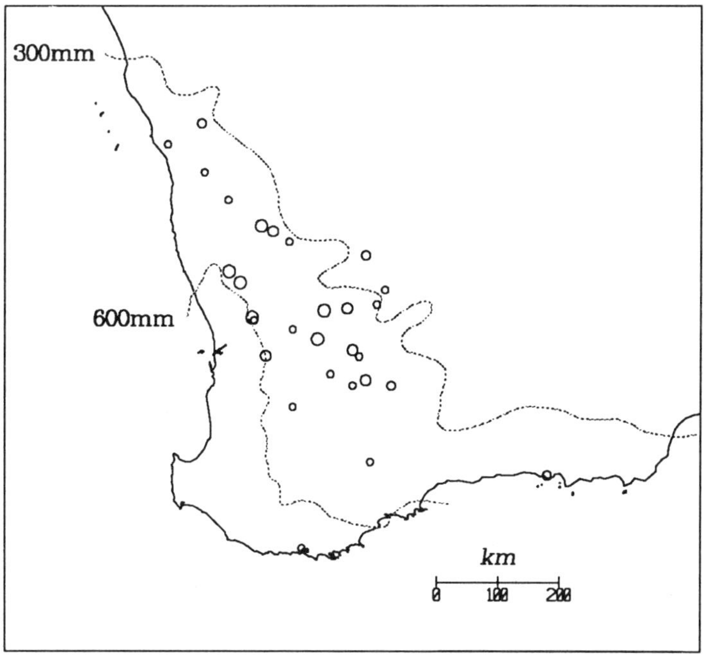

	1900/37	1987	1988	1989	1990
% of observers recording species	70	5.5	11	6.0	5.8
% of recording period species seen	R-C	2.6	3.2	0.3	1.4

Fairy martin

Cecropis ariel

In the wheatbelt, the fairy martin is found in woodland and farmland, often near water due to mud being available for building nests (Serventy and Whittell 1962) (Fig. 131). Its status is unclear, but it may be migratory (Readers' Digest 1976). It feeds on mayflies, flying termites, water boatmen, beetles, flies and ants. It breeds in spring and early summer. It nests colonially in constructions of mud and vegetation built on old buildings, culverts, etc, and lays 4–5 eggs.

The fairy martin may have decreased in abundance over the last 50 years. For example, in the Kellerberrin district Ford and Stone (1957) noted that it was not as plentiful as the tree martin. The fairy martin was not recorded there between 1985 and 1994, despite extensive field work. It has not been resident in the Avon Valley over the past 60 years. However, numbers vary enormously, and it was absent over some periods, as between 1990 to 1993 (J. Masters, in litt.).

Richard's pipit

Anthus novaeseelandiae

Richard's pipit is ubiquitous in Australia, and a common resident on farmland throughout the wheatbelt (Fig. 132). It feeds on seed, millipedes, cockroaches, earwigs, grasshoppers, bugs, beetles, flies, caterpillars, moths, ants and spiders. It breeds in winter and spring in a nest on the ground. It lays 3–5 eggs.

Richard's pipit is a species of open country. It noticeably increased its distribution as clearing for agriculture progressed (Serventy and Whittell 1962). For example, this species colonized freshly-cleared farmland in the eastern wheatbelt within one year of the first clearing (Sedgwick 1974).

Fig. 132. Distribution 1900–1937

Richard's pipit

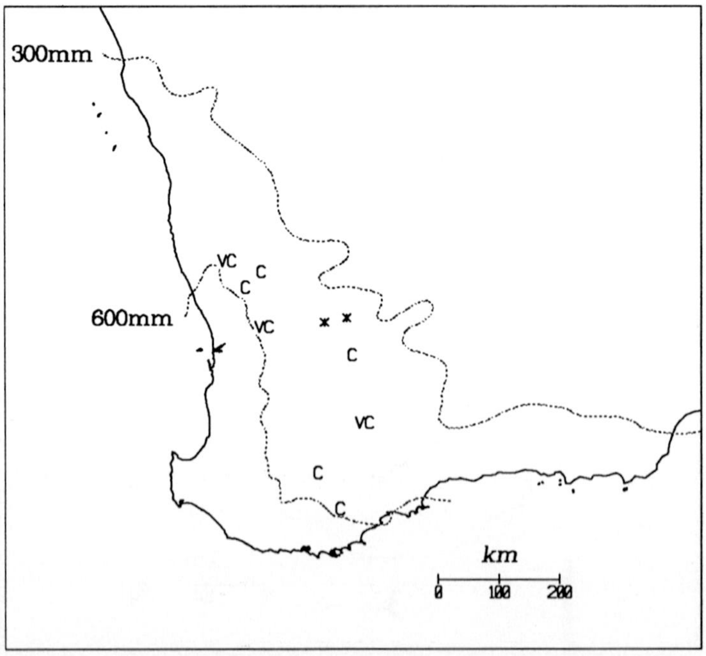

Fig. 132. Distribution 1987–1990

Richard's pipit

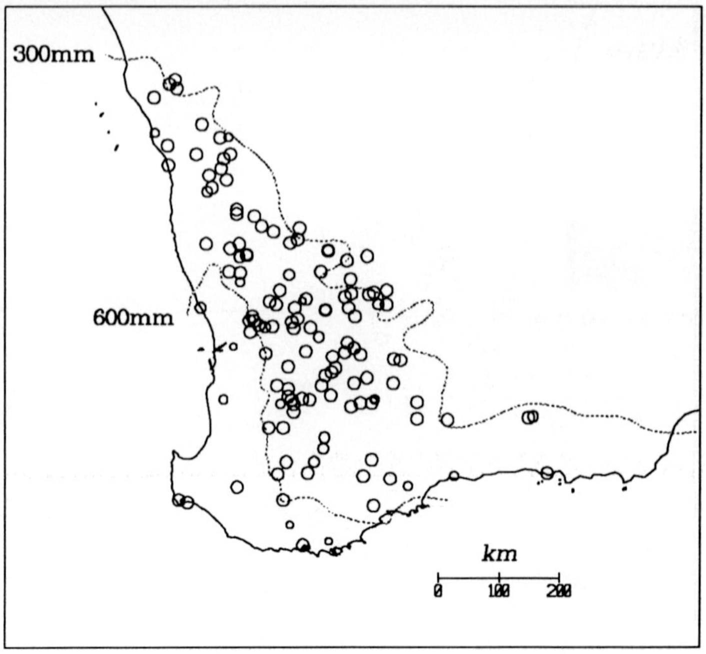

	1900/37	1987	1988	1989	1990
% of observers recording species	100	71	73	74	71
% of recording period species seen	C-VC	57	57	58	60

Fig. 133. Distribution 1900–1937

Black-faced cuckoo-shrike

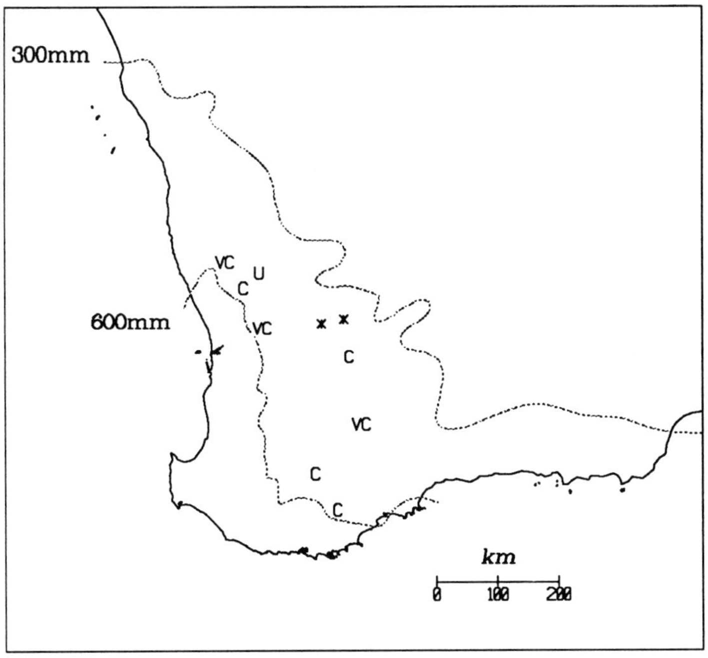

Fig. 133. Distribution 1987–1990

Black-faced cuckoo-shrike

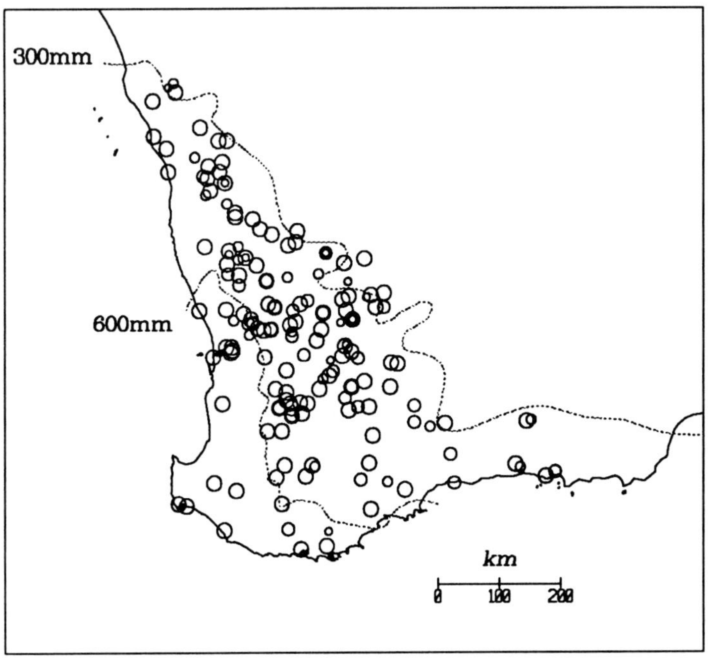

	1900/37	1987	1988	1989	1990
% of observers recording species	100	82	82	88	90
% of recording period species seen	U-VC	50	53	50	57

Black-faced cuckoo-shrike

Coracina novaehollandiae

The black-faced cuckoo-shrike is ubiquitous in Australia, and common in woodland, mallee, shrubland, heath, towns and farmland throughout the wheatbelt (Fig. 133). It feeds on seed, fruit, worms, dragonflies, cockroaches, flying termites, mantids, grasshoppers, crickets, locusts, lerps, beetles, caterpillars, moths, wasps, ants and spiders. The black-faced cuckoo-shrike breeds in spring and summer. It nests on a saucer-shaped nest on a large horizontal branch in a tree and lays 2–3 eggs.

There is no evidence to suggest that the black-faced cuckoo-shrike has changed in status.

Ground cuckoo-shrike

Coracina maxima

In Western Australia, the ground cuckoo-shrike is predominantly a bird of the arid zone. It is a rare nomad in woodland and shrubland of the northern wheatbelt (Fig. 134). It feeds on mantids, grasshoppers, locusts, ants, spiders and small birds. It breeds in spring on a bowl-shaped nest on a horizontal branch of a tree. It lays 2–3 eggs.

The ground cuckoo-shrike has declined in abundance and range in the wheatbelt. For example, it was "generally rather common" around Broome Hill (Carter 1923). Leake (1962) noted that it "did not like the country being settled, as it has not been seen" around Kellerberrin since the late 1910s.

Fig. 134. Distribution 1900–1937

Ground cuckoo-shrike

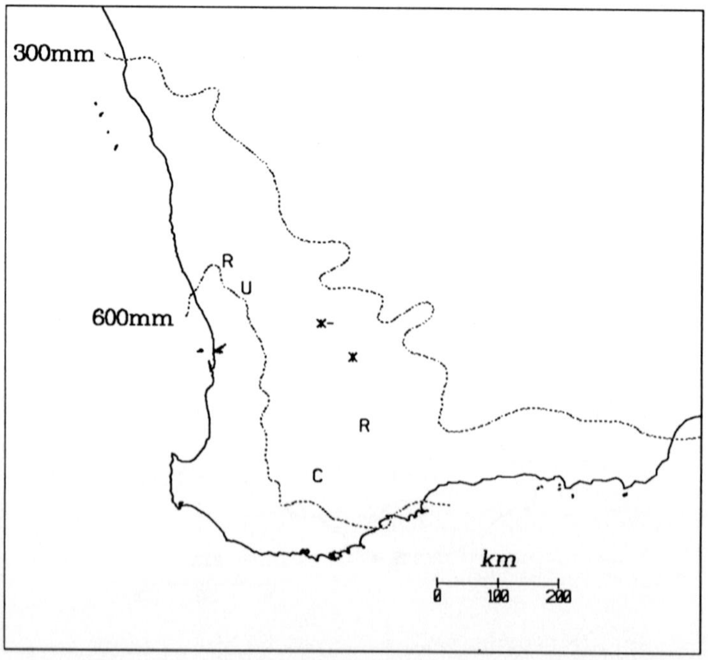

Fig. 134. Distribution 1987–1990

Ground cuckoo-shrike

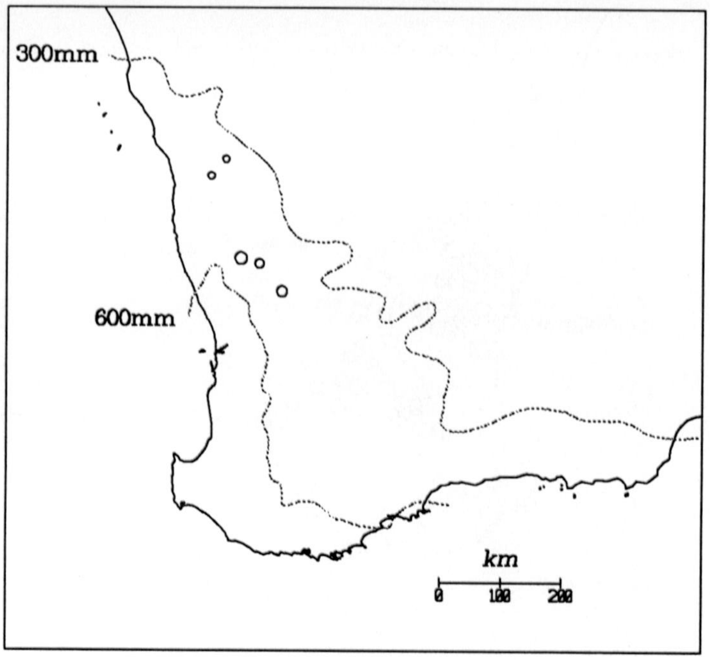

	1900/37	1987	1988	1989	1990
% of observers recording species	60	2.4	1.4	0.9	0
% of recording period species seen	R-C	1.1	0.2	0.02	0

(Top photo) Development of the wheatbelt has provided opportunities for some species of native birds. For example, removal of gravel and sand has created banks which provide potential nest sites for species which excavate burrows (photo by D. A. Saunders).

(Bottom photo) Entrances to burrows constructed by white-backed swallows and rainbow bee-eaters (photo by D. A. Saunders).

White-winged triller

Lalage sueurii

The white-winged triller is an uncommon, seasonal migrant in woodland, mallee, shrubland and heath throughout the wheatbelt (Fig. 135). It feeds on seed, nectar, cockroaches, mantids, grasshoppers, crickets, locusts, bugs, lerps, beetles, flies, moths, caterpillars, bees, ants and spiders. The white-winged triller breeds in spring and summer. It nests on a saucer-shaped nest on a horizontal or sloping branch of a tree or shrub and lays 2–3 eggs.

The seasonal pattern of occurrence was noted around Moora (plentiful from September to January with the occasional bird remaining until April) (Orton and Sandland 1913), Broome Hill (Carter 1923), Northam (earliest record 10 July, although the birds usually arrived in late August or early September and left in late January or February) (Jenkins 1931; Masters and Milhinch 1974), the Lake Grace area (August to December) (Carnaby 1933), and around Kellerberrin (Leake 1962). This pattern was also shown by arrival and departure dates from four locations in the wheatbelt during this survey (Table below and Fig. on facing page).

	1988	1989	1990
Tardun	Jun. Wk 4 Nov. Wk 4	Aug. Wk 4 Jan. Wk 2	Jun. Wk 2 Dec. Wk 3
Arrino	Jul. Wk 3 Nov. Wk 3	Aug. Wk 1 Oct. Wk 4	Jun. Wk 2 Dec. Wk 3
Avon River between Northam and Toodyay	Aug. Wk 2 Dec. Wk 2	Sep. Wk 4* Oct. Wk 2	Jun. Wk 4 Dec. Wk 3
Pingrup	Aug. Wk 1 Jan. Wk 4	Aug. Wk 3 Feb. Wk 1	Sep. Wk 2 Nov. Wk 1

*This observer noted that the white-winged triller was scarce in 1989.

The white-winged triller was common to very common throughout the wheatbelt (Orton and Sandland 1913; Milligan 1904), but is no longer common.

Fig. 135. Distribution 1900–1937

White-winged triller

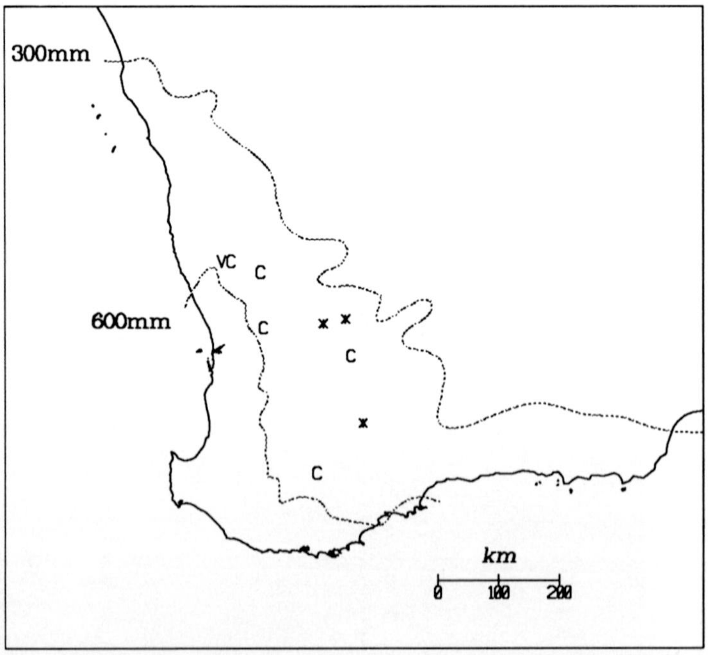

Fig. 135. Distribution 1987–1990

White-winged triller

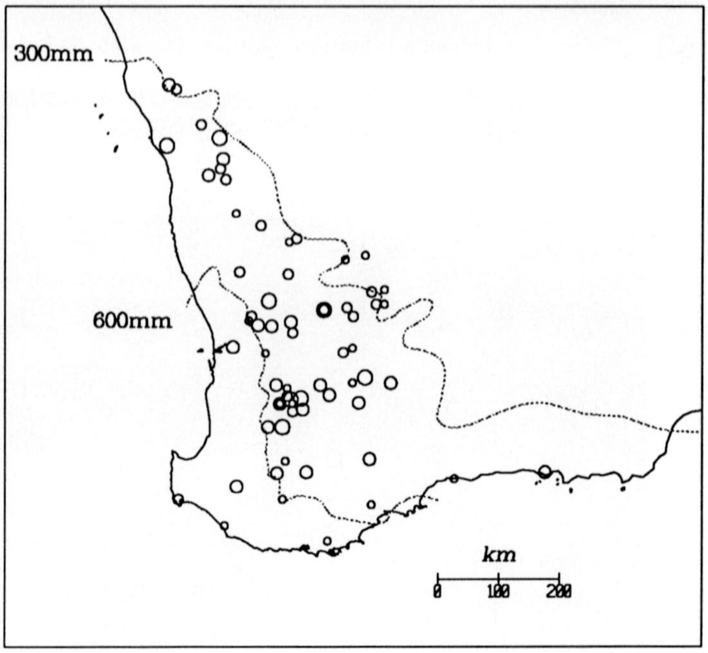

	1900/37	1987	1988	1989	1990
% of observers recording species	80	34	36	33	35
% of recording period species seen	C-VC	7.9	6.9	6.5	6.8

Fig. 135. Distribution 1987–1990 — February–June

White-winged triller

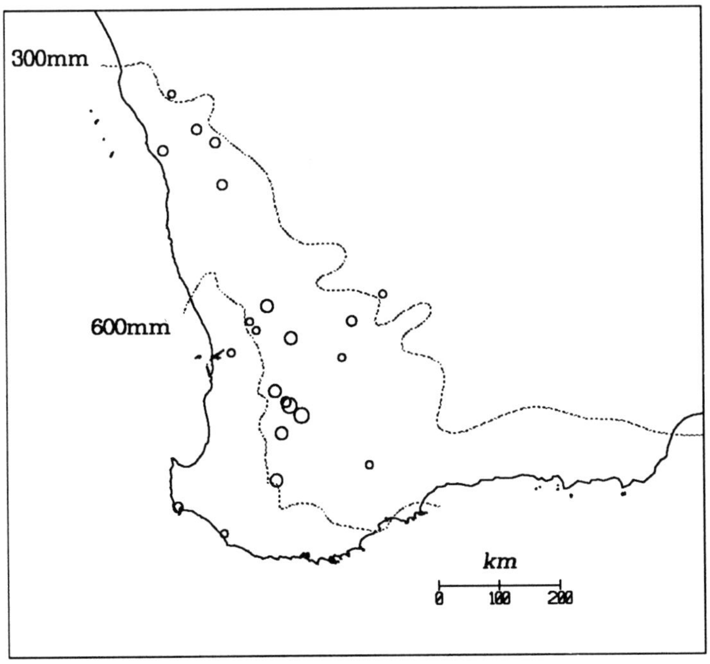

Fig. 135. Distribution 1987–1990 — August–January

White-winged triller

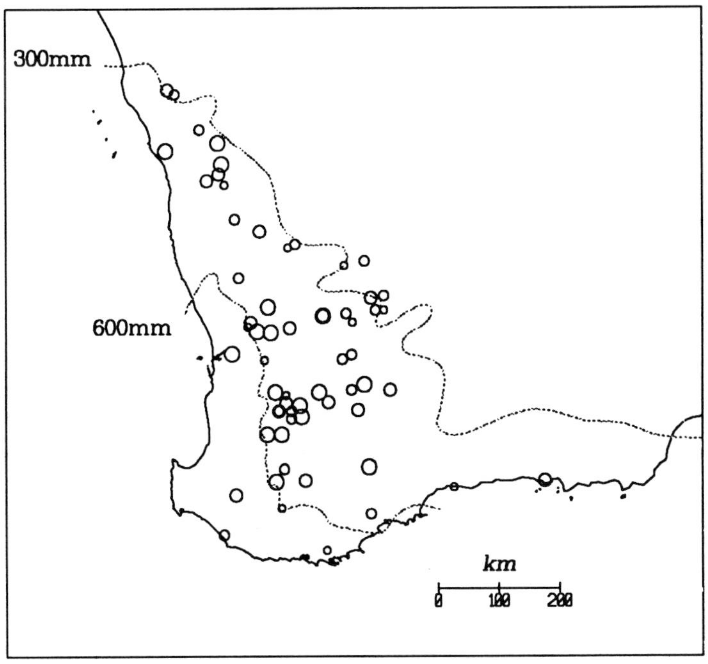

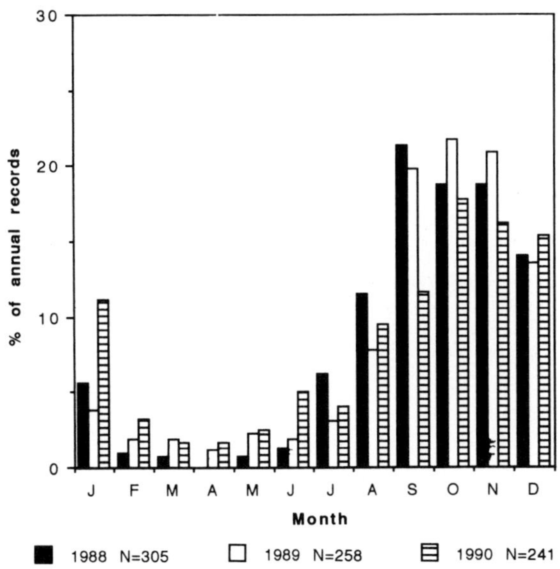

Southern scrub robin

Drymodes brunneopygia

The southern scrub robin is a rare resident of remnant patches of mallee and shrubland in much of the wheatbelt (Fig. 136). It feeds on seed, grasshoppers, crickets, bugs, beetles, caterpillars, wasps, ants and spiders. It breeds in late winter and spring. It nests on a shallow cup-shaped structure on or near the ground and lays 1 egg.

This species is dependent on native vegetation and it must have declined in abundance, and possibly range over the last 80 years. For example, Sedgwick (1974) noted that clearing native vegetation drove this species from a newly-cleared farm in the eastern wheatbelt.

Although still found in remnant patches, many of these populations may be isolated. They may also be too small to be viable and populations are still being lost from individual remnants. For example, the southern scrub robin was scarce on East Yorkrakine Nature Reserve, north of Kellerberrin, in the mid-1970s (Dell 1980), but it had disappeared by the mid-1980s (Saunders 1989). As these populations disappear, the range of this species will contract.

Fig. 136. Distribution 1900–1937

Southern scrub robin

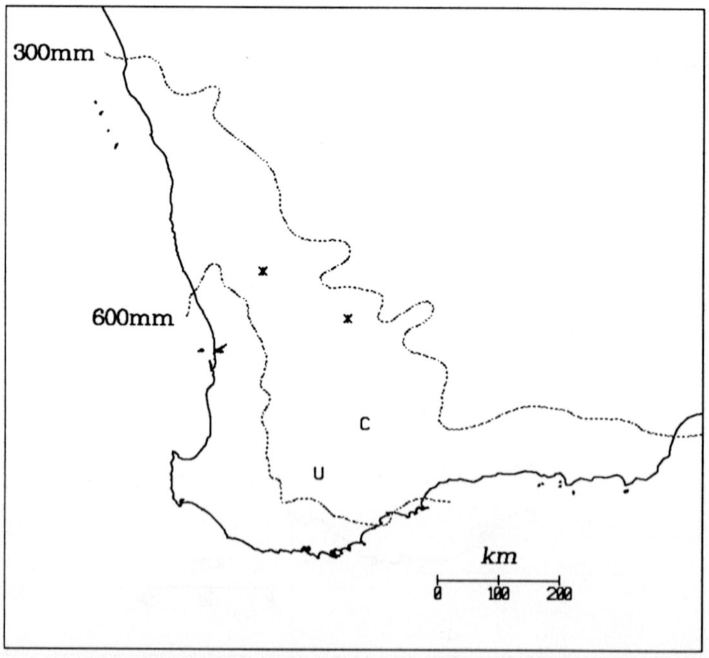

Fig. 136. Distribution 1987–1990

Southern scrub robin

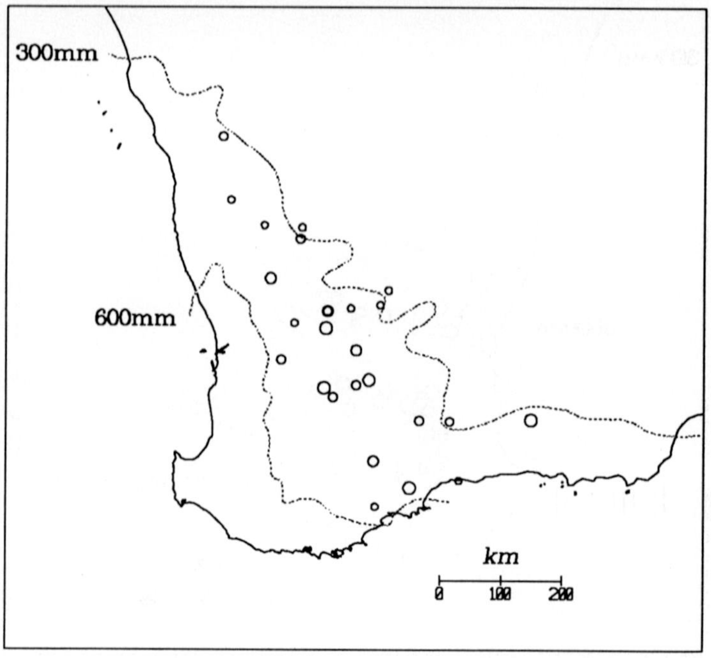

	1900/37	1987	1988	1989	1990
% of observers recording species	40	13	9.2	7.7	11
% of recording period species seen	U-C	2.4	1.1	2.9	3.4

Fig. 137. Distribution 1900–1937

Scarlet robin

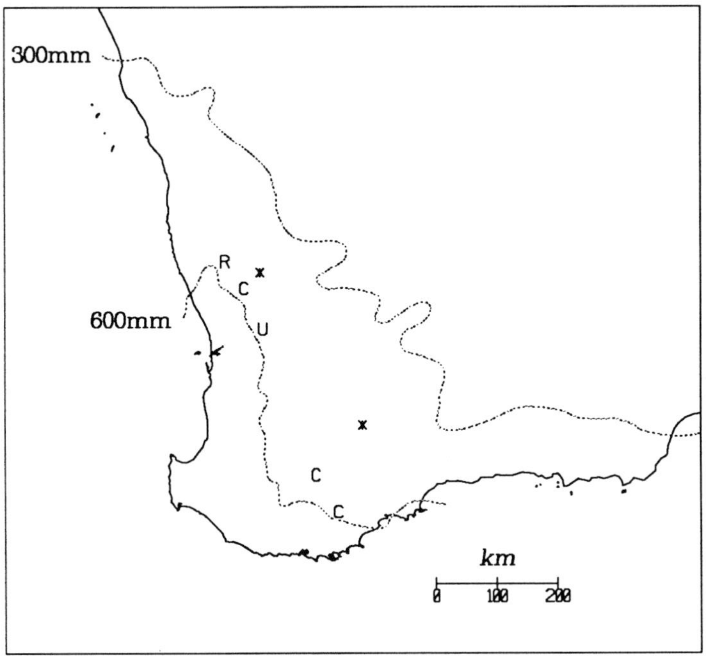

Fig. 137. Distribution 1987–1990

Scarlet robin

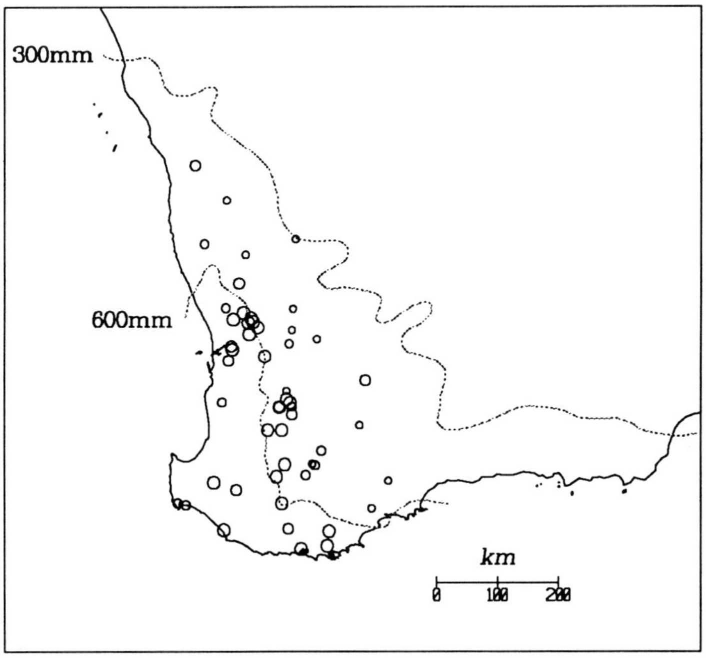

	1900/37	1987	1988	1989	1990
% of observers recording species	70	18	25	24	27
% of recording period species seen	R-C	8.5	12	13	14

Scarlet robin

Petroica multicolor

The scarlet robin is an uncommon resident of woodland in the higher rainfall districts of the wheatbelt (Fig. 137). It feeds on millipedes, centipedes, termites, earwigs, grasshoppers, bugs, cicadas, beetles, flies, moths, caterpillars, wasps, bees, ants and spiders. It breeds in late winter and spring in a cup-shaped nest in a tree fork or hollow stump. It lays 3 eggs.

This species is dependent on native vegetation and has decreased in abundance and possibly range over the last 70 years. It formerly occurred in the Kellerberrin/Kwolyin districts and was last recorded in 1941 (Ford and Stone 1957). It was not recorded in the Kellerberrin district between 1985 and 1994 and must be regarded as extinct there. Serventy and Whittell (1962) noted that the scarlet robin was becoming scarce around Perth and was extremely scarce in Kings Park (a large patch of remnant woodland in the centre of Perth) where it had been regularly present formerly.

Red-capped robin

Petroica goodenovii

The red-capped robin is an uncommon resident of woodland and shrubland throughout the wheatbelt (Fig. 138). It may be locally common in some patches of native vegetation. It feeds on worms, mantids, earwigs, flying termites, grasshoppers, bugs, beetles, flies, moths, caterpillars, wasps, bees, ants and spiders. It breeds in late winter and spring in a cup-shaped nest in a tree or shrub and lays 2–3 eggs.

Recording rates from 1988 to 1990 showed a cyclic pattern with the highest recording rates between April and July (see Fig. below). Cale (1990) believes this cyclic pattern is the result of the red-capped robin using road reserves as corridors for dispersal, and consequently, being seen more often.

The dependence of the red-capped robin on remnant vegetation has meant that it has probably decreased in abundance as a result of the loss of suitable habitat.

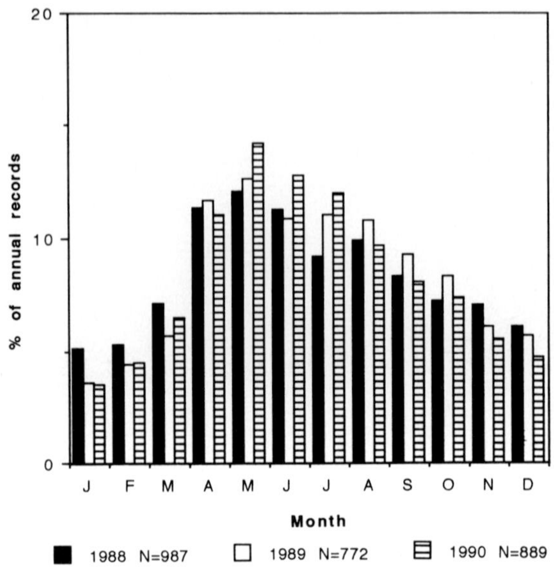

Fig. 138. Distribution 1900–1937

Red-capped robin

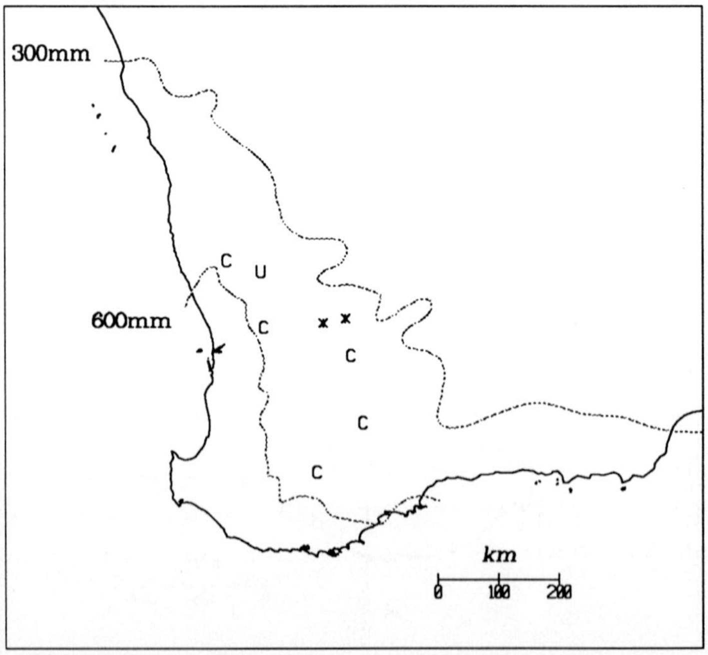

Fig. 138. Distribution 1987–1990

Red-capped robin

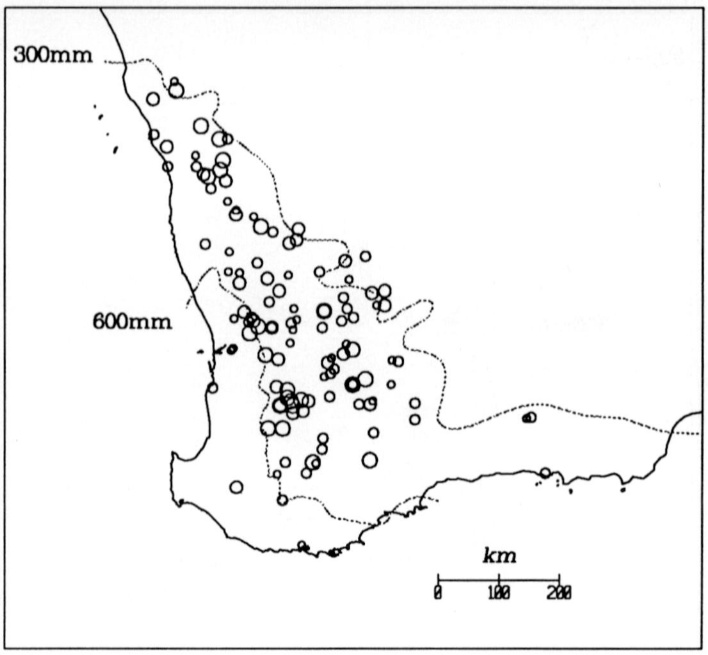

	1900/37	1987	1988	1989	1990
% of observers recording species	80	53	56	53	63
% of recording period species seen	U-C	17	19	17	21

Hooded robin

Melanodryas cucullata

The hooded robin is an uncommon resident in woodland, mallee and shrubland in parts of the central and southern wheatbelt (Fig. 139). It feeds on seed, millipedes, centipedes, termites, grasshoppers, crickets, bugs, beetles, flies, moths, caterpillars, wasps, ants and spiders. The hooded robin breeds in late winter and spring on a cup-shaped structure in a tree or shrub. It lays 2 eggs.

Regarded as common in several places in the wheatbelt before 1937, the hooded robin is no longer common and has disappeared from many parts of its former range. For example, it disappeared from Kellerberrin and Northam as a result of clearing for agriculture (Leake 1962; Masters and Milhinch 1974; Saunders 1989).

Fig. 139. Distribution 1900–1937

Hooded robin

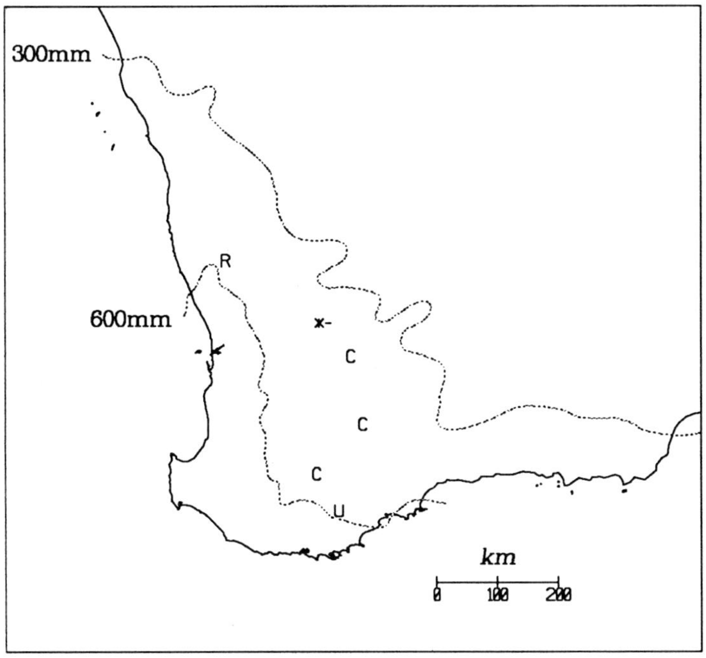

Fig. 139. Distribution 1987–1990

Hooded robin

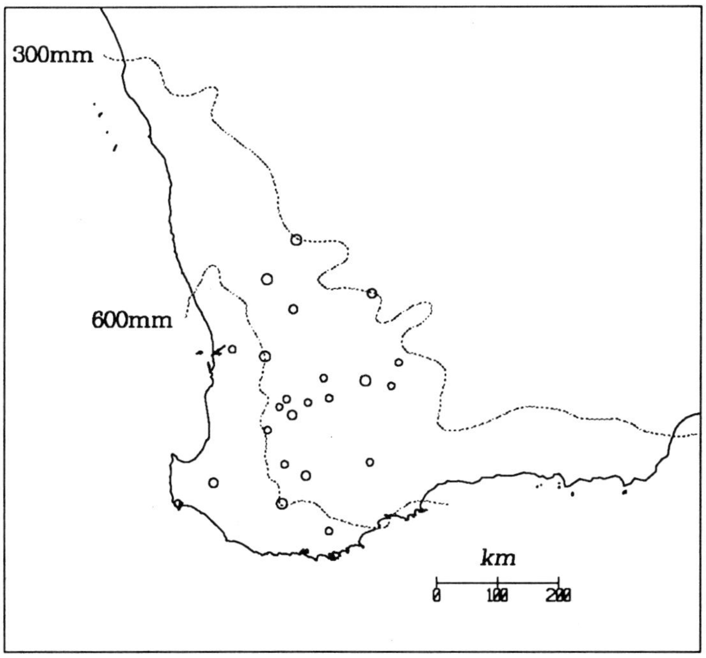

	1900/37	1987	1988	1989	1990
% of observers recording species	60	5.5	8.5	6.0	7.7
% of recording period species seen	R-C	1.2	1.2	0.5	0.5

Western yellow robin

Eopsaltria griseogularis

OTHER NAME:

Yellow robin

The western yellow robin is an uncommon resident in woodland, mallee and shrubland throughout the wheatbelt (Fig. 140). It feeds on cockroaches, beetles, caterpillars, wasps and ants, foraging mainly on the ground. It breeds in late winter and spring in a cup-shaped nest in a tree and lays 2 eggs.

The western yellow robin may move locally. For example, one individual moved 7.5 km between two nature reserves north of Kellerberrin. However, it moved along a 40 m wide well-vegetated road verge that connected the two reserves (Saunders and de Rebeira 1991).

The western yellow robin became scarce around the Perth area and has not been seen in Kings Park, in the centre of Perth, since the winter of 1935 (Serventy and Whittell 1962). It is dependent on native vegetation, particularly woodland (Masters and Milhinch 1974) which explains its decline in abundance and range. It is still decreasing in abundance and being lost from individual remnants of native vegetation (Saunders 1989).

Fig. 140. Distribution 1900–1937

Western yellow robin

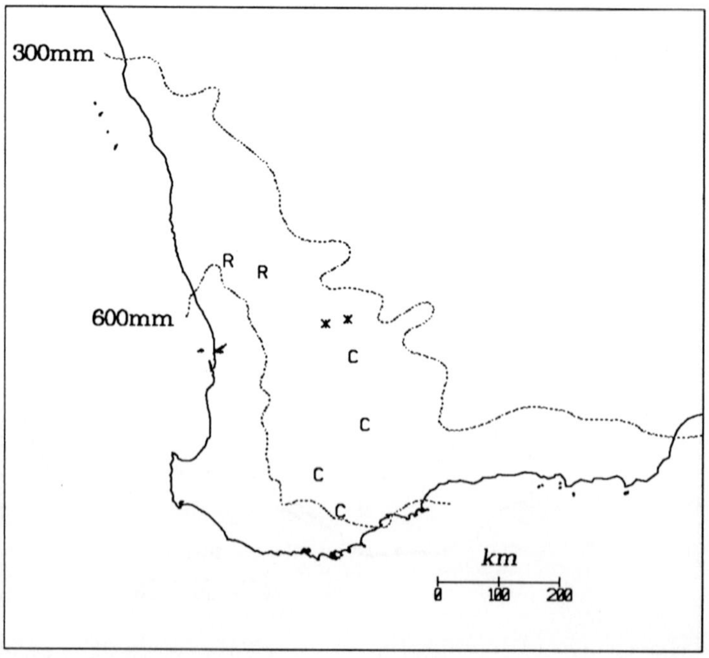

Fig. 140. Distribution 1987–1990

Western yellow robin

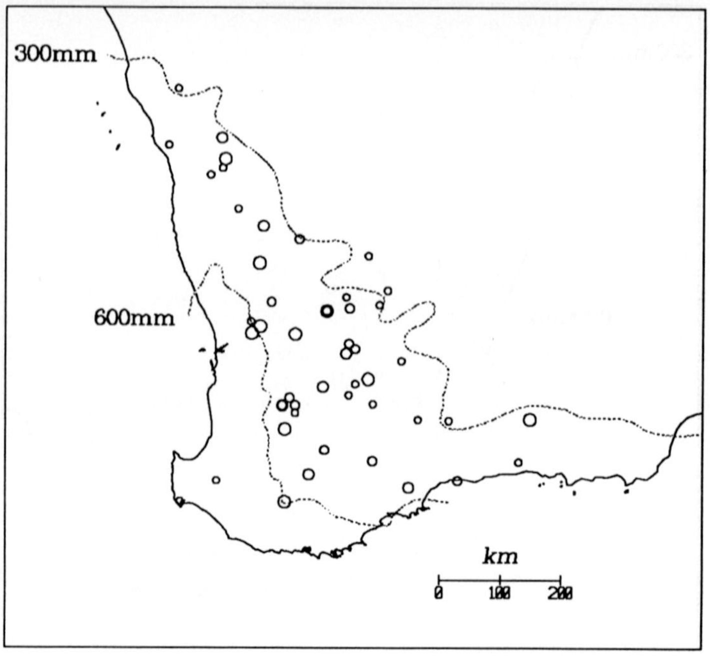

	1900/37	1987	1988	1989	1990
% of observers recording species	80	19	22	17	17
% of recording period species seen	R-C	3.8	5.7	4.8	6.9

Fig. 141. Distribution 1900–1937

Jacky winter

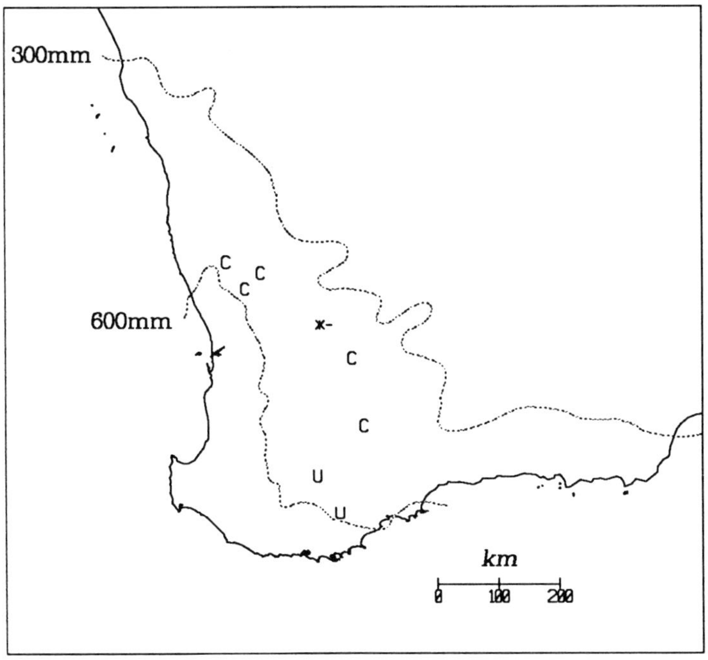

Fig. 141. Distribution 1987–1990

Jacky winter

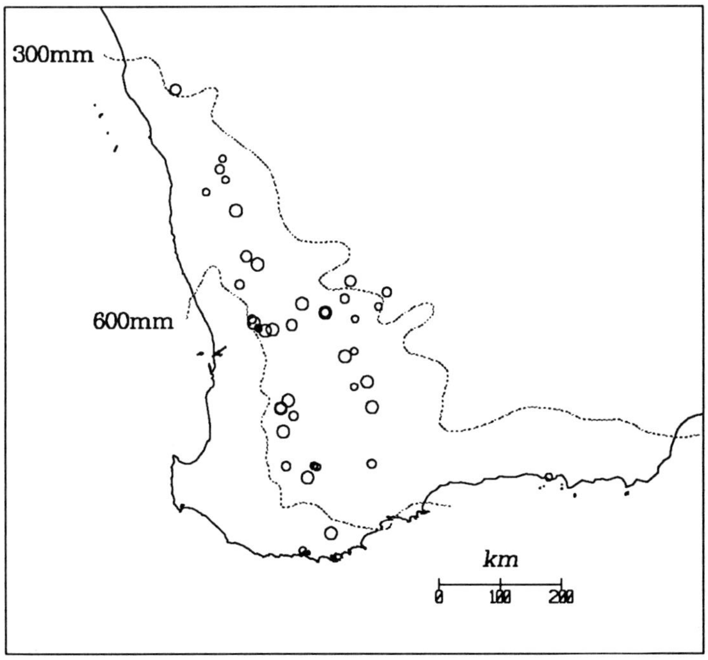

	1900/37	1987	1988	1989	1990
% of observers recording species	80	17	20	19	16
% of recording period species seen	U-C	7.4	4.6	6.8	5.3

Jacky winter

Microeca leucophaea

The jacky winter is an uncommon resident in woodland and mallee throughout the wheatbelt (Fig. 141). It feeds on termites, earwigs, grasshoppers, cicadas, bugs, beetles, flies, butterflies, moths, caterpillars, wasps, bees, ants and spiders. The jacky winter breeds in late winter and early summer on a saucer-shaped nest in a tree and lays 2 eggs.

The jacky winter was common at a number of locations throughout the wheatbelt (Fig. 141), but has declined in abundance. For example, it was numerous between Mogumber and the Wongan Hills (Milligan 1904), but was not recorded in the 1970s around the Wongan Hills (de Rebeira and de Rebeira 1977). Similarly, it was fairly common along the Moore River near Mogumber (Lawson 1905), but rare in the same area in August 1950 (Loaring and Serventy 1952). In the mid-1970s, the jacky winter was moderately common in Durokoppin and Kodj Kodjin Nature Reserves north of Kellerberrin (Dell 1978); by the mid-1980s it was uncommon to rare.

Crested shrike-tit

Falcunculus frontatus

The crested shrike-tit is a rare resident of woodland in the wheatbelt (Fig. 142). It feeds on seed, berries, grasshoppers, crickets, bugs, beetles, moths, caterpillars, wasps and spiders, particularly under the bark of trees. It breeds in spring in a goblet-shaped nest in a tree and lays 2–3 eggs.

Serventy and Whittell (1962) stated that the distribution of the crested shrike-tit was southwest from Morawa, Moora, Kellerberrin and Norseman, but not in the forested extreme southwest. It was rare to common in various locations in the wheatbelt before 1937 (Fig. 142), but has since declined in abundance and range. For example, it was scarce in Kodj Kodjin Nature Reserve north of Kellerberrin in the mid-1970s (Dell 1978); by the mid-1980s it had disappeared.

Fig. 142. Distribution 1900–1937

Crested shrike-tit

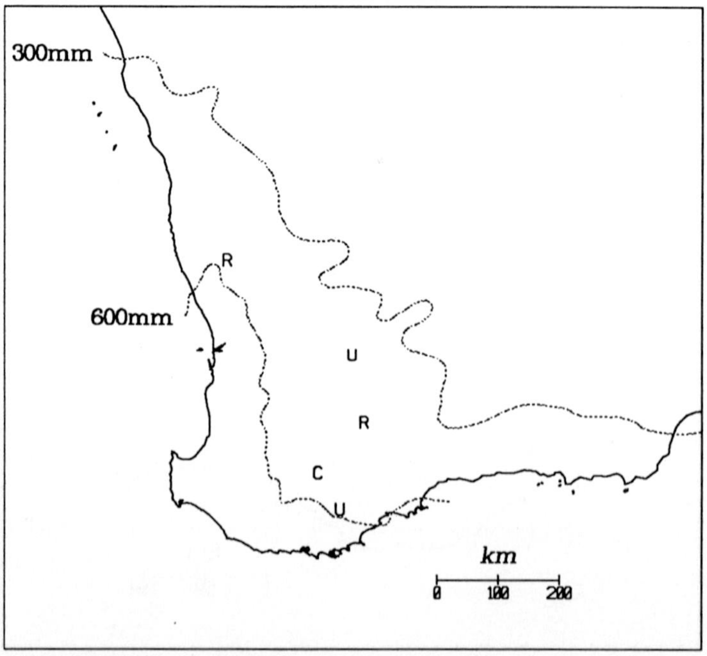

Fig. 142. Distribution 1987–1990

Crested shrike-tit

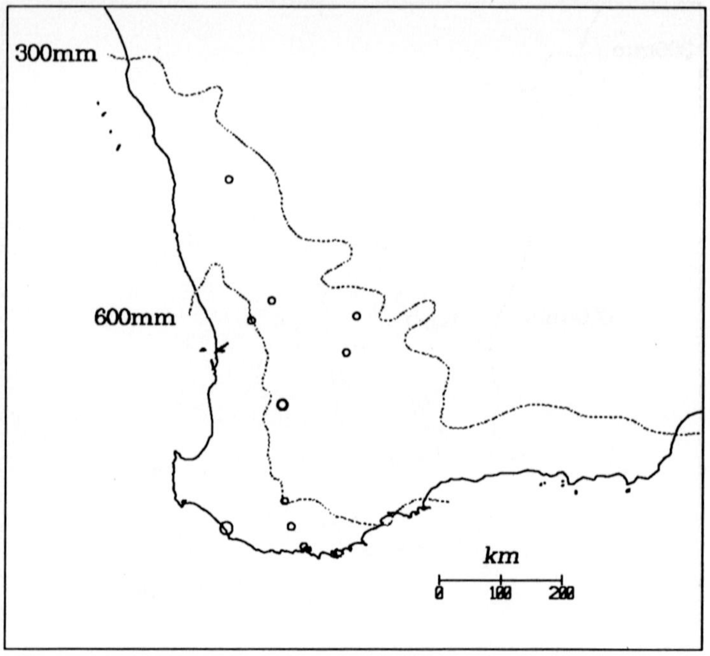

	1900/37	1987	1988	1989	1990
% of observers recording species	50	4.7	2.8	3.4	2.9
% of recording period species seen	R-C	0.5	0.6	0.7	0.7

Gilbert's whistler

Pachycephala inornata

Gilbert's whistler is a rare nomad in woodland and mallee in the drier margins of the wheatbelt (Fig. 143) adjacent to large areas of uncleared native vegetation. It feeds on seed and other plant material, grasshoppers, stick insects, bugs, beetles, moths, caterpillars, ants and spiders, foraging often on the ground. Gilbert's whistler breeds in spring in a cup-shaped nest in a shrub and lays 2–3 eggs.

Gilbert's whistler was not widely distributed, nor common in the wheatbelt before clearing began, but has declined in abundance and range. For example, it was present at Kellerberrin during the first half of this century (Leake 1962; Ford and Stone 1957), recorded near Bruce Rock (Crossman 1909), common at Broome Hill (Carter 1923), and rare around Lake Grace (Carnaby 1933). It no longer occurs in these areas and has disappeared from most of the wheatbelt.

Fig. 143. Distribution 1900–1937

Gilbert's whistler

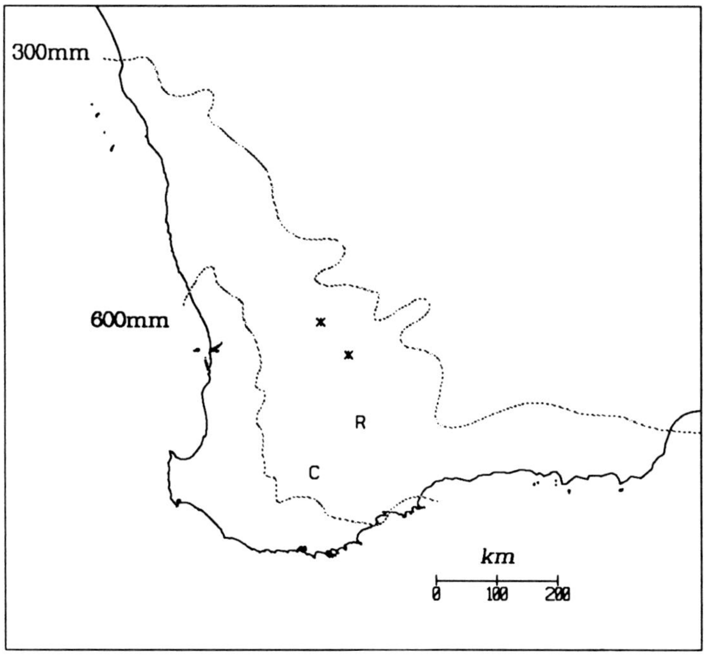

Fig. 143. Distribution 1987–1990

Gilbert's whistler

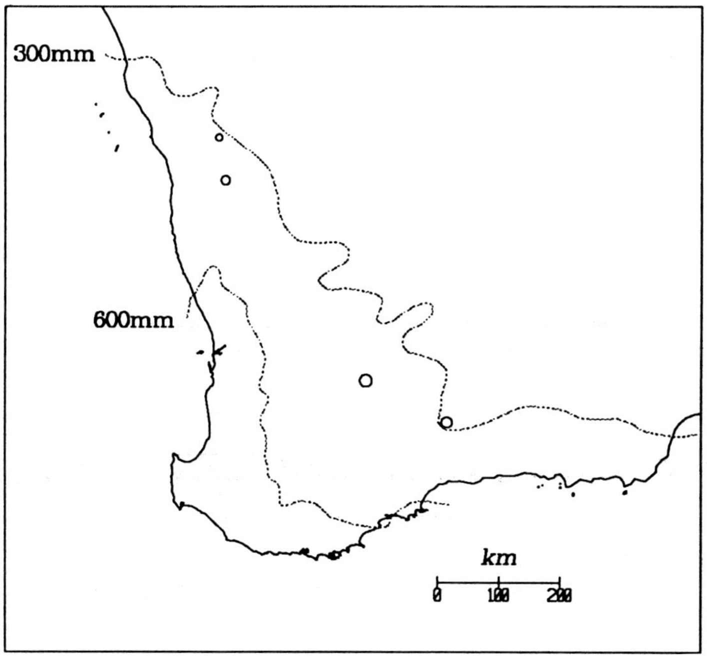

	1900/37	1987	1988	1989	1990
% of observers recording species	40	0	0.7	1.7	2.9
% of recording period species seen	R-C	0	0.5	0.7	0.5

Golden whistler

Pachycephala pectoralis

The golden whistler is an uncommon resident in woodland, mallee and shrubland throughout the wheatbelt (Fig. 144). It feeds on seed, berries, cockroaches, earwigs, grasshoppers, bugs, lerps, beetles, flies, moths, caterpillars, wasps, bees, ants and spiders. It breeds in spring and early summer in a cup-shaped nest in a tree or shrub and lays 2 eggs.

Recording rates for the period 1988 to 1990 showed a cyclic pattern with rates doubling in late winter and spring (see Fig. below). Little is known of the movements of the golden whistler, but birds are believed to wander locally (Readers' Digest 1976).

The golden whistler has declined in abundance in the wheatbelt. It is resident in large patches of remnant vegetation in the Kellerberrin district (Saunders and Curry 1990), but has disappeared from smaller remnants. For example, it was uncommon on East Yorkrakine Nature Reserve in the mid-1970s (Dell 1980), but disappeared between 1986 and 1987 (Saunders 1989). As populations isolated on remnants die out, the range of this species will contract.

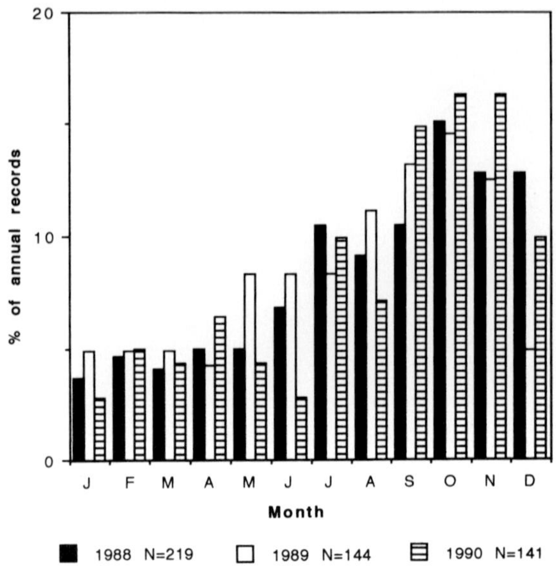

Fig. 144. Distribution 1900–1937

Golden whistler

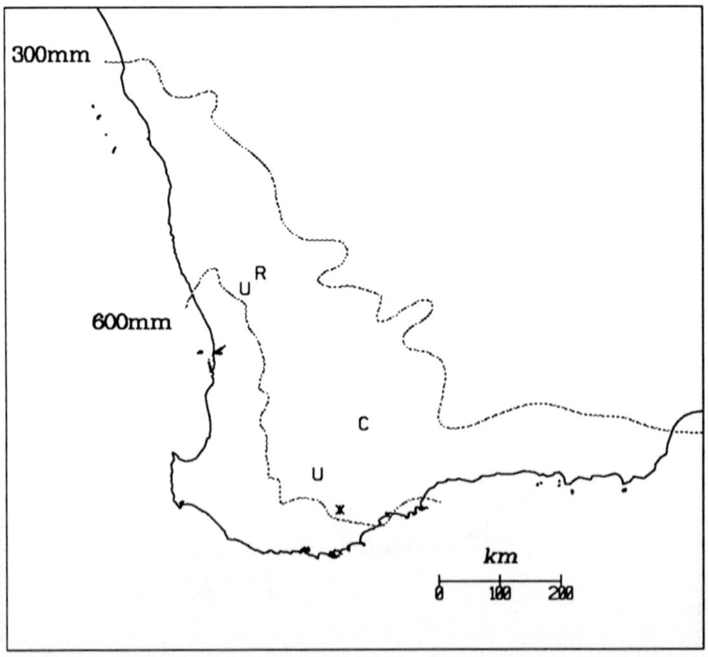

Fig. 144. Distribution 1987–1990

Golden whistler

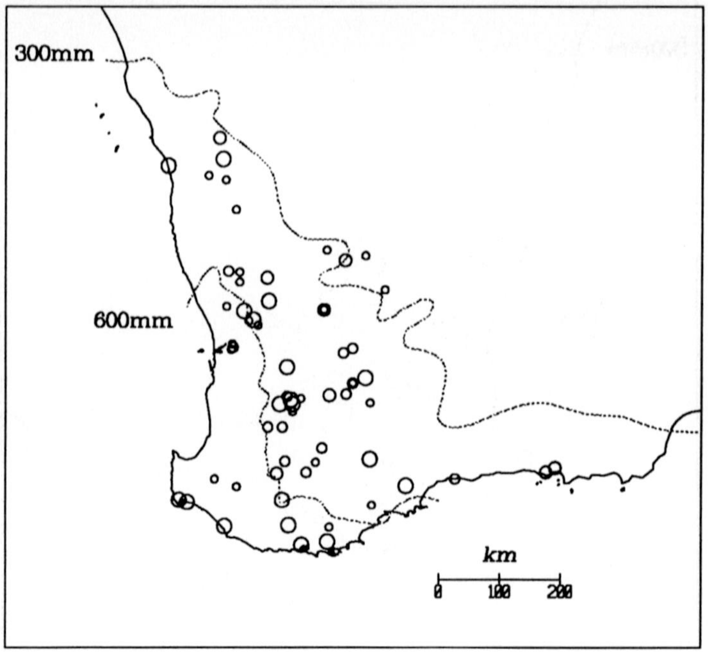

	1900/37	1987	1988	1989	1990
% of observers recording species	50	23	29	22	24
% of recording period species seen	R-C	4.5	11	9.5	8.1

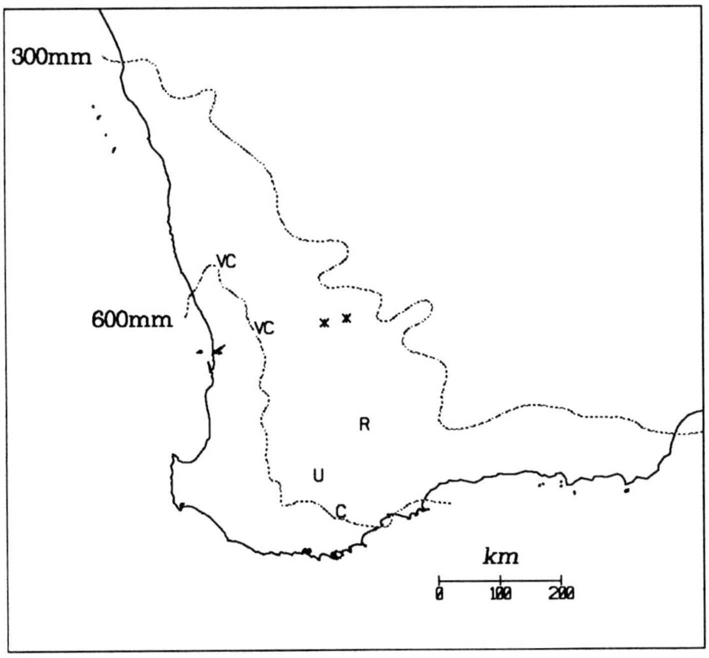

Fig. 145. Distribution 1900–1937

Rufous whistler

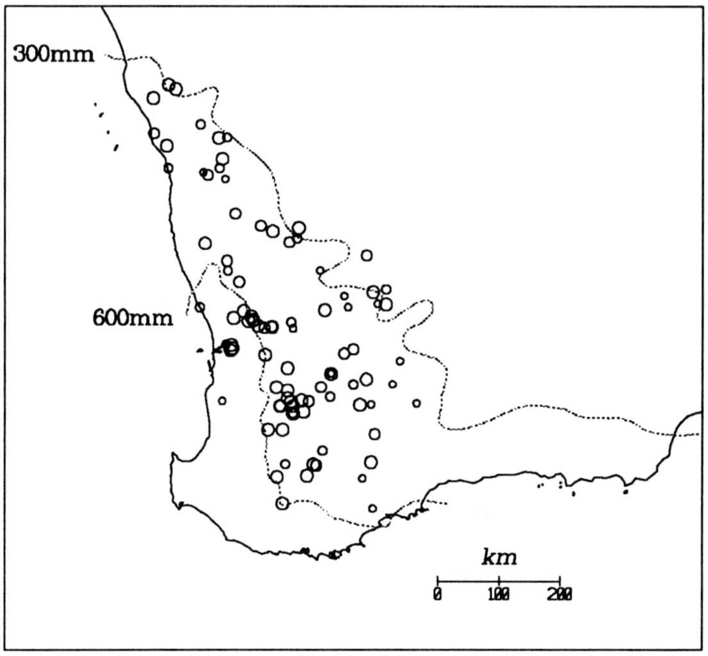

Fig. 145. Distribution 1987–1990

Rufous whistler

	1900/37	1987	1988	1989	1990
% of observers recording species	70	41	49	54	50
% of recording period species seen	R-VC	22	24	26	25

Rufous whistler

Pachycephala rufiventris

The rufous whistler is a resident which may be locally common in woodland, mallee and shrubland throughout the wheatbelt (Fig. 145). It feeds on seed, grasshoppers, bugs, lerps, beetles, flies, moths, caterpillars, wasps, bees, ants and spiders. It breeds in spring and summer in a cup-shaped nest in a tree or shrub and lays 2–3 eggs.

The rufous whistler is sedentary. Banding studies north of Kellerberrin have shown that most individuals were recaptured at the site they were originally banded, some up to 7 years later. However, individuals do wander locally as demonstrated by data obtained from one individual which moved 3.6 km along a vegetated road verge leading from a nature reserve (Saunders and de Rebeira 1991).

Although common in remnant vegetation and along well-vegetated road verges, the fact that the rufous whistler is dependent on native vegetation means that it must have declined in abundance as a result of the loss of suitable habitat.

Grey shrike-thrush

Colluricincla harmonica

The grey shrike-thrush is a common resident in woodland, mallee and shrubland throughout the wheatbelt (Fig. 146). It feeds on seed and other plant material, snails, millipedes, centipedes, damselflies, cockroaches, mantids, earwigs, grasshoppers, crickets, stick insects, cicadas, lerps, bugs, beetles, lacewings, moths, caterpillars, wasps, bees, ants, spiders, frogs, lizards, nestling birds, and small mammals. It breeds in spring in a bowl-shaped nest in a tree or shrub and lays 3–4 eggs.

Banding studies north of Kellerberrin have shown that the grey shrike-thrush is sedentary. However, it does move locally as revealed by data from one individual which moved 1.2 km along a well-vegetated road verge leading from a large nature reserve (Saunders and de Rebeira 1991).

Although widespread and locally common, the grey shrike-thrush is dependent on native vegetation, and does not occur in heavily disturbed and degraded remnant vegetation. This species has declined in abundance as a result of loss of suitable habitat. For example, it disappeared from around Perth (Serventy and Whittell 1962) and underwent a reduction in numbers around Kellerberrin as a result of clearing native vegetation (Leake 1962).

Fig. 146. Distribution 1900–1937

Grey shrike-thrush

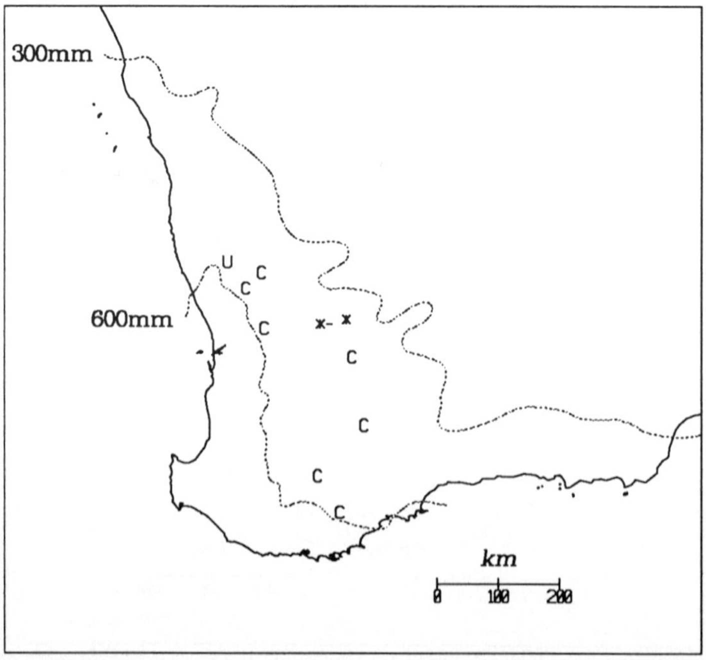

Fig. 146. Distribution 1987–1990

Grey shrike-thrush

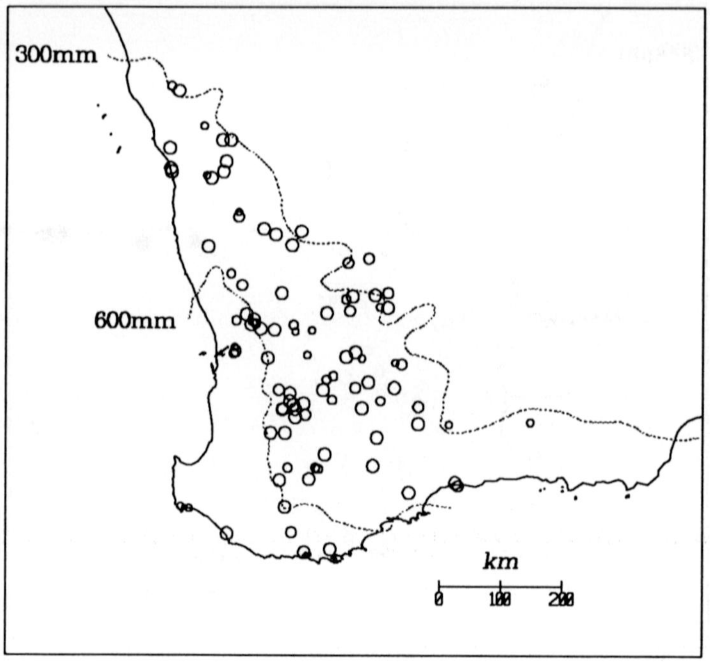

	1900/37	1987	1988	1989	1990
% of observers recording species	100	45	53	51	49
% of recording period species seen	U-C	22	29	31	29

Fig. 147. Distribution 1900–1937

Crested bellbird

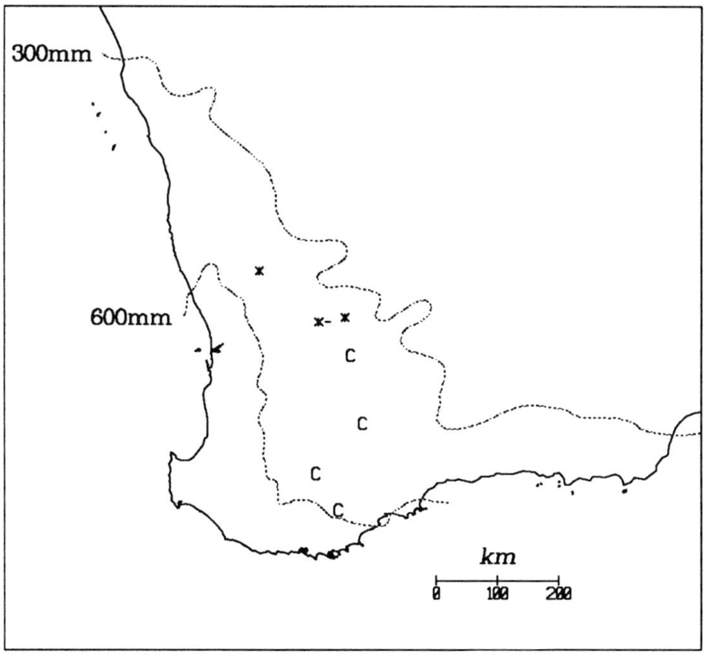

Fig. 147. Distribution 1987–1990

Crested bellbird

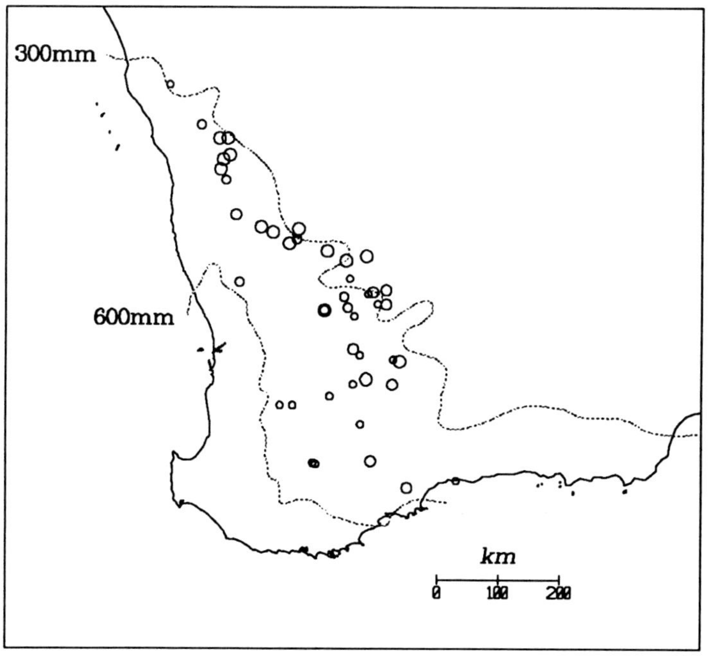

	1900/37	1987	1988	1989	1990
% of observers recording species	70	24	23	20	20
% of recording period species seen	C	8.7	10	8.8	9.1

Crested bellbird

Oreoica gutturalis

The crested bellbird is an uncommon resident of woodland, mallee and shrubland throughout most of the wheatbelt (Fig. 147). It feeds on plant material, insects and spiders. It breeds in spring in a cup-shaped nest in a shrub or tree and lays 3–4 eggs.

Although widely regarded as common early this century in the wheatbelt, the crested bellbird has declined in abundance. For example, Leake (1962) noted that this species was declining in the Kellerberrin area, because of clearing for agriculture.

Restless flycatcher

Myiagra inquieta

The restless flycatcher is an uncommon resident in woodland and mallee of the higher rainfall districts of the wheatbelt (Fig. 148), and a rare nomad in others, particularly the central and northern wheatbelt. It feeds on damselflies, grasshoppers, crickets, bugs, cicadas, beetles, flies, moths, wasps, ants and spiders. The restless flycatcher breeds in spring and early summer in a cup-shaped nest in a tree and lays 3–4 eggs.

Over the past 80 years the restless flycatcher has declined in range and abundance. Early this century, it was numerous on good soils between Mogumber (Lawson 1905) and the Wongan Hills (Milligan 1904). It had declined around Mogumber by the early 1950s (Loaring and Serventy 1952) and disappeared from the Wongan Hills by the early 1970s (de Rebeira and de Rebeira 1977). Leake (1962) noted that it was declining around Kellerberrin. It no longer occurs there, except as a rare nomad (Saunders 1989), being seen there twice in nine years of intensive field work (1985–94). Similarly, it was common near Bruce Rock (Crossman 1909); now it is a rare nomad.

Fig. 148. Distribution 1900–1937

Restless flycatcher

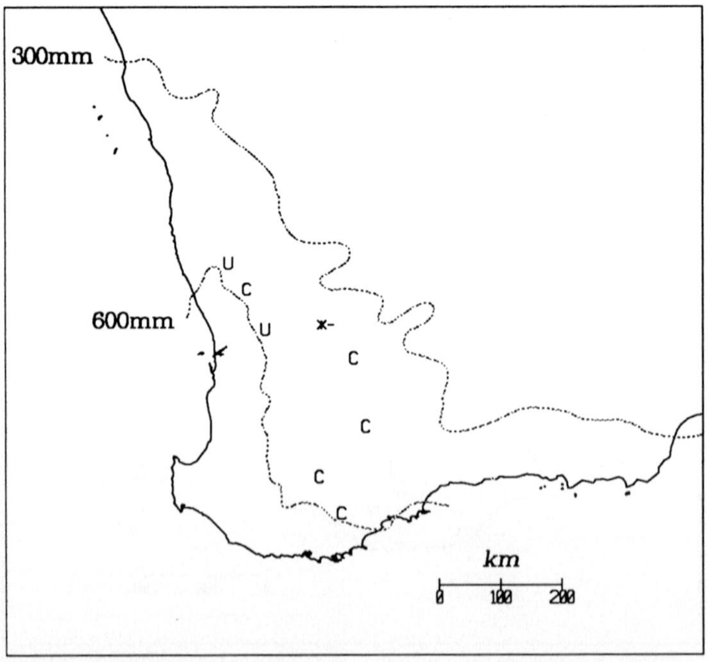

Fig. 148. Distribution 1987–1990

Restless flycatcher

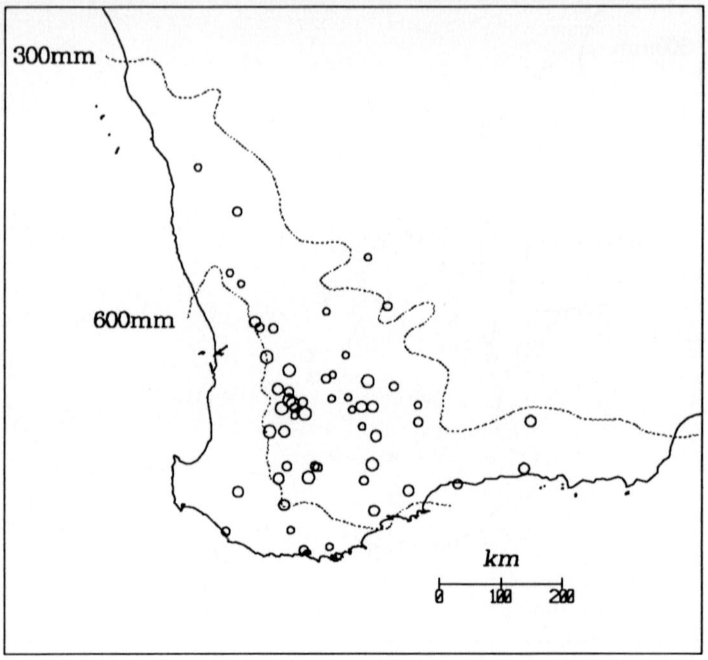

	1900/37	1987	1988	1989	1990
% of observers recording species	80	26	26	24	27
% of recording period species seen	U-C	5.3	7.5	6.8	5.8

(Top photo) Wide, well-vegetated road verges that link large patches of native vegetation like this strip between Kodj Kodjin and Durokoppin Nature Reserves in the central wheatbelt foster the movements of species like the western yellow robin which has disappeared from large areas of its former range. Note that the road is on one side of the verge, not down the middle as is the norm (photo by G. Chapman).

(Bottom photo) Narrow, sparsely-vegetated road verges with little understorey may not aid the movements of many species of birds which are dependent upon native vegetation (photo by D. E. Steven).

Grey fantail

Rhipidura fuliginosa

The grey fantail is a common resident in woodland, mallee, shrubland, towns and farm gardens of the higher rainfall districts of the wheatbelt. It is a common winter and spring breeding migrant in the remainder (Fig. 149). It feeds on mantids, earwigs, grasshoppers, bugs, beetles, flies, mosquitoes, moths, native bees, ants and spiders. It breeds in spring and early summer in a cup-shaped nest in a tree and lays 2–4 eggs.

Historically, the seasonal movement pattern of the grey fantail has been described for several locations: Moora (appeared with the first rains and departed during August) (Orton and Sandland 1913); the Kellerberrin/Kwolyin districts (arrived April to June and departed in November or December) (Ford and Stone 1957; Leake 1962); and around the Wongan Hills (April to July with one sighting in September) (de Rebeira and de Rebeira 1977). This pattern is also shown by arrival and departure dates from five wheatbelt locations during this survey (Table below and Fig. on facing page).

	1988	1989	1990
Tardun	May Wk 1 Sep. Wk 2	Apr. Wk 2 Oct. Wk 1	Apr. Wk 4 Oct. Wk 1
Arrino	May Wk 2 Aug. Wk 3	Apr. Wk 4 Sep. Wk 3	Apr. Wk 4 Sep. Wk 4
Avon River between Northam and Toodyay	Resident throughout period		
Bruce Rock	Jun. Wk 1 Sep. Wk 2	May Wk 1 Sep. Wk 1	May Wk 4 Sep. Wk 1
Pingrup	Arrived first week in April 1988 and resident from then.		

Limited data from banding studies north of Kellerberrin where the grey fantail is a seasonal migrant, revealed that an individual banded in the area was recaptured at the same site 14 months later (Saunders and de Rebeira 1991).

There are no data to suggest any change in status of this species.

Fig. 149. Distribution 1900–1937

Grey fantail

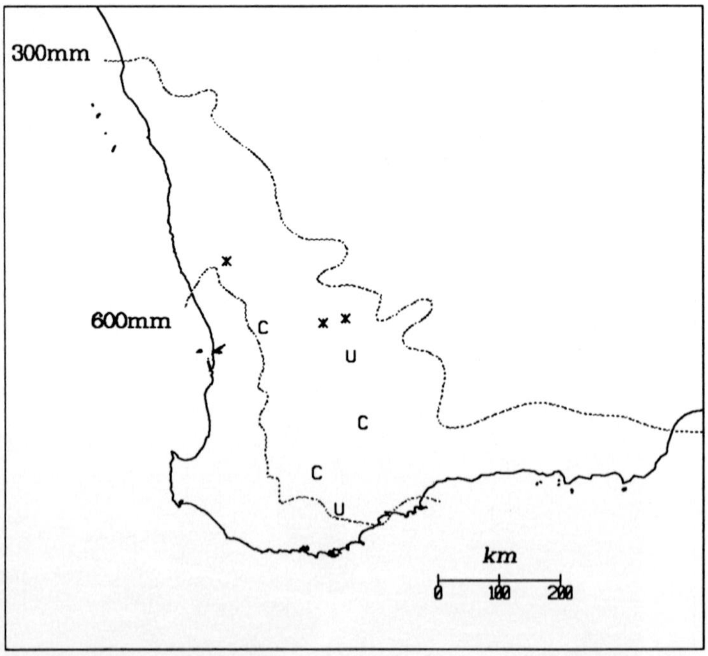

Fig. 149. Distribution 1987–1990

Grey fantail

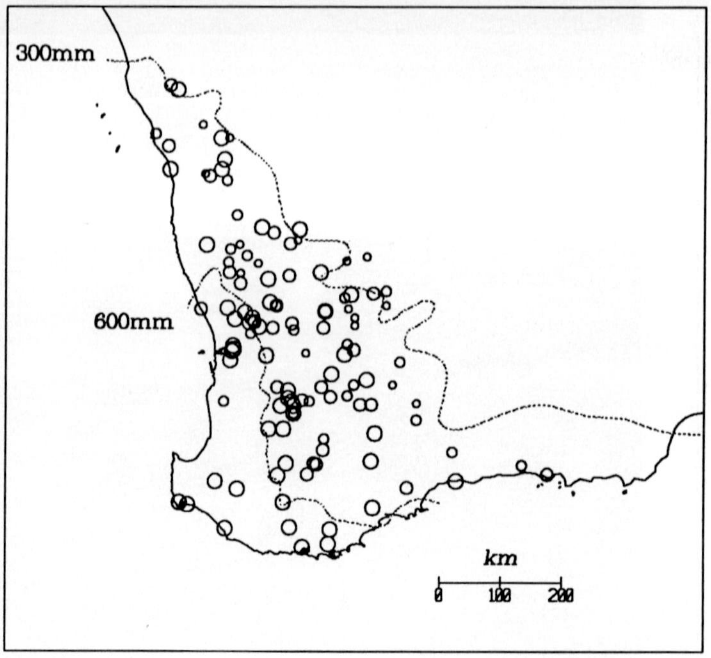

	1900/37	1987	1988	1989	1990
% of observers recording species	80	54	63	63	67
% of recording period species seen	U-C	24	29	32	33

Fig. 149. Distribution 1987–1990 — April–October

Grey fantail

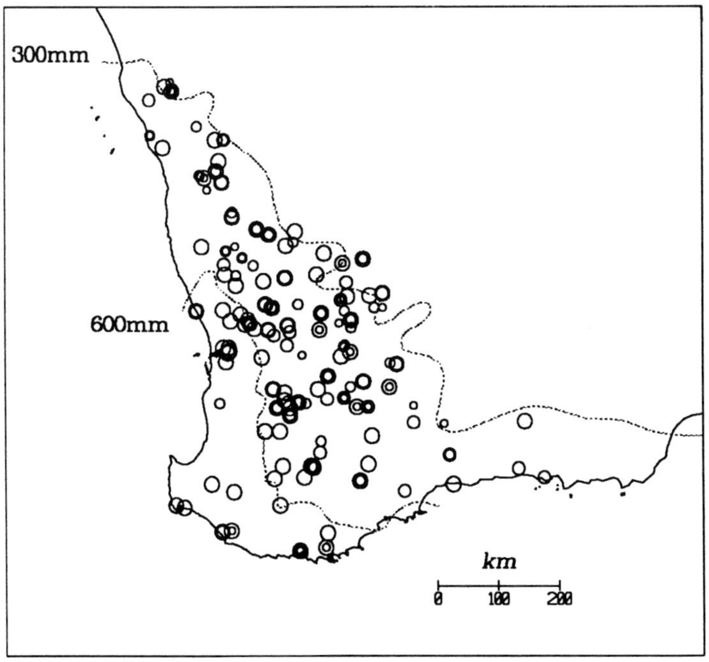

Fig. 149. Distribution 1987–1990 — December–February

Grey fantail

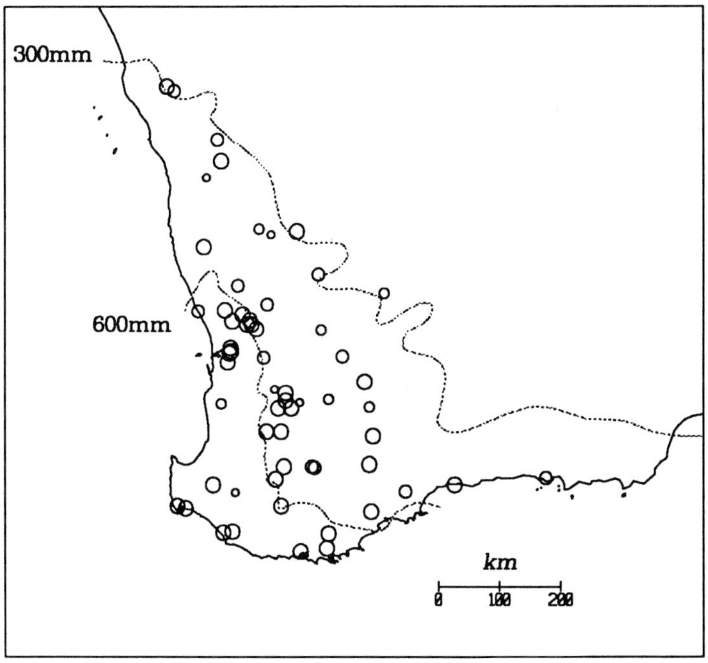

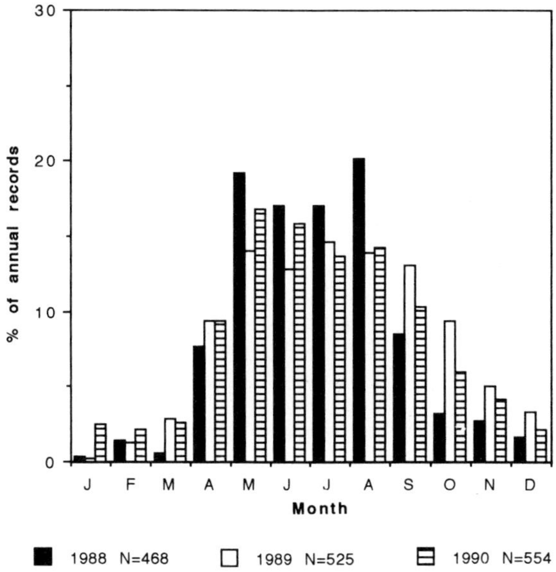

Willie wagtail

Rhipidura leucophrys

The willie wagtail is ubiquitous on mainland Australia and a common resident of woodland, mallee, shrubland, heath, towns and farmland throughout the wheatbelt (Fig. 150). It feeds on seed, damselflies, cockroaches, grasshoppers, bugs, cicadas, water boatmen, beetles, flies, moths, caterpillars, butterflies, wasps, ants and spiders. The willie wagtail breeds from winter to summer in a cup-shaped nest in a tree, shrub, shed or verandah. It lays 3–4 eggs.

The willie wagtail is sedentary, however, one individual did move 7 km from the site where it was banded north of Kellerberrin (Saunders and de Rebeira 1991).

The willie wagtail is a bird of open country and it has benefited greatly from development for agriculture (Serventy and Whittell 1962). Early this century it was not seen around the Wongan Hills (Milligan 1904), but it became established following clearing for agriculture (de Rebeira and de Rebeira 1977). It is common around human settlements and can often be seen perched on the back of domestic livestock as it seeks out insect prey.

Fig. 150. Distribution 1900–1937

Willie wagtail

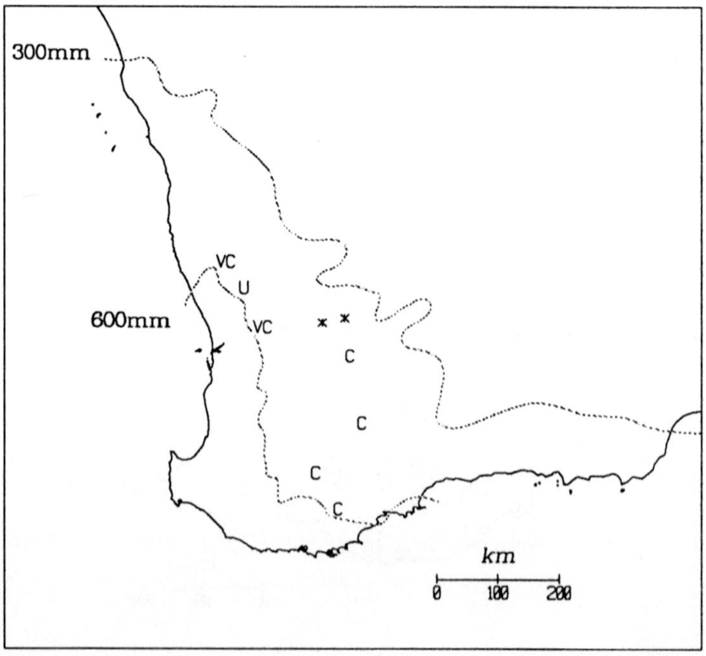

Fig. 150. Distribution 1987–1990

Willie wagtail

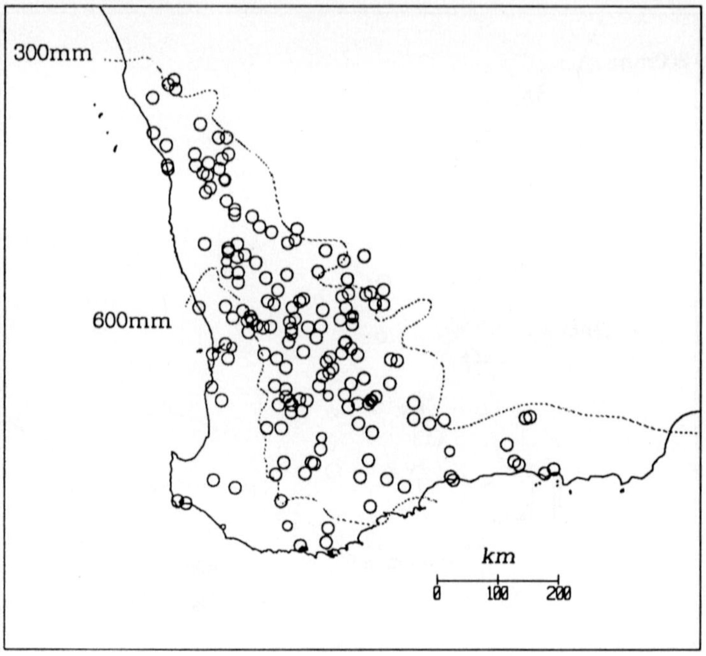

	1900/37	1987	1988	1989	1990
% of observers recording species	90	98	96	98	96
% of recording period species seen	U-VC	85	85	84	86

Fig. 151. Distribution 1900–1937

Chestnut quail-thrush

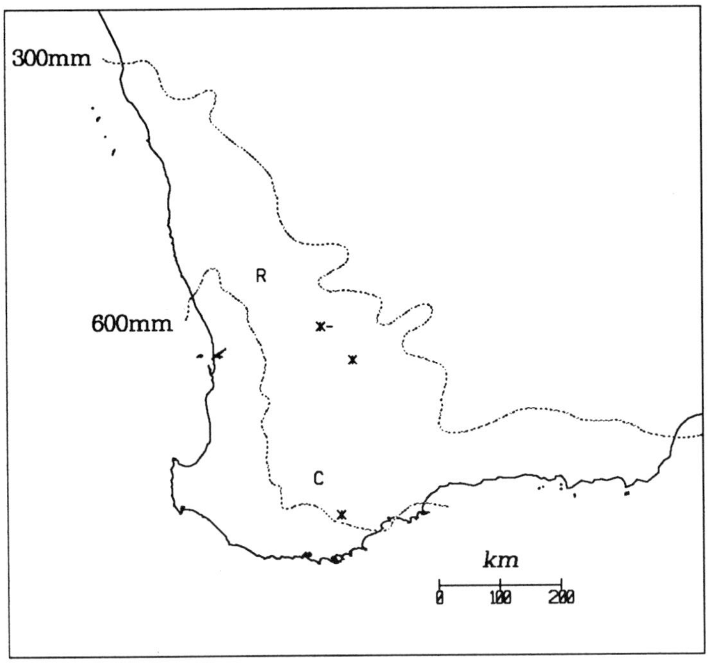

Fig. 151. Distribution 1987–1990

Chestnut quail-thrush

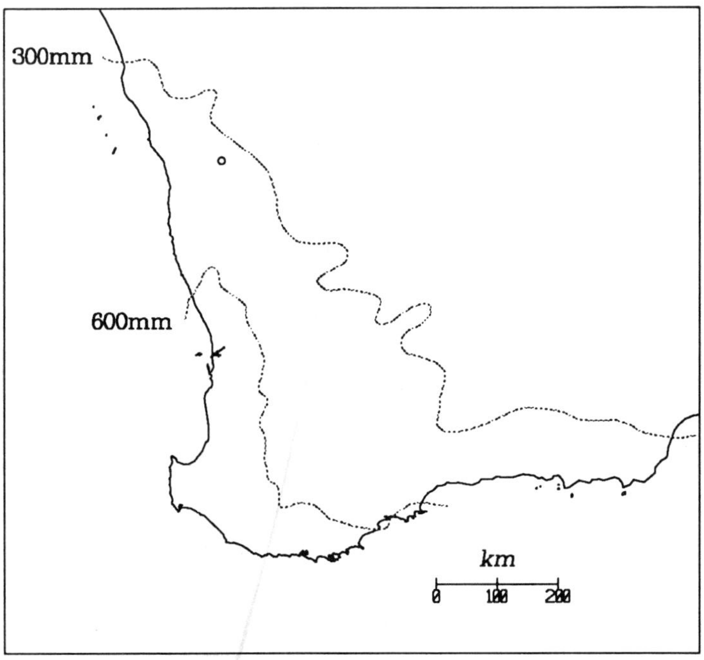

	1900/37	1987	1988	1989	1990
% of observers recording species	50	0.8	0	0	0
% of recording period species seen	R-C	0.03	0	0	0

Chestnut quail-thrush

Cinclosoma castanotum

Historically, the chestnut quail-thrush was common around Tenterden (Milligan 1903) and at Broome Hill (Carter 1923), was rare around the Wongan Hills (Milligan 1904), was declining around Kellerberrin (Leake 1962), and was recorded at Cumminin Station near Bruce Rock (Crossman 1909) and in the Stirling Range (Milligan 1903) (Fig. 151). It no longer occurs in these areas and is a rare nomad in the drier districts of the wheatbelt. It feeds on seed, berries, centipedes, grasshoppers, bugs, beetles, flies, caterpillars and ants. It breeds in spring on the ground and lays 2 eggs.

Loss of habitat, degradation of understory by grazing, and predation by foxes and cats have all probably contributed to its decline. The wheatbelt may have represented the moister margins of the distribution of the chestnut quail-thrush. It is present in the uncleared areas on the eastern edge of the wheatbelt (Blakers *et al.* 1984; Lynch and Saunders 1991).

White-browed babbler

Pomatostomus superciliosus

The white-browed babbler is a resident which may be locally common in woodland, mallee, shrubland and heath throughout the wheatbelt (Fig. 152). It forages on the ground for seed, berries, millipedes, cockroaches, earwigs, termites, crickets, grasshoppers, bugs, cicadas, beetles, caterpillars, wasps, ants, spiders and predates nests of small birds (e.g., red-capped robin (P. Cale, pers. comm.)). It breeds in late winter and spring in a solidly constructed, enclosed nest in a shrub and lays 2–3 eggs.

Banding studies carried out north of Kellerberrin reveal that the white-browed babbler is dependent on native vegetation, and sedentary. The longest movement was for an individual which was retrapped 1.5 km along a vegetated road verge that was connected to the nature reserve in which it had been originally caught (Saunders and de Rebeira 1991).

This species has probably declined in abundance in the extensively cleared districts of the wheatbelt as a result of loss of suitable habitat.

Fig. 152. Distribution 1900–1937

White-browed babbler

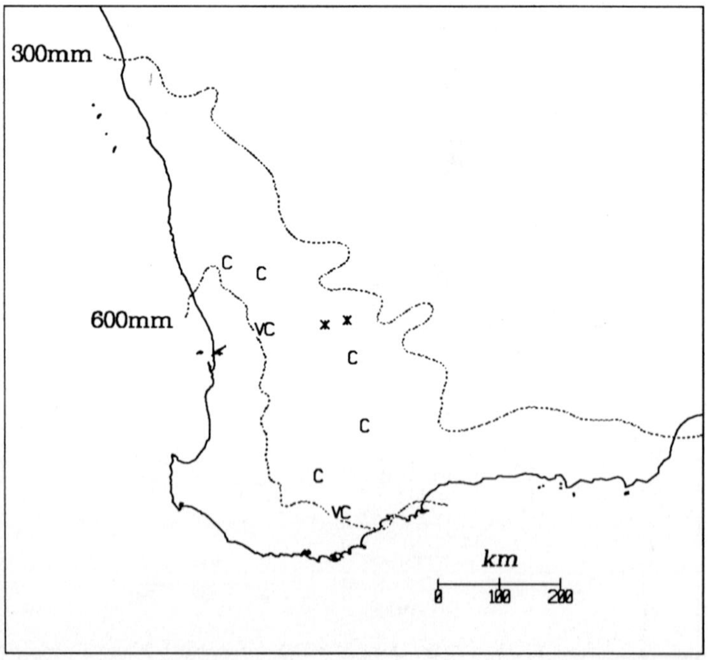

Fig. 152. Distribution 1987–1990

White-browed babbler

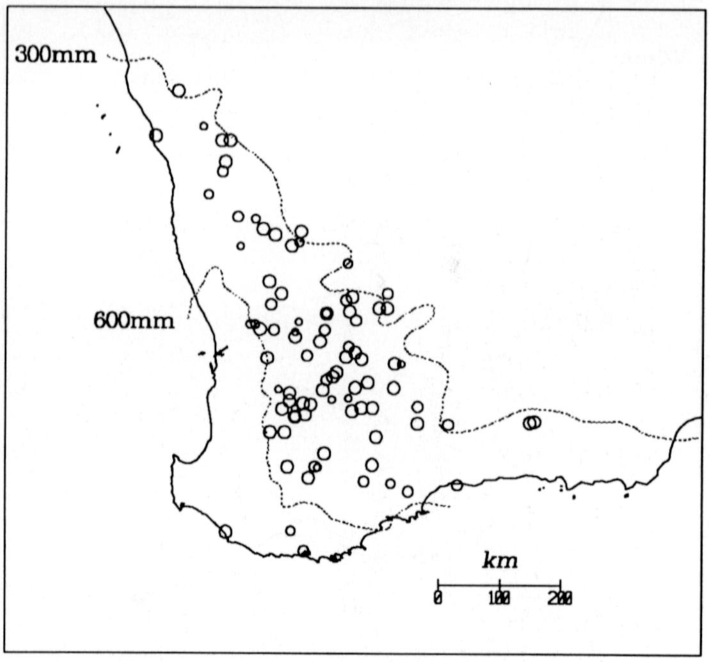

	1900/37	1987	1988	1989	1990
% of observers recording species	90	45	49	47	47
% of recording period species seen	C-VC	26	26	29	30

Much native vegetation on farmland *(above)* and along roads *(below)* has been grazed and/or burnt regularly. As a result, the understorey and litter layers have been destroyed. This structural simplification degrades the quality of the habitat, leading to the elimination of species dependent on these structural layers. There is little or no regeneration of native vegetation and these patches of remnant vegetation will disappear over time (photos by D. A. Saunders).

Clamorous reed warbler

Acrocephalus stentoreus

The clamorous reed warbler is a rare breeding migrant (spring and early summer) in reed-beds in suitable aquatic habitats in the wheatbelt (Fig. 153). For example, in 1988 it arrived along the Avon River between Northam and Toodyay during the second week in October, and left during the last week in January. In 1989 it arrived during the second week in September and departed the last week in January. It was not recorded there during the spring of 1990.

The clamorous reed warbler feeds on seed, damselflies, dragonflies, grasshoppers, bugs, cicadas, beetles, flies, moths, ants and spiders. It breeds in spring and summer in a cup-shaped nest in reeds or a shrub. It lays 2–4 eggs.

There is no evidence to suggest any change in the status of this species.

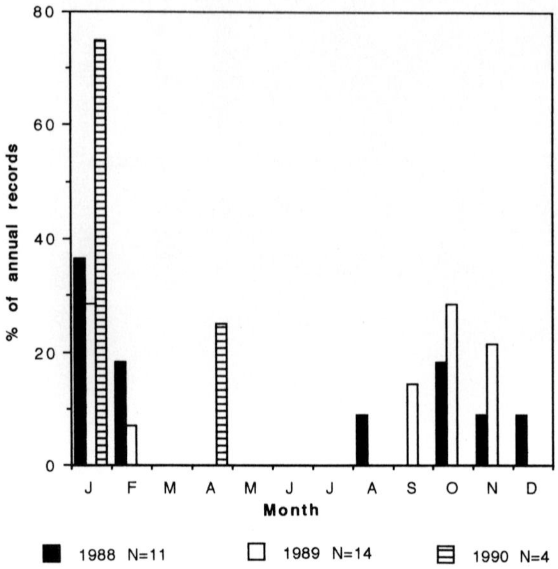

Fig. 153. Distribution 1900–1937

Clamorous reed warbler

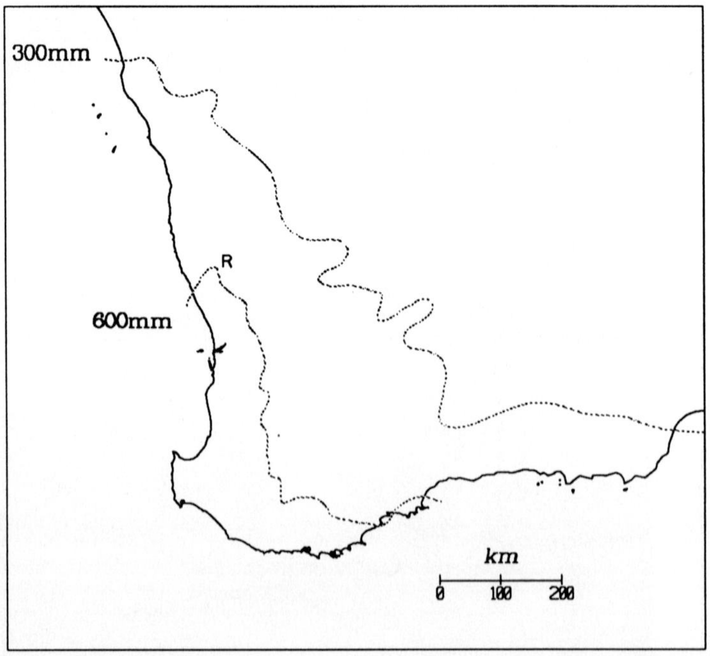

Fig. 153. Distribution 1987–1990

Clamorous reed warbler

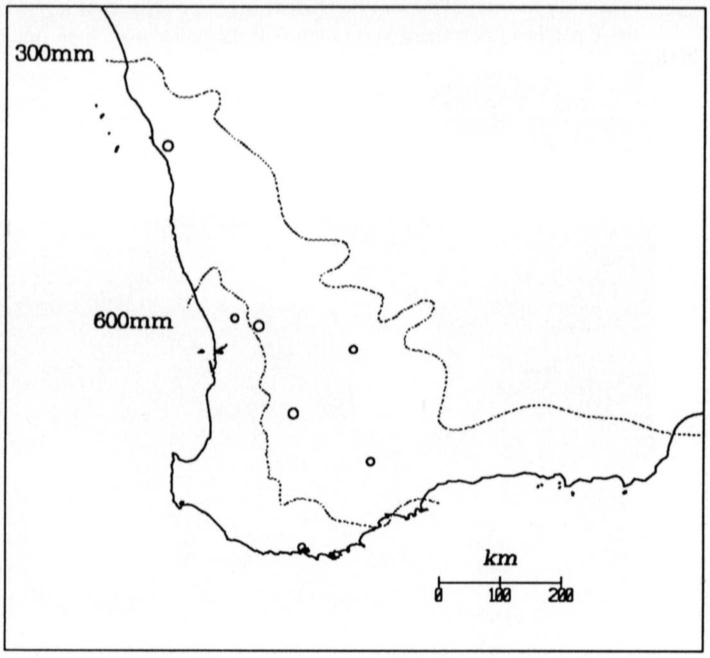

	1900/37	1987	1988	1989	1990
% of observers recording species	10	0.8	1.4	4.3	1.0
% of recording period species seen	R	0.1	0.2	0.3	0.1

Fig. 153. Distribution 1987–1990 — March–July

Clamorous reed warbler

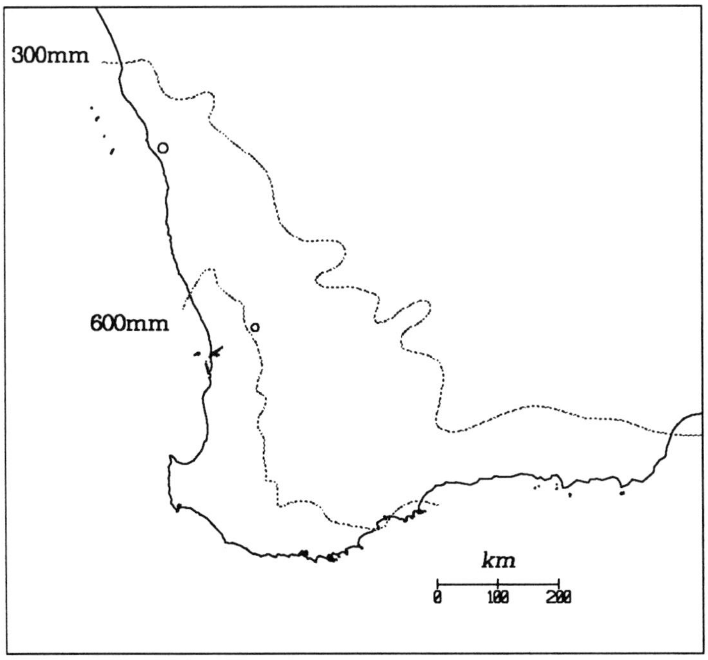

Fig. 153. Distribution 1987–1990 — August–February

Clamorous reed warbler

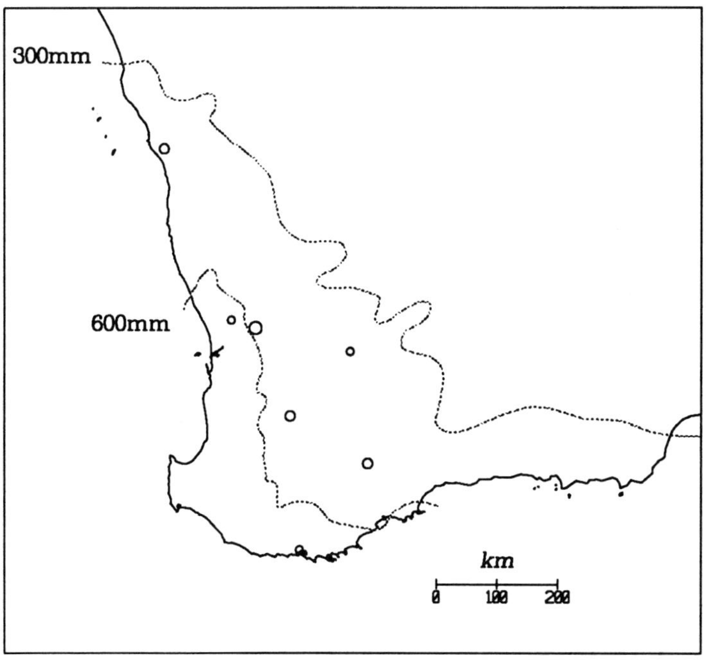

Little grassbird

Megalurus gramineus

The little grassbird is a rare nomad in wetlands in the higher rainfall districts of the wheatbelt (Fig. 154). It feeds on freshwater snails, damselflies, grasshoppers, bugs, beetles, flies, caterpillars, moths, wasps, ants and spiders. It breeds in spring and early summer in a cup-shaped nest in rushes or a clump of grasses. It lays 3–5 eggs.

There is no evidence to suggest that this species has changed in status.

Fig. 154. Distribution 1987–1990

Little grassbird

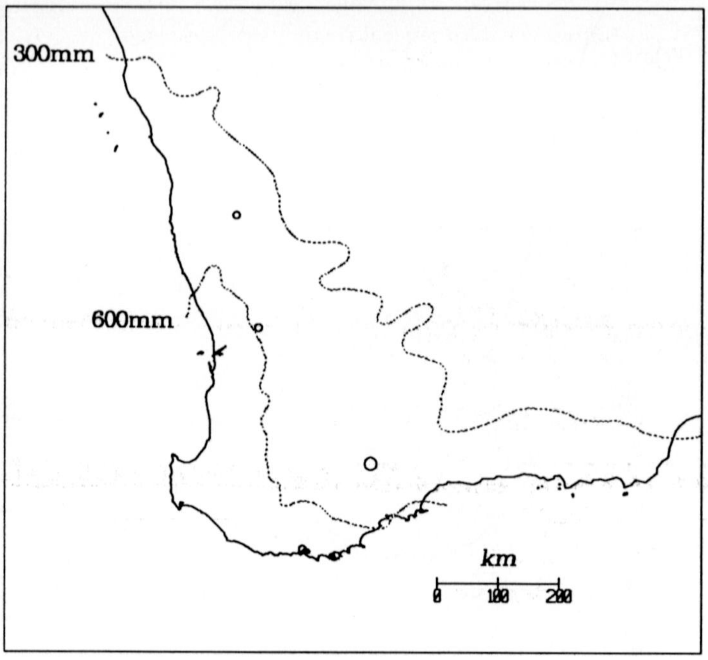

	1900/37	1987	1988	1989	1990
% of observers recording species	0	0	2.1	0.9	1.9
% of recording period species seen	NR	0	0.1	0.1	0.8

(Top photo) Many patches of remnant vegetation are isolated from other patches. Species dependent on native vegetation may be isolated on these patches because they are unable to cross extensive areas of cleared land. Most of these isolated populations are too small to be viable and as they die out their ranges contract (photo by D. A. Saunders).

(Bottom photo) Road verges may provide the skeleton on which to construct linkages to aid the movement of species dependent on native vegetation (photo by G. Chapman).

Rufous song lark

Cinclorhamphus mathewsi

The rufous song lark is an uncommon breeding migrant in open woodland throughout much of the wheatbelt (Fig. 155). It feeds on seed, mantids, grasshoppers, crickets, locusts, bugs, beetles, caterpillars, moths, wasps, bees, ants and spiders. It breeds in spring and summer in a cup-shaped nest on the ground and lays 3–4 eggs.

Historically, the seasonal movement pattern of the rufous song lark has been noted for a number of locations: Moora (numerous from September until January or February) (Orton and Sandland 1913); Northam (earliest record late June and latest record first week in February) (Jenkins 1931; Masters and Milhinch 1974); and Kellerberrin/Kwolyin (August to end of December) (Ford and Stone 1957). This pattern is also shown in Figure on facing page and by arrival and departure dates from four locations in the wheatbelt during this survey (Table below).

	1988	1989	1990
Tardun	Not seen	Aug. Wk 3 Oct. Wk 1	Not seen
Arrino	Sep. Wk 3 Dec. Wk 4	Jun. Wk 1 Dec. Wk 2	Aug. Wk 4 Dec. Wk 2
Avon River between Northam and Toodyay	Sep. Wk 4 Oct. Wk 4	Sep. Wk 3 Dec. Wk 3	Jul. Wk 3 Dec. Wk 1
Pingrup	Oct. Wk 2 Dec. Wk 3	Aug. Wk 4 Oct. Wk 2	Jun. Wk 3 Oct. Wk 3

This species may have declined in abundance.

Fig. 155. Distribution 1900–1937

Rufous song lark

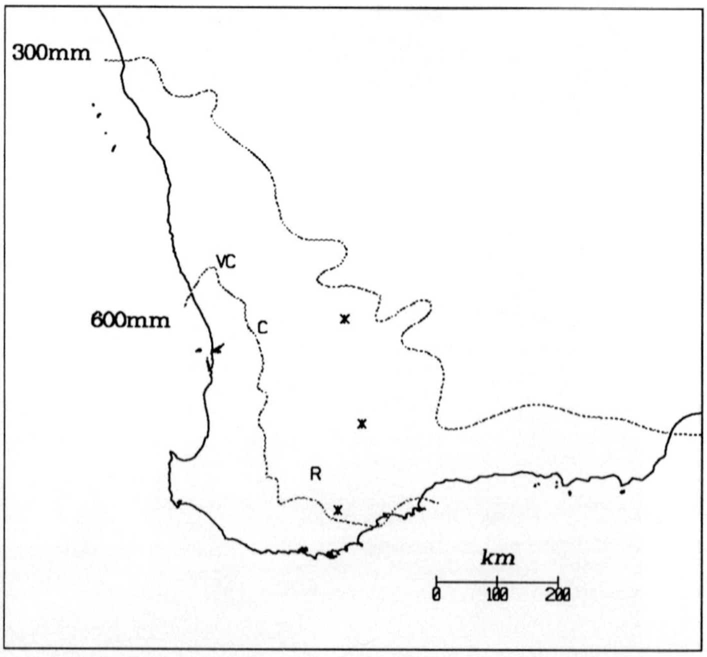

Fig. 155. Distribution 1987–1990

Rufous song lark

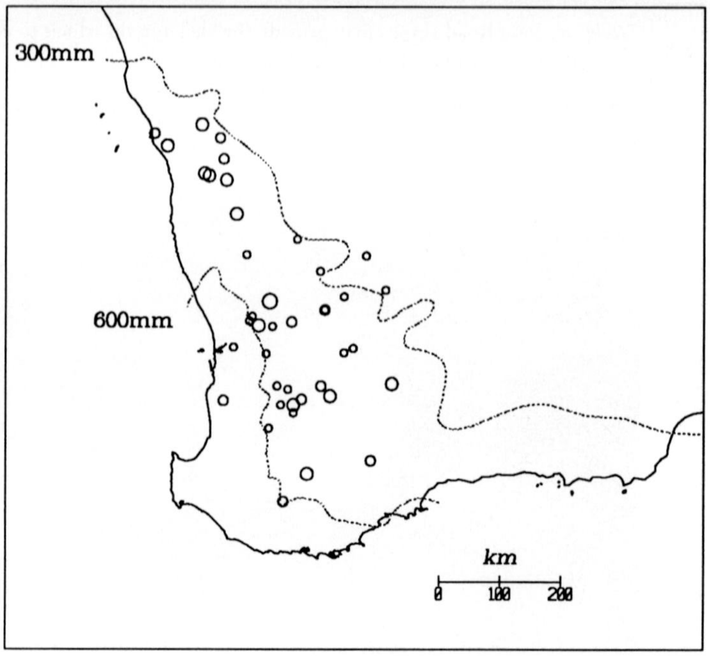

	1900/37	1987	1988	1989	1990
% of observers recording species	60	13	18	15	9.6
% of recording period species seen	R-VC	3.1	2.7	1.9	2.1

Fig. 155. Distribution 1987–1990 — February–May

Rufous song lark

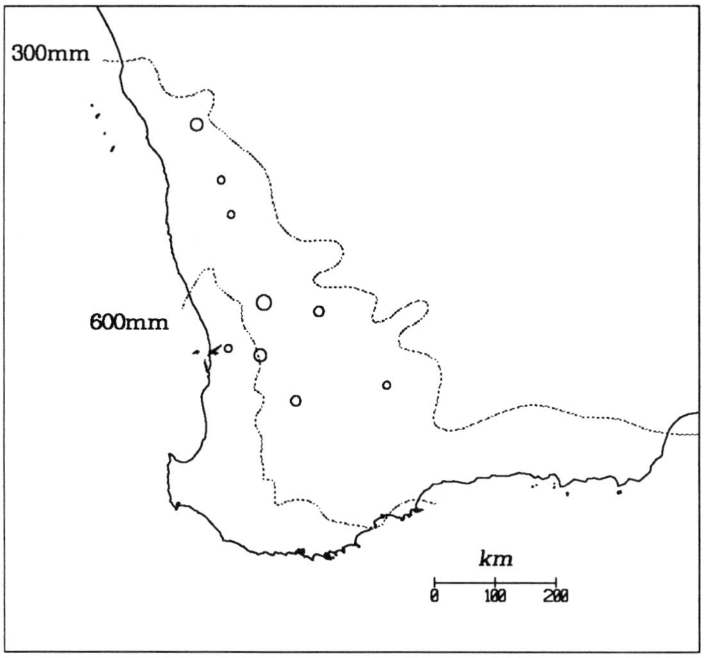

Fig. 155. Distribution 1987–1990 — June–January

Rufous song lark

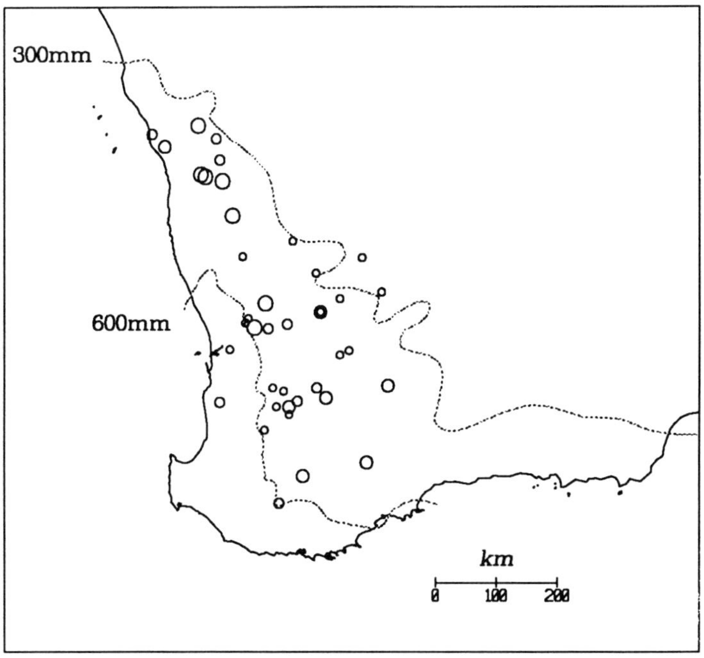

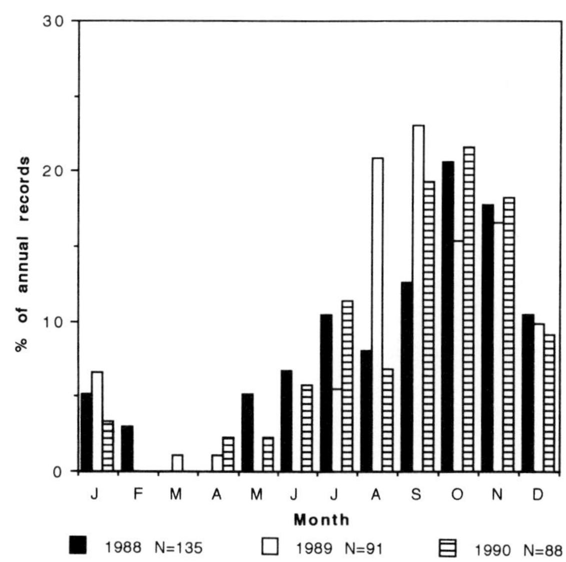

Brown song lark

Cinclorhamphus cruralis

The brown song lark is an uncommon breeding migrant in farmland throughout the wheatbelt (Fig. 156). It feeds on seed, cockroaches, grasshoppers, crickets, bugs, beetles, flies, caterpillars, wasps, bees, ants and spiders. It breeds in spring and summer in a cup-shaped nest on the ground and lays 3–4 eggs.

Historically, the seasonal movement pattern of the brown song lark has been noted for several locations: Northam (normally July to January, but may be June to February) (Jenkins 1931; Masters and Milhinch 1974); Lake Grace (Carnaby 1933); and Kellerberrin/Kwolyin (June or July to December or January) (Ford and Stone 1957; Leake 1962). This pattern is also shown in Figure on facing page and by arrival and departure dates from five wheatbelt locations during this survey (Table below).

	1988	1989	1990
Tardun	Seen Aug. Wks 3 & 4	Jun. Wk 4 Feb. Wk 1	Aug. Wk 1 Dec. Wk 1
Arrino	Jun. Wk 1 Nov. Wk 3	Not seen	Aug. Wk 2 *
Avon River between Northam and Toodyay	Seen Oct. Wk 2	Not seen	Seen Oct. Wk 4 and Nov. Wk 1
Bruce Rock	Jul. Wk 4 Oct. Wk 4	May Wk 2 Sep. Wk 3	Jul. Wk 1 Dec. Wk 2
Pingrup	Sep. Wk. 2 Feb. Wk 2	Jun. Wk 1 Remained all through 1990	

*No data for departure time as the survey ended 31 December, 1990.

Numbers may vary between years, as during 1990 when recording rates for the brown song lark were nearly twice those of the previous two years. In some areas they may remain throughout the year. There is no evidence to indicate that this species has changed in abundance, however, it is possible that it has increased as a result of development for agriculture.

Fig. 156. Distribution 1900–1937

Brown song lark

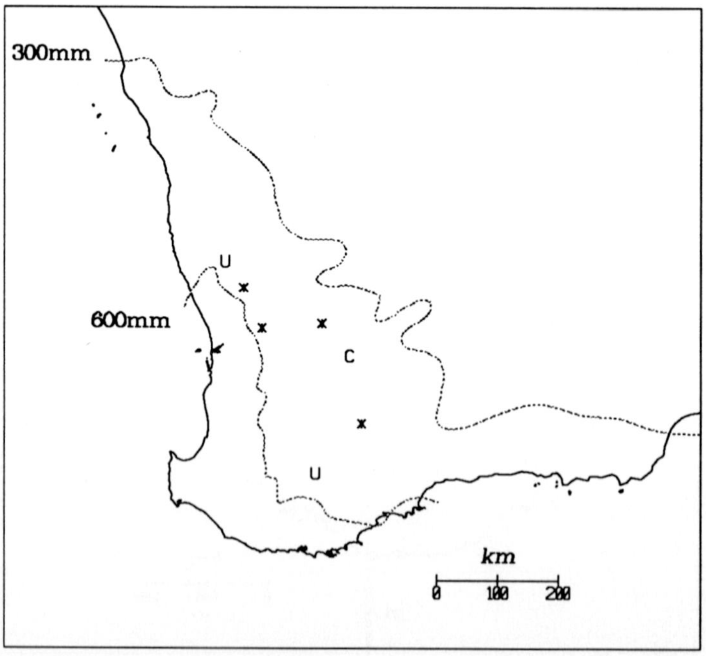

Fig. 156. Distribution 1987–1990

Brown song lark

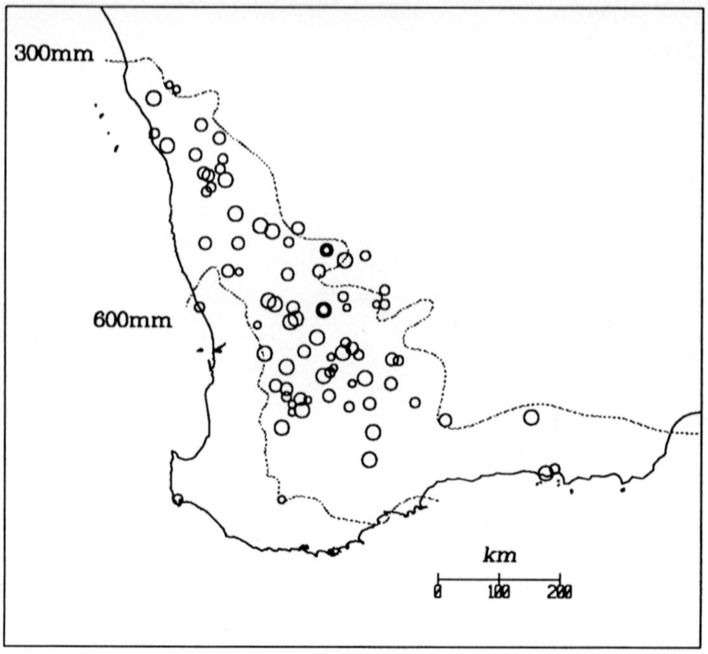

	1900/37	1987	1988	1989	1990
% of observers recording species	70	42	39	32	43
% of recording period species seen	U-C	13	10	10	19

Fig. 156. Distribution 1987–1990 — March–May

Brown song lark

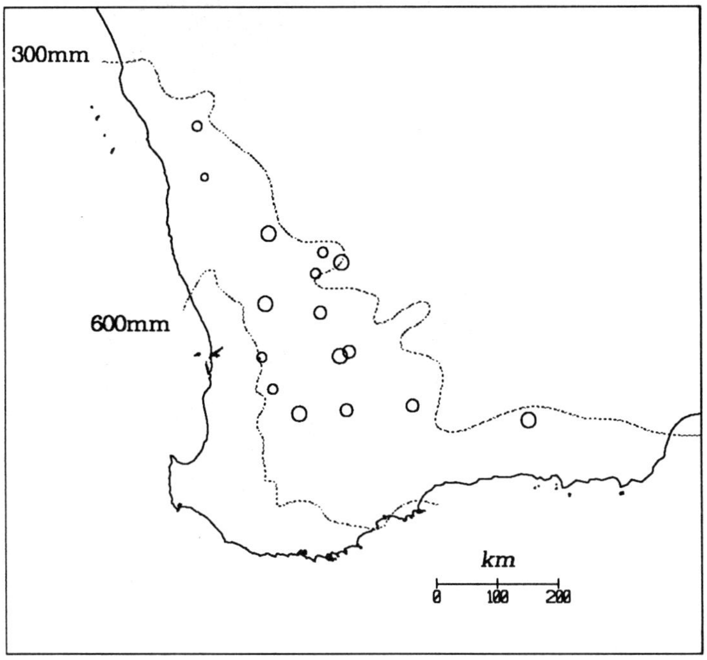

Fig. 156. Distribution 1987–1990 — June–February

Brown song lark

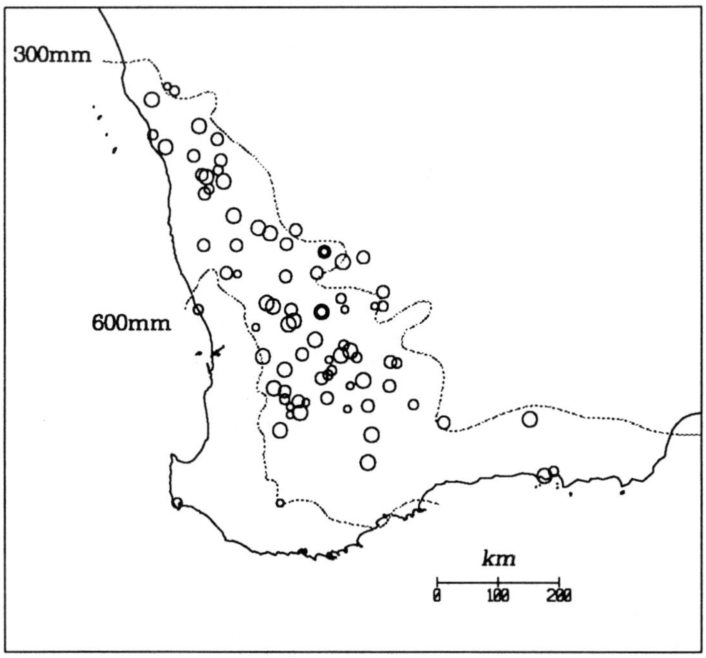

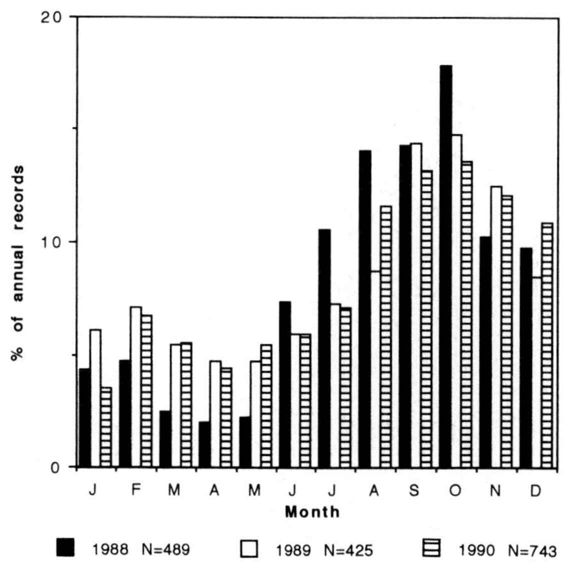

Splendid wren

Malurus splendens

OTHER NAME:

Splendid fairy-wren

The splendid wren is an uncommon resident in woodland, mallee and shrubland of the northern wheatbelt and of the higher rainfall districts of the southern wheatbelt (Fig. 157). It feeds on seed, termites, crickets, grasshoppers, aphids, bugs, beetles, caterpillars, moths, ants and spiders. It breeds in spring and summer in a ball-shaped nest in a small shrub and lays 2–4 eggs.

It is possible that the splendid wren, which is dependent on native vegetation, has declined in abundance in the wheatbelt because of loss of suitable habitat. For example, in the Northam district, the splendid wren was once common, but numbers had fallen severely since about 1950 (Masters and Milhinch 1974).

Fig. 157. Distribution 1900–1937

Splendid wren

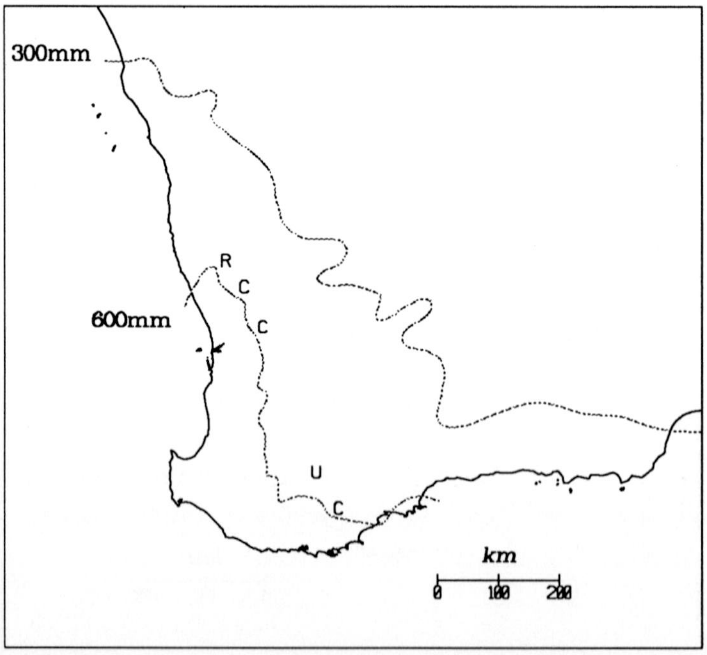

Fig. 157. Distribution 1987–1990

Splendid wren

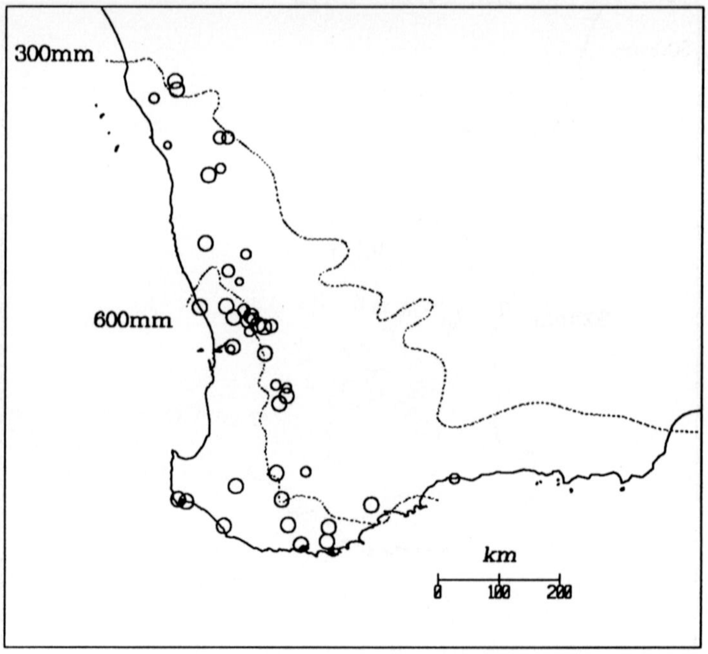

	1900/37	1987	1988	1989	1990
% of observers recording species	50	15	27	28	27
% of recording period species seen	R-C	8.1	14	18	16

Variegated wren

Malurus lamberti

OTHER NAME:

Variegated fairy-wren

The variegated wren is an uncommon resident in shrubland and heath of the northern wheatbelt (Fig. 158). This represents the southern limits of its range in Western Australia. It feeds on seed, cockroaches, earwigs, grasshoppers, bugs, lacewings, beetles, flies, caterpillars, moths, wasps, bees, ants and spiders. It breeds in spring in an enclosed, oval-shaped nest in a small shrub and lays 3–4 eggs.

The variegated wren is dependent on native vegetation and has probably declined in those areas which have been extensively cleared.

Fig. 158. Distribution 1987–1990

Variegated wren

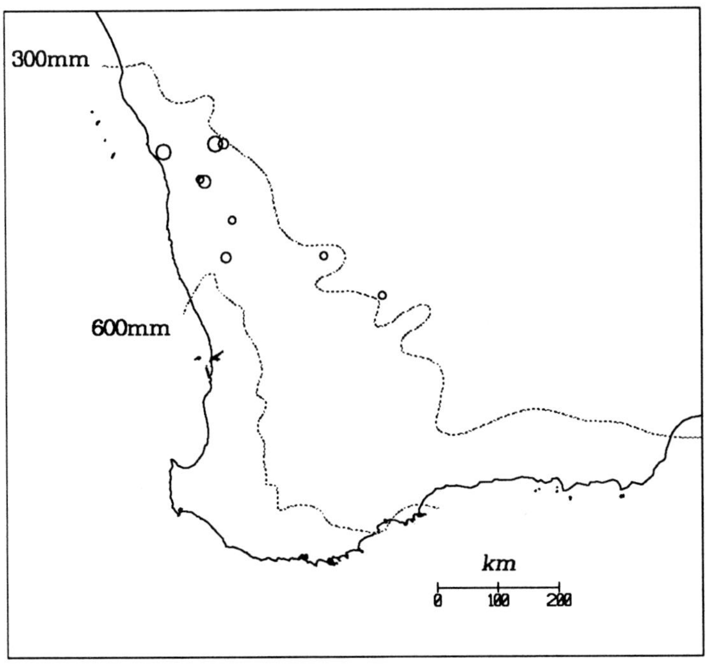

	1900/37	1987	1988	1989	1990
% of observers recording species	0	1.6	3.5	4.3	7.7
% of recording period species seen	NR	0.8	1.1	1.6	1.6

Red-winged wren

Malurus elegans

OTHER NAME:

Red-winged fairy-wren

The red-winged wren occurs close to water, including streams and swamps (Serventy and Whittell 1962) and was not found in the wheatbelt (Fig. 159) during this survey. It feeds on insects, breeds in spring in a globular-shaped nest close to the ground and lays 2–3 eggs.

The red-winged wren was formerly recorded from two localities in the wheatbelt (Fig. 159) and has declined in range over the past 80 years as result of loss of habitat.

Fig. 159. Distribution 1900–1937

Red-winged wren

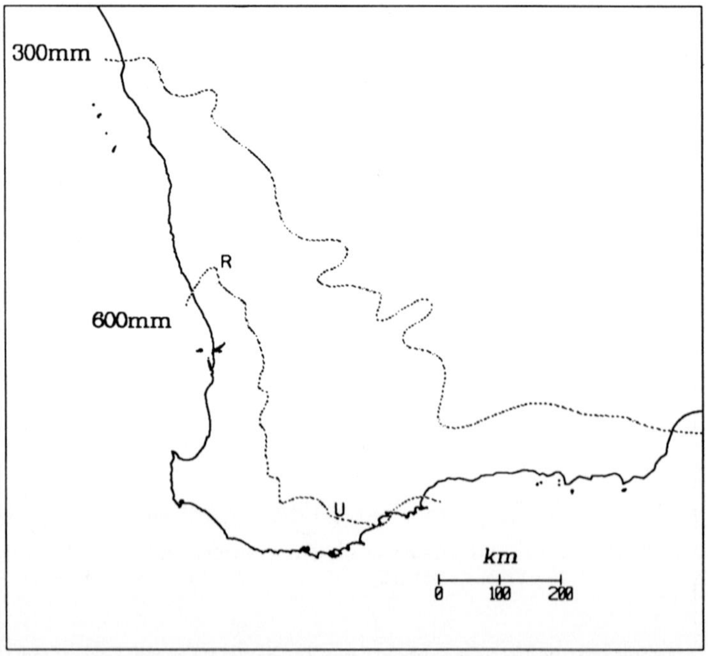

Fig. 159. Distribution 1987–1990

Red-winged wren

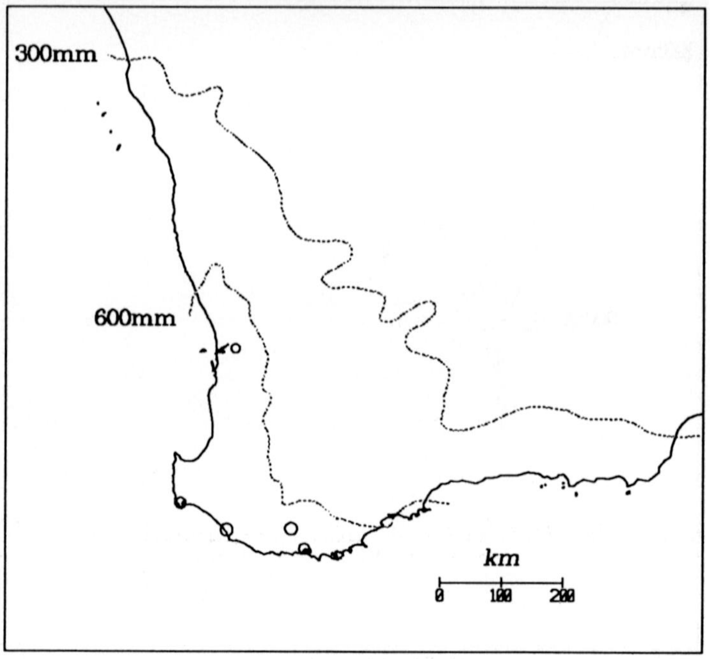

	1900/37	1987	1988	1989	1990
% of observers recording species	20	0.8	3.5	3.4	3.8
% of recording period species seen	R-U	0.2	1.7	1.7	1.5

Fig. 160. Distribution 1900–1937

Blue-breasted wren

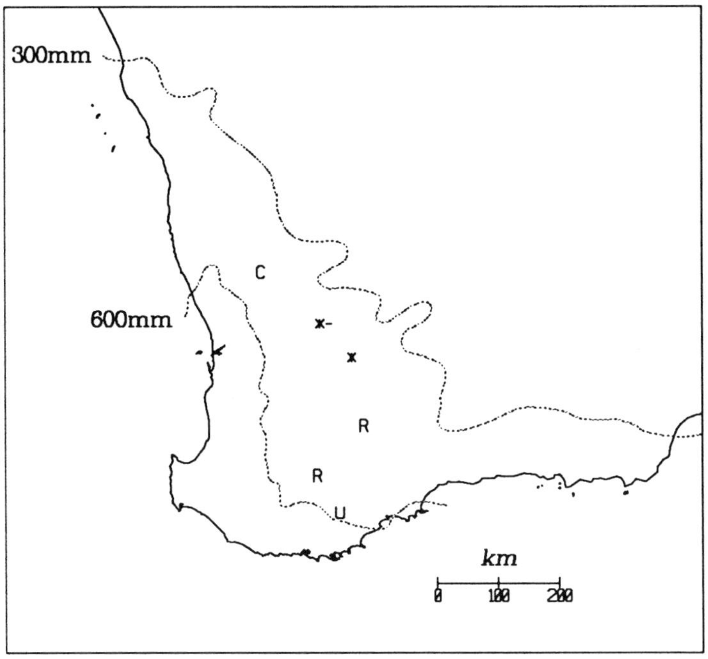

Fig. 160. Distribution 1987–1990

Blue-breasted wren

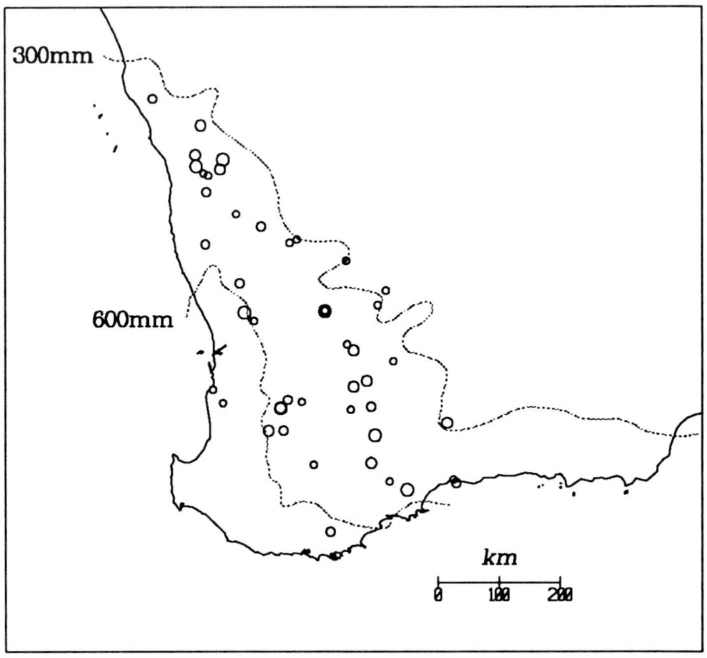

	1900/37	1987	1988	1989	1990
% of observers recording species	60	19	15	16	23
% of recording period species seen	R-C	4.0	3.8	3.8	3.1

Blue-breasted wren

Malurus pulcherrimus

OTHER NAME:

Blue-breasted fairy-wren

The blue-breasted wren is an uncommon resident in mallee, shrubland and heath throughout the wheatbelt (Fig. 160). It feeds on insects, breeds in spring in an enclosed, oval-shaped nest near the ground and lays 3 eggs.

Large-scale clearing has seen much suitable habitat disappear and the blue-breasted wren has declined in abundance as a result. It is dependent on native vegetation and restricted to remnants. As most of these are small, there are many scattered isolated populations of this species in the wheatbelt. Many of these are too small to be viable. If these die out the range of the blue-breasted wren will contract.

White-winged wren

Malurus leucopterus

OTHER NAME:

White-winged fairy-wren

The white-winged wren is an uncommon resident of shrubland, heath and samphire flats around salt lakes of the northern half of the wheatbelt (Fig. 161). This is the southern limit of its distribution in Western Australia. It feeds on insects, breeds in spring in an enclosed, oval-shaped nest near the ground and lays 3–4 eggs.

There is no evidence to suggest that the status of this species has changed.

Fig. 161. Distribution 1900–1937

White-winged wren

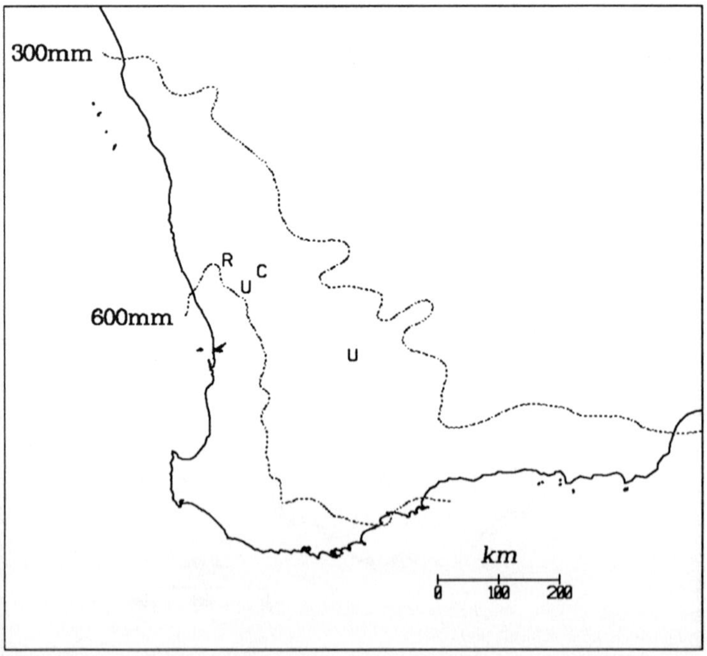

Fig. 161. Distribution 1987–1990

White-winged wren

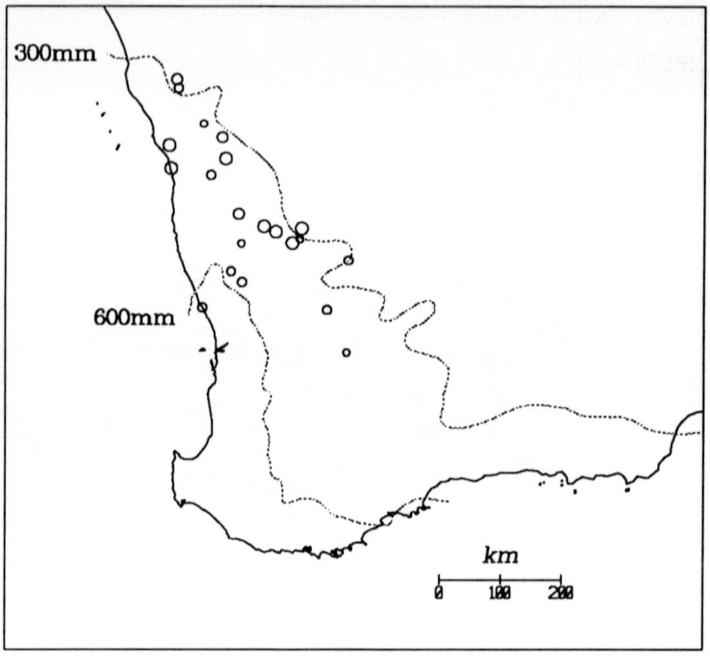

	1900/37	1987	1988	1989	1990
% of observers recording species	40	6.3	11	9.4	11
% of recording period species seen	R-C	3.0	3.6	5.0	4.4

Fig. 162. Distribution 1900–1937

Southern emu wren

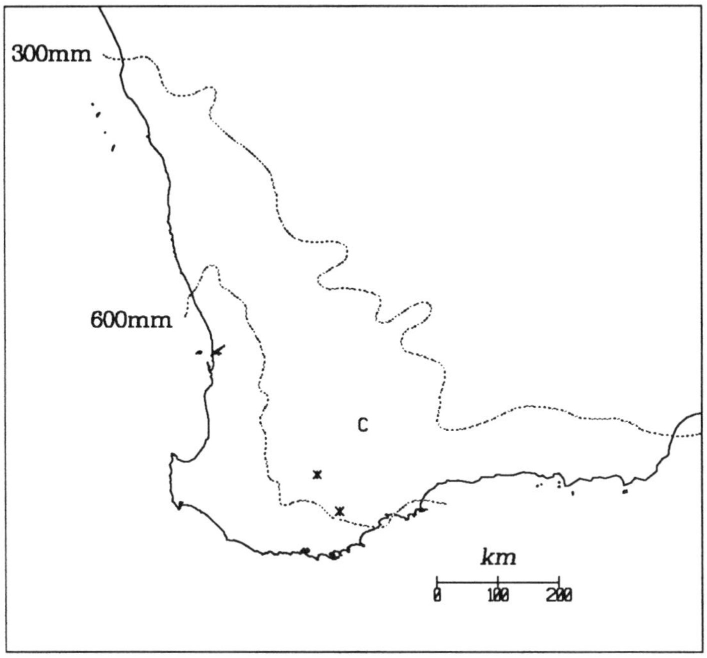

Fig. 162. Distribution 1987–1990

Southern emu wren

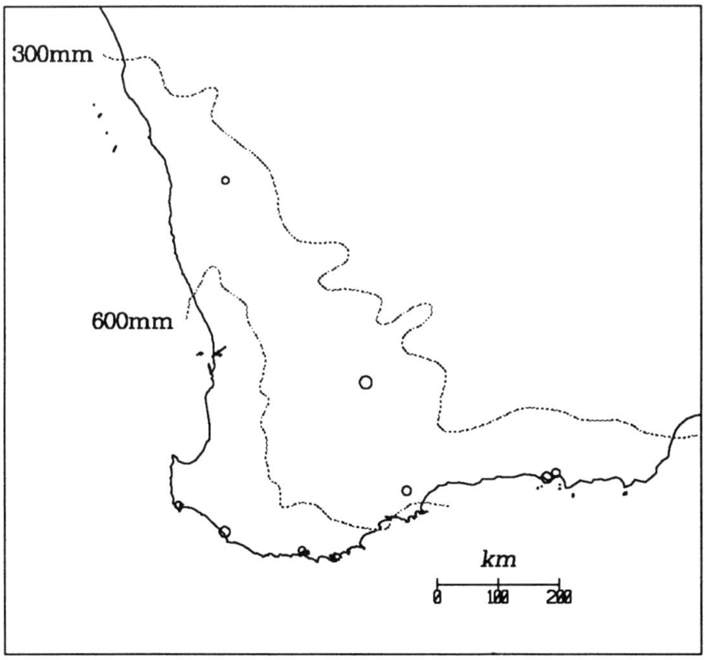

	1900/37	1987	1988	1989	1990
% of observers recording species	30	0	2.8	4.3	4.8
% of recording period species seen	C	0	0.4	1.6	1.4

Southern emu wren

Stipiturus malachurus

The southern emu wren is a bird of the coast which occurs in dense shrubland and heath around swamps and on coastal dunes (Serventy and Whittell 1962). It is rare in the wheatbelt (Fig. 162). It feeds on seed, mantids, grasshoppers, bugs, lacewings, beetles, flies, caterpillars, bees, wasps, ants and spiders. It breeds in spring in an enclosed, oval-shaped nest near the ground and lays 3 eggs.

Formerly, the southern emu wren was plentiful at Lake Grace (Carnaby 1933), and was recorded with no comment on abundance at Broome Hill (Carter 1923), in the Stirling Range (Milligan 1903) and near Tammin (J. Masters, in litt.). The southern emu wren has decreased its range in the wheatbelt, almost certainly as the result of loss of habitat.

Thick-billed grass wren

Amytornis textilis

The thick-billed grass wren was collected by Carter (1923) at Broome Hill in 1908, and seen in marlock scrub east of Broome Hill in 1910. Carter considered them rare in the district (Fig. 163). It was not recorded in the wheatbelt during this survey. In Western Australia it only occurs near Shark Bay (Blakers et al. 1984), and is extinct in the wheatbelt.

The thick-billed grass wren is classified by the Royal Australasian Ornithologists Union as vulnerable (Garnett 1992). Its disappearance is considered to be the result of loss of habitat, degradation of remnant vegetation by domestic stock and rabbits, which damaged or removed the understory vegetation on which the bird depended, and predation by cats (Garnett 1992).

Fig. 163. Distribution 1900–1937

Thick-billed grass wren

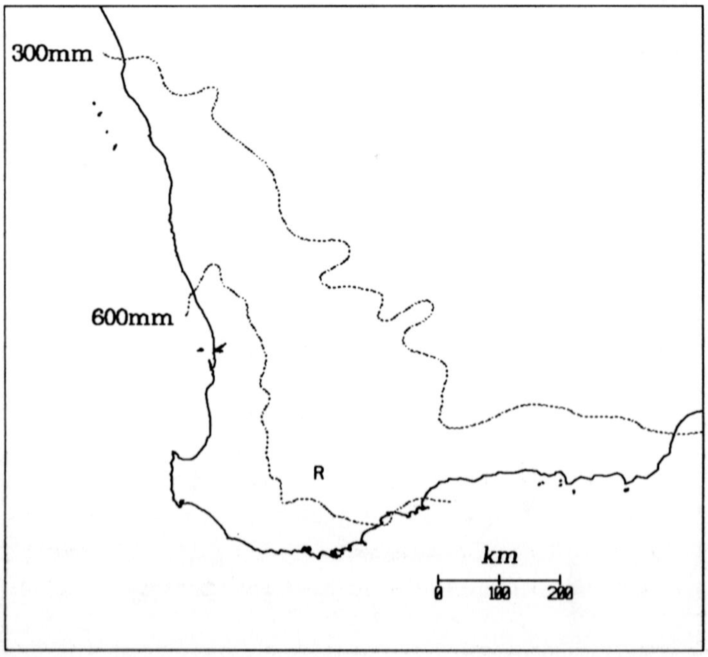

	1900/37	1987	1988	1989	1990
% of observers recording species	10	0	0	0	0
% of recording period species seen	R	0	0	0	0

Fig. 164. Distribution 1900–1937

White-browed scrub wren

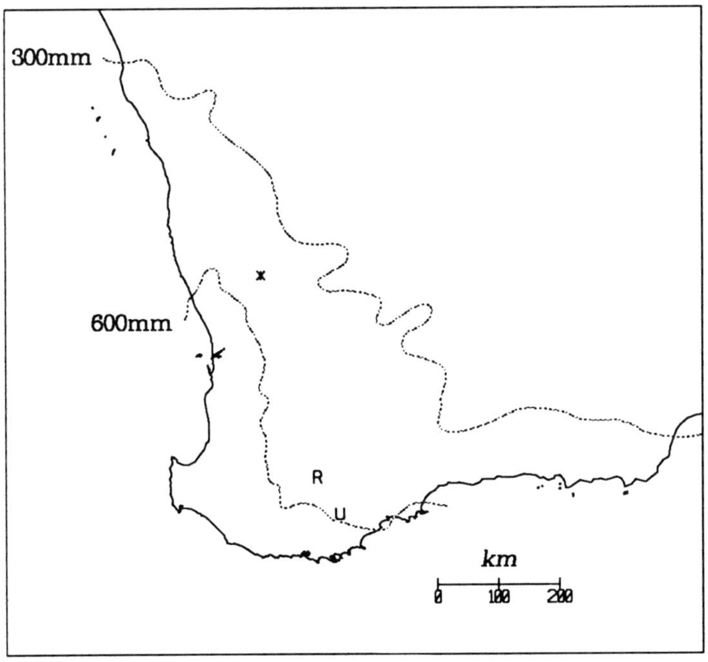

Fig. 164. Distribution 1987–1990

White-browed scrub wren

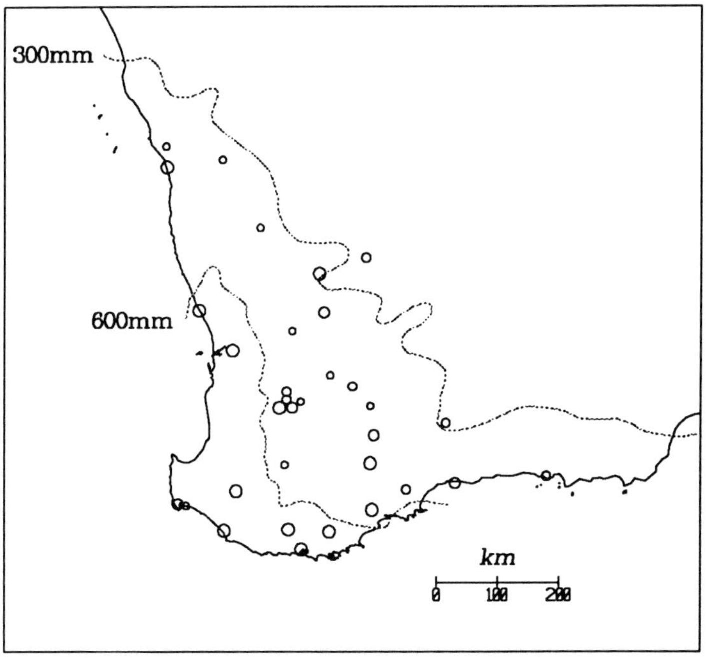

	1900/37	1987	1988	1989	1990
% of observers recording species	30	9.4	15	13	15
% of recording period species seen	R-U	3.0	5.7	0.6	7.3

White-browed scrub wren

Sericornis frontalis

The white-browed scrub wren is an uncommon resident in mallee, shrubland and heath throughout the wheatbelt (Fig. 164). It feeds on seed, centipedes, termites, grasshoppers, bugs, lacewings, beetles, flies, moths, caterpillars, wasps, ants, scorpions and spiders. It breeds from winter to mid-summer in a dome-shaped nest on or near the ground and lays 2–3 eggs.

The white-browed scrub wren has undergone a significant decline in abundance in the wheatbelt because of loss of habitat. Degradation of the understory of remnant patches by domestic stock and grazing by rabbits would have also adversely affected it. It was one of the most common small birds in the Bunketch district (Reid 1951), yet by the 1970s it was scarce in an equivalent area 45 km to the northwest (Dell 1979). It is possible that species like this that remain in patches of native vegetation, are cryptic and uncommon, may be under-sampled in surveys such as the one reported here.

Shy hylacola

Sericornis cautus

The shy hylacola is a rare resident in remnant patches of mallee, shrubland and heath in the southern half of the wheatbelt (Fig. 165). It feeds on seed, cockroaches, grasshoppers, bugs, lerps, beetles, caterpillars, moths, ants and spiders. It breeds in spring in a dome-shaped nest on or near the ground and lays 2–3 eggs.

This species is dependent on remnant native vegetation, and forages on the ground. It has declined in abundance throughout its wheatbelt range because of loss of habitat and degradation of much of the remainder. This is another species that may have been under-sampled during this survey.

Fig. 165. Distribution 1900–1937

Shy hylacola

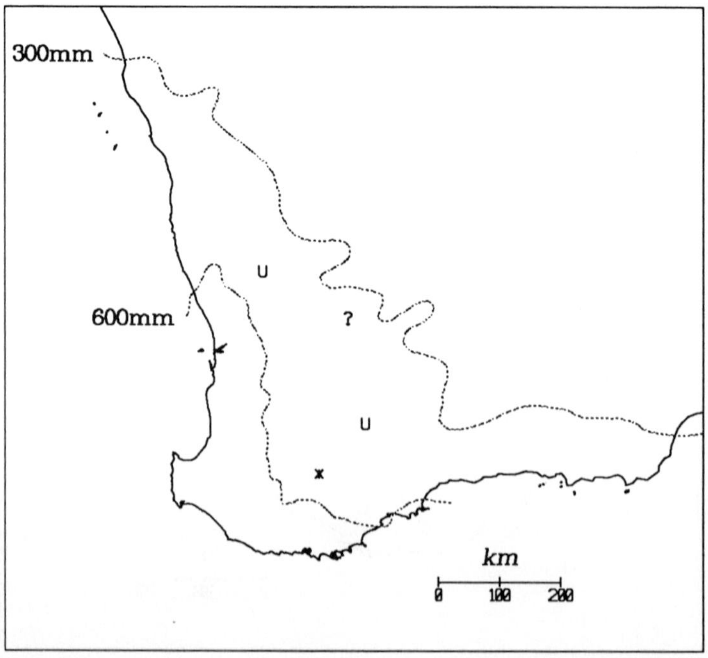

Fig. 165. Distribution 1987–1990

Shy hylacola

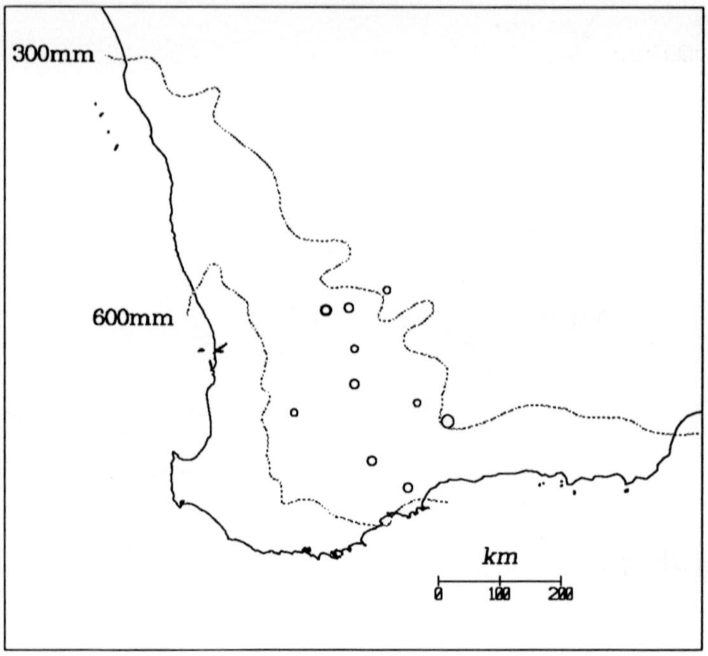

	1900/37	1987	1988	1989	1990
% of observers recording species	40	3.9	4.9	4.3	7.7
% of recording period species seen	U	0.4	0.4	0.6	1.1

Fig. 166. Distribution 1900–1937

Redthroat

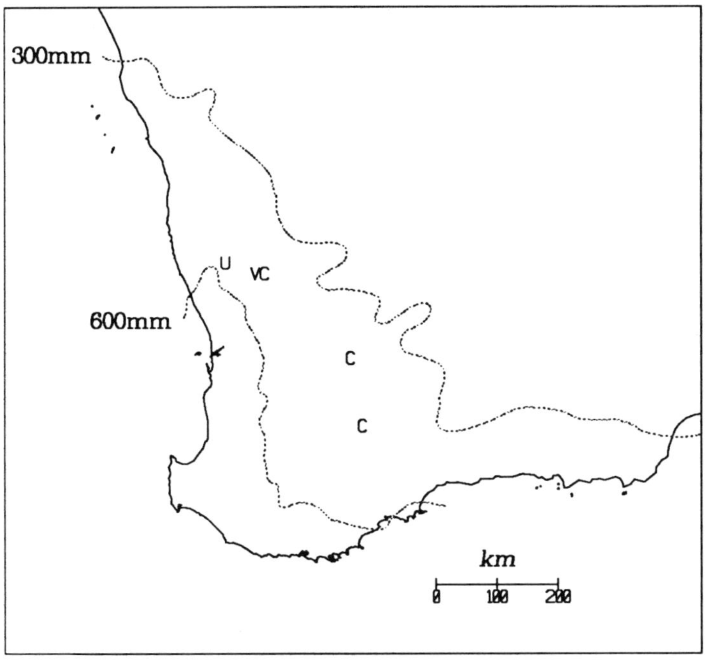

Fig. 166. Distribution 1987–1990

Redthroat

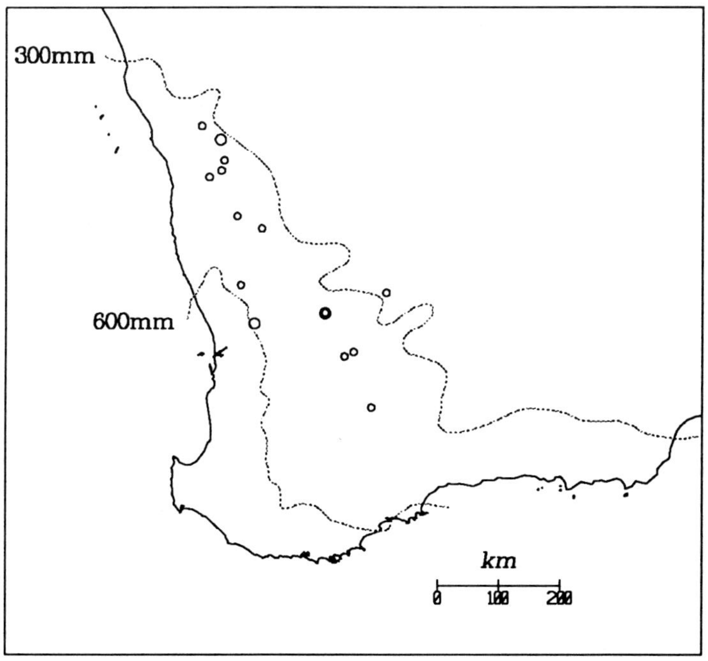

	1900/37	1987	1988	1989	1990
% of observers recording species	40	4.7	7.0	5.1	6.7
% of recording period species seen	U-VC	0.5	0.4	0.6	1.0

Redthroat

Sericornis brunneus

Formerly, the redthroat was common in several wheatbelt localities. It is now a rare resident in woodland, mallee and shrubland in the northern half of the wheatbelt (Fig. 166). It feeds on seed, termites, earwigs, grasshoppers, bugs, beetles, maggots, caterpillars, butterflies, moths, wasps, ants and spiders. It breeds in spring in a globe-shaped nest close to the ground and lays 3–4 eggs.

The redthroat has declined in abundance as a result of loss of habitat, and degradation of much of the remainder. This species is dependent on the maintenance of remnant vegetation in good condition for its continued survival in the wheatbelt. This is another species that may have been under-sampled during this survey.

Field wren

Sericornis fuliginosus

OTHER NAME:
Calamanthus

The field wren is a rare resident of remnant patches of shrubland and heath in the wheatbelt (Fig. 167). It feeds on seed, grasshoppers, crickets, bugs, beetles, caterpillars, moths, wasps, ants and spiders. It breeds in spring in a globular-shaped nest on or near the ground and lays 3–4 eggs.

During the early part of this century, the field wren was regarded as common over much of its wheatbelt range. It has declined markedly in abundance over the past 50 years. For example, Leake (1962) noted it was declining in the Kellerberrin area because of clearing for agriculture. It is rare there now (Saunders and Curry 1990). The same trends have occurred around the Wongan Hills (Milligan 1904; de Rebeira and de Rebeira 1977) and Tammin (J. Masters, in litt.). Sedgwick (1974) noted that the field wren was the most frequent species on a newly taken up farm block in the eastern wheatbelt, however, it disappeared when the land was cleared. Its future survival in the wheatbelt will depend on the maintenance and persistance of remnant vegetation, and possibly the planting of vegetation to link isolated remnants.

Fig. 167. Distribution 1900–1937

Field wren

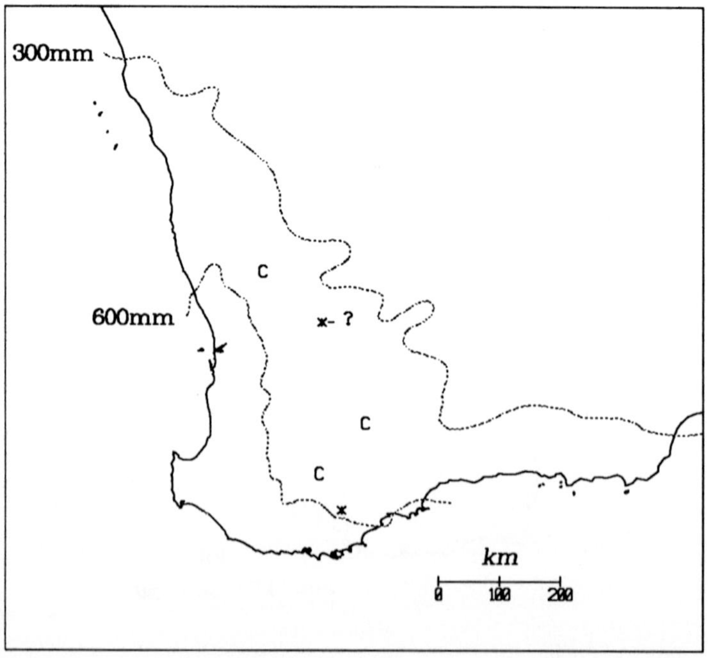

Fig. 167. Distribution 1987–1990

Field wren

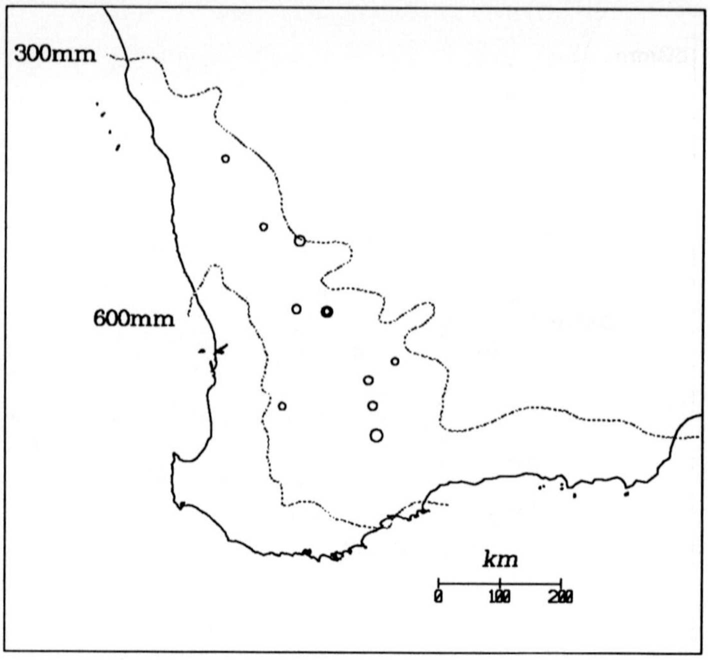

	1900/37	1987	1988	1989	1990
% of observers recording species	60	3.1	4.2	4.3	2.9
% of recording period species seen	C	0.6	0.3	0.8	0.1

Southern whiteface

Aphelocephala leucopsis

The southern whiteface is a rare resident in woodland of the drier districts of the northern half of the wheatbelt (Fig. 168) adjacent to the pastoral zone. It forages on the ground for seed, earwigs, grasshoppers, crickets, bugs, water boatmen, beetles, flies, caterpillars, ants and spiders. It breeds in winter and spring in a dome-shaped nest in a shrub or tree. It lays 2–5 eggs.

There is no evidence that this species has changed in status.

Fig. 168. Distribution 1987–1990

Southern whiteface

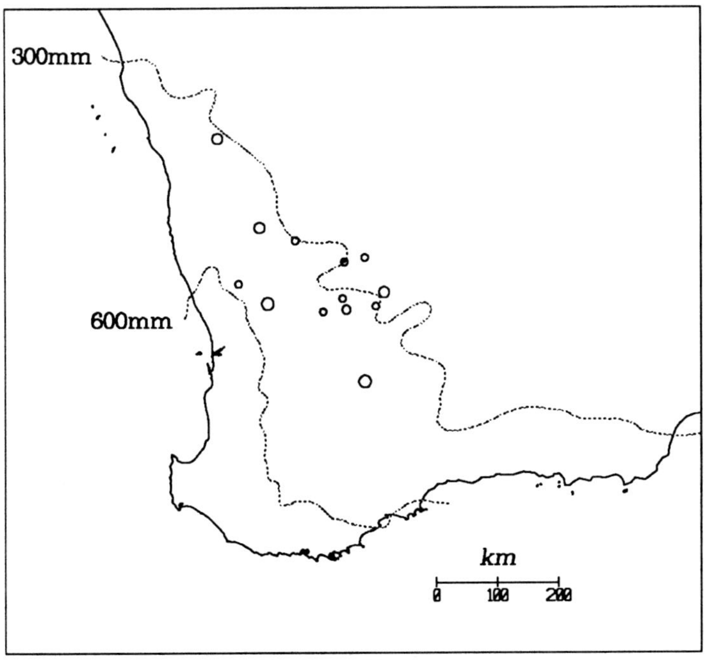

	1900/37	1987	1988	1989	1990
% of observers recording species	0	6.3	4.9	2.6	4.8
% of recording period species seen	NR	2.3	2.2	1.5	1.5

Weebill

Smicrornis brevirostris

The weebill is Australia's smallest bird averaging only 5 gm (Saunders and de Rebeira 1991). It is an uncommon resident in woodland and mallee throughout the wheatbelt, although it may be locally common in some remnants of native vegetation (Fig. 169). It feeds on seed, grasshoppers, bugs, lerps, beetles, flies, midges, caterpillars, moths, wasps, ants and spiders. It breeds in spring and summer in a pear-shaped, domed nest in a tree or shrub and lays 2 eggs.

This species has declined in abundance in the wheatbelt as a result of destruction of suitable habitat. For example, Leake (1962) noted that the weebill had undergone a reduction in numbers as a result of clearing for agriculture in the Kellerberrin district.

Fig. 169. Distribution 1900–1937

Weebill

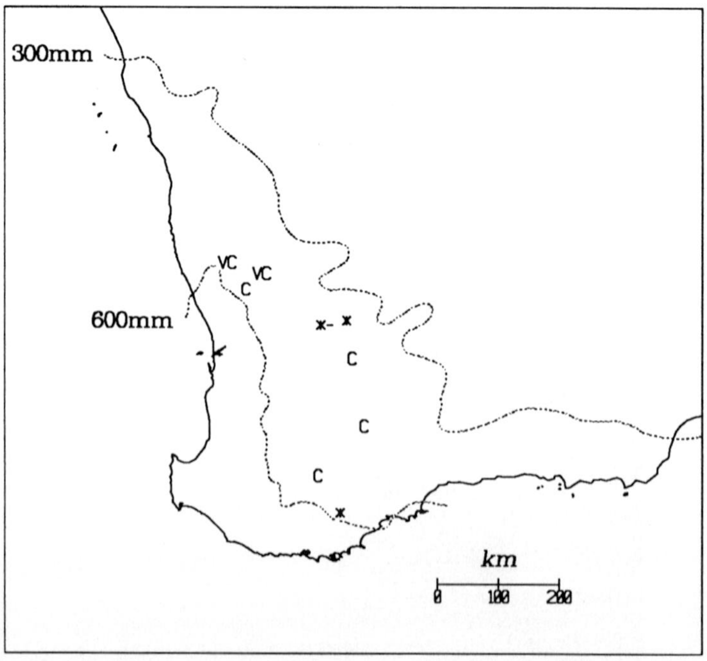

Fig. 169. Distribution 1987–1990

Weebill

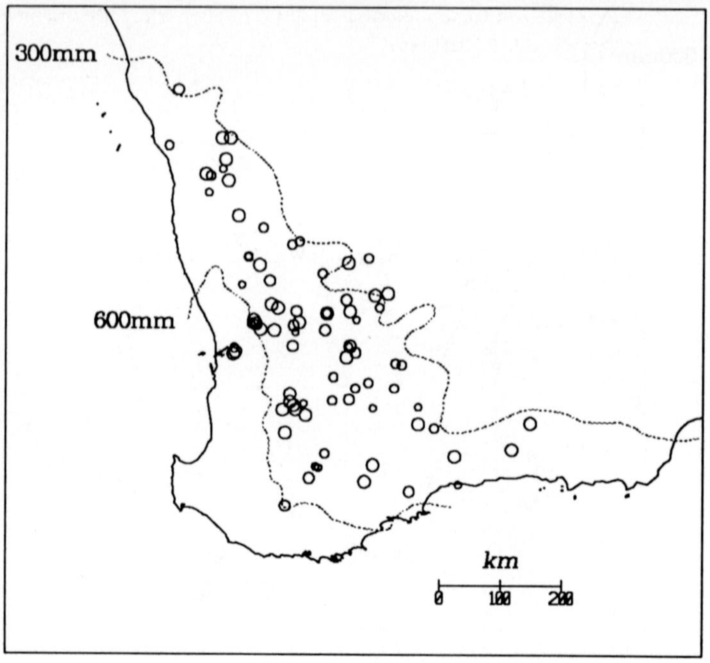

	1900/37	1987	1988	1989	1990
% of observers recording species	90	40	43	36	38
% of recording period species seen	C-VC	21	19	19	17

Some species of honeyeater depend on a range of plant species which flower at different times of the year. These honeyeaters move around the landscape to satisfy their food requirements. Plans for revegetation should include patches made up of species known to be important as food sources locally. *Banksia prionites* which grows on deep yellow sands is one such species (photo by D. A. Saunders).

Western warbler

Gerygone fusca

OTHER NAMES:

Western gerygone, Western flyeater

The western warbler is a common resident in woodland and mallee of the higher rainfall districts of the wheatbelt (Fig. 170) and a seasonal (April to October) migrant in the drier portions of its range. This seasonal pattern is illustrated by the percentage of annual recordings for each month of the period 1988 to 1990 for those locations where the species was not resident (see Fig. below). It feeds on termites, bugs, aphids, beetles, flies, wasps, ants and spiders. It breeds in spring or early summer in an enclosed, oval-shaped nest in a tree or shrub and lays 2–3 eggs.

The western warbler is dependent on remnant vegetation. The extensive loss of woodland and mallee throughout its range in the wheatbelt has caused a decline in abundance of this species.

Fig. 170. Distribution 1900–1937

Western warbler

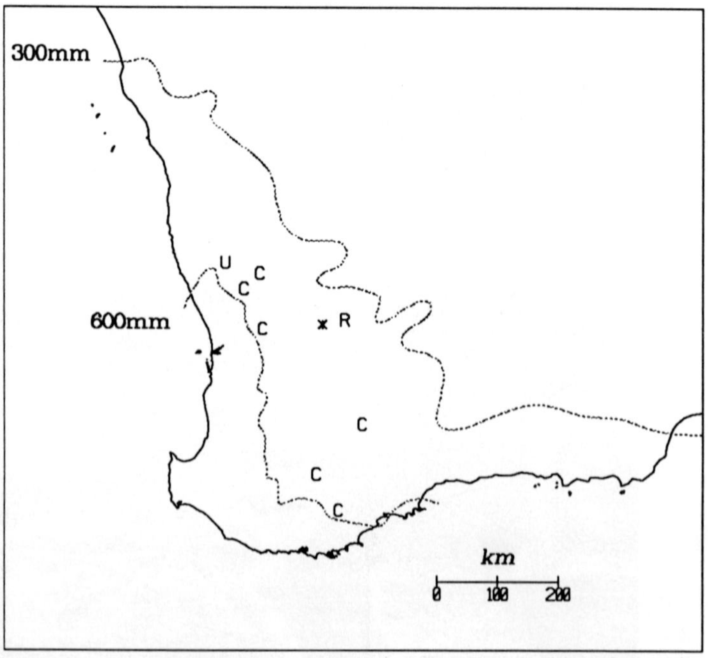

Fig. 170. Distribution 1987–1990

Western warbler

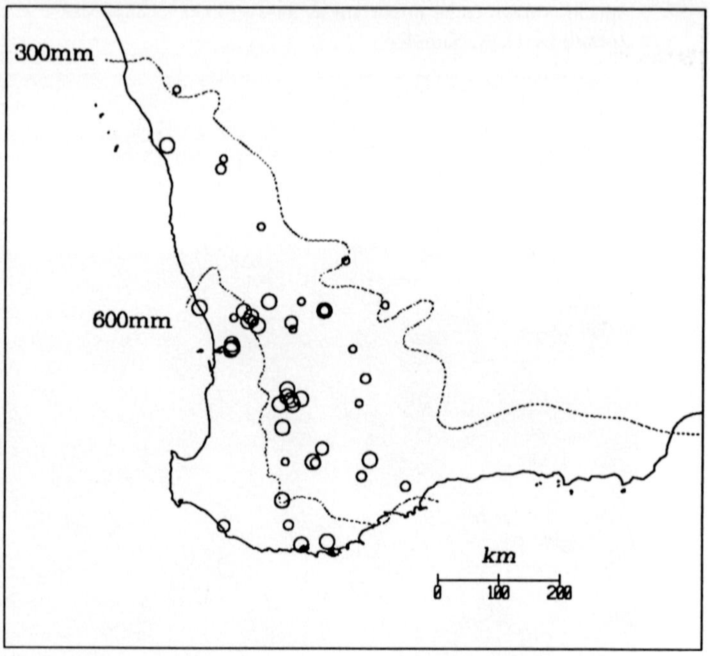

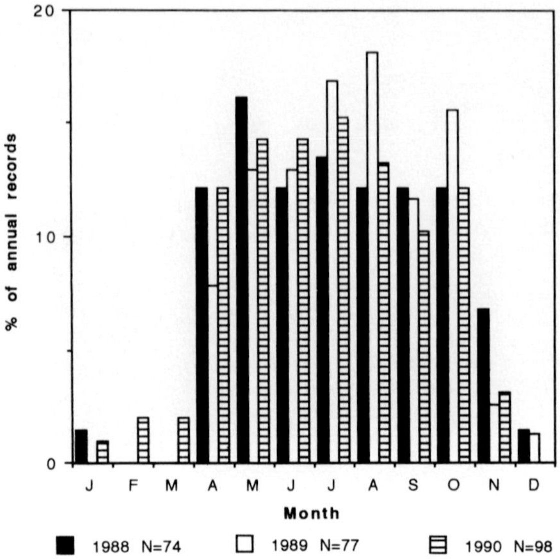

	1900/37	1987	1988	1989	1990
% of observers recording species	90	19	26	22	27
% of recording period species seen	R-C	11	14	13	14

Fig. 170. Distribution 1987–1990 — April–November

Western warbler

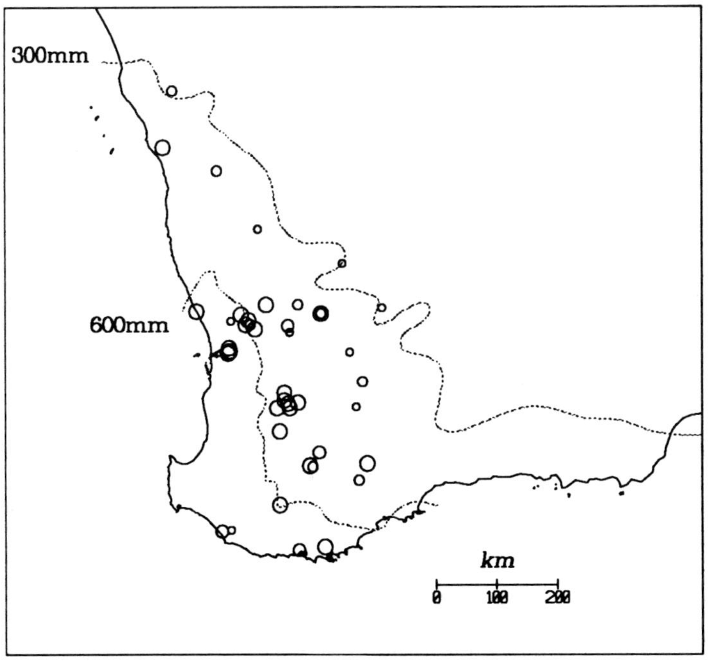

Fig. 170. Distribution 1987–1990 — December–March

Western warbler

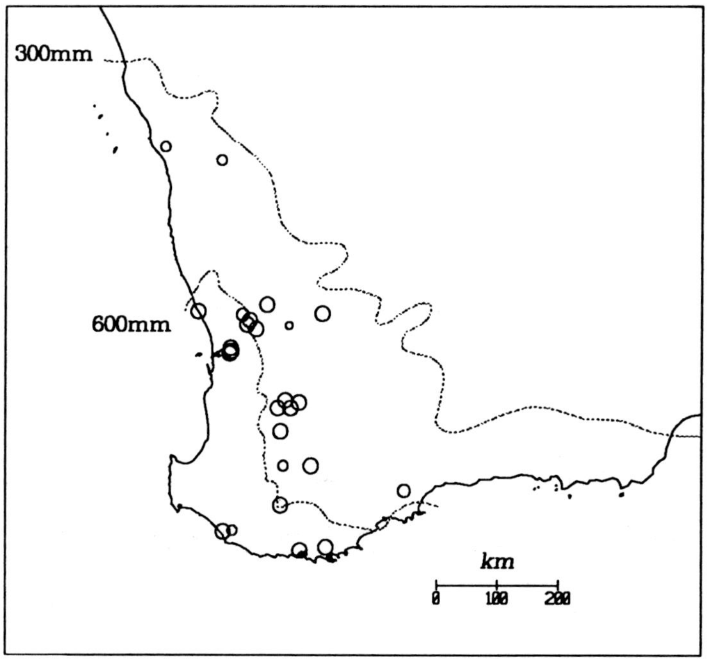

Broad-tailed thornbill

Acanthiza apicalis

OTHER NAME:
Inland thornbill

The broad-tailed thornbill is an uncommon resident of woodland, mallee and shrubland throughout the wheatbelt (Fig. 171). It forages in foliage for seed, berries, termites, grass-hoppers, bugs, aphids, beetles, flies, blowflies, caterpillars, moths, bees, wasps, ants and spiders. It breeds in late winter and spring in a dome-shaped nest in a low shrub and lays 3 eggs.

Banding studies north of Kellerberrin have shown that the broad-tailed thornbill is sedentary. One individual was retrapped 1.1 km from its banding site. However, it was still within the nature reserve in which it was banded (Saunders and de Rebeira 1991).

Orton and Sandland (1913) stated that around Moora the broad-tailed thornbill was nearly as plentiful as the yellow-rumped thornbill, which they listed as extremely plentiful. This is not the case now and this species has declined in abundance due to clearing of much of its former habitat (Leake 1962).

Fig. 171. Distribution 1900–1937

Broad-tailed thornbill

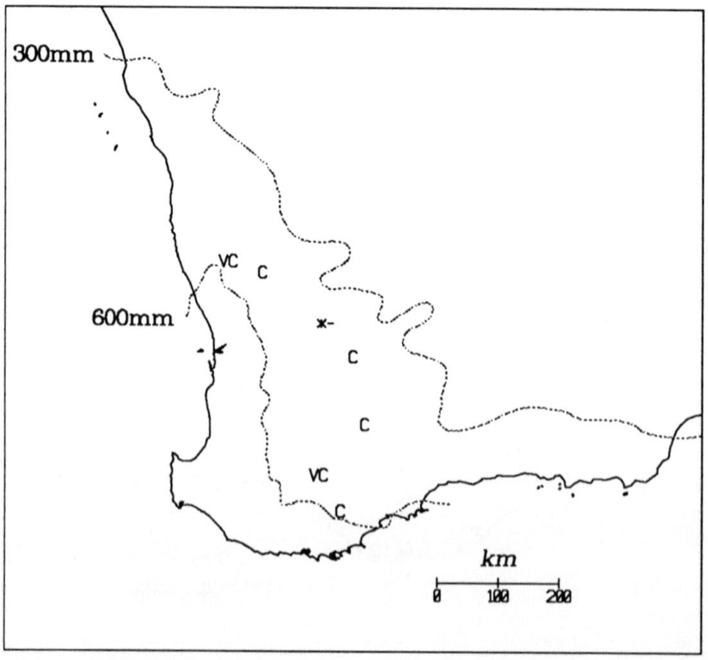

Fig. 171. Distribution 1987–1990

Broad-tailed thornbill

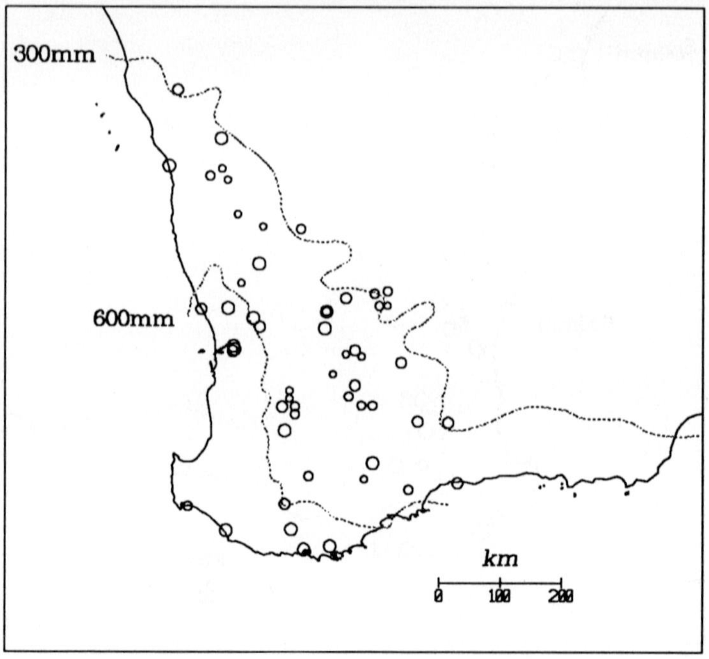

	1900/37	1987	1988	1989	1990
% of observers recording species	70	21	30	28	29
% of recording period species seen	C-VC	7.1	10	11	12

Fig. 172. Distribution 1900–1937

Chestnut-rumped thornbill

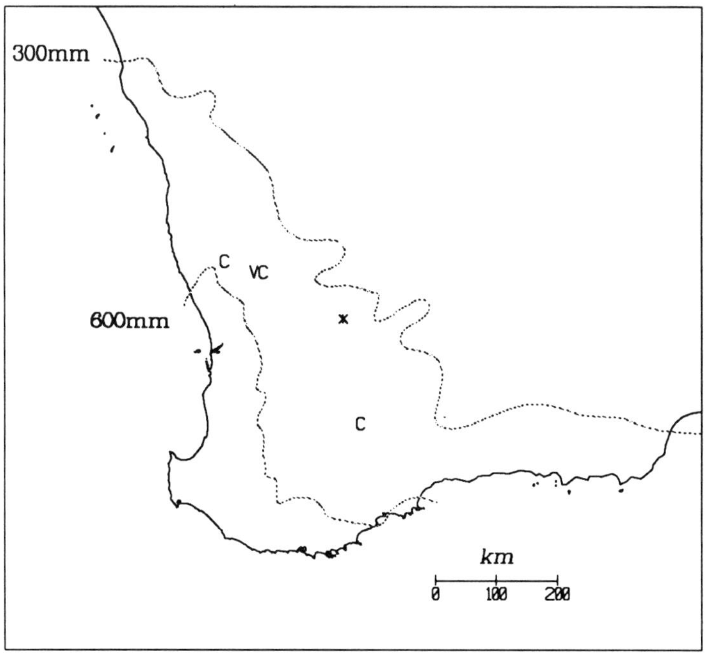

Fig. 172. Distribution 1987–1990

Chestnut-rumped thornbill

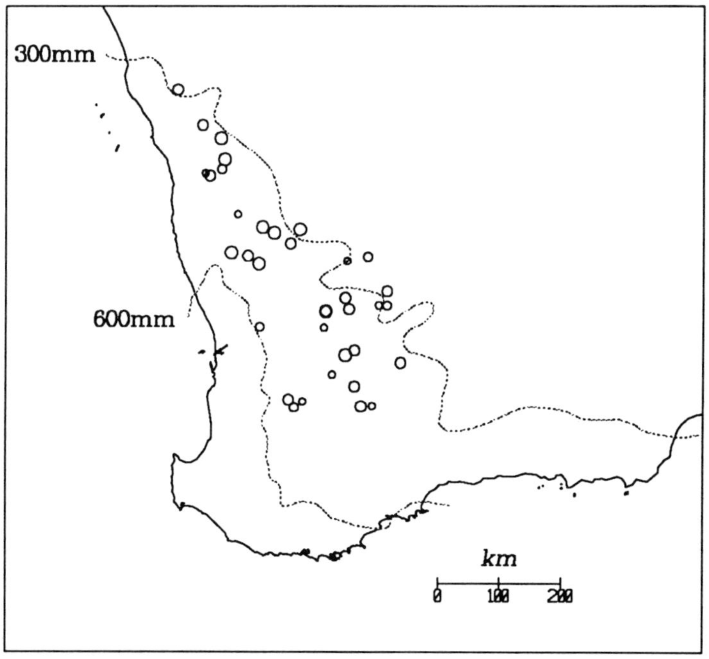

	1900/37	1987	1988	1989	1990
% of observers recording species	40	20	18	16	20
% of recording period species seen	C-VC	6.0	4.8	5.2	8.9

Chestnut-rumped thornbill

Acanthiza uropygialis

The chestnut-rumped thornbill is an uncommon resident of woodland, mallee and shrubland of the northern two-thirds of the wheatbelt (Fig. 172). It forages on the ground and in shrubs for seed, grasshoppers, bugs, cicadas, thrips, lace bugs, beetles, flies, caterpillars, ants and spiders. It breeds in late winter and spring in a dome-shaped nest in a tree hollow and lays 3 eggs.

Banding studies north of Kellerberrin have shown that this species is sedentary with 95 per cent of birds retrapped at their original banding site. The birds do wander locally, with individuals being caught 2.5 km from their banding site, having moved along a vegetated road verge connected to a nature reserve in which they were resident (Saunders and de Rebeira 1991).

The chestnut-rumped thornbill has declined in abundance in the wheatbelt because of the loss of suitable habitat.

Yellow-rumped thornbill

Acanthiza chrysorrhoa

Throughout the wheatbelt, the yellow-rumped thornbill is a common resident (Fig. 173) of woodland, mallee, shrubland, and farmland where there are remnants of native vegetation, even if badly degraded. It is often seen where patches of native vegetation abut farmland. It forages in trees, shrubs and on the ground for seed, buds, grasshoppers, crickets, bugs, beetles, flies, caterpillars, moths, wasps, bees, ants and spiders. It breeds in late winter and spring in a dome-shaped nest in a tree or shrub. It lays 3–4 eggs.

Banding studies north of Kellerberrin have shown that the yellow-rumped thornbill is sedentary, but individuals have been retrapped up to 1 km from the original banding site (Saunders and de Rebeira 1991).

The yellow-rumped thornbill has declined in abundance throughout the wheatbelt where extensive clearing of native vegetation has taken place. For example, Leake (1962) noted it was plentiful sixty to seventy years previously, but had almost disappeared from the Kellerberrin district because of clearing for agriculture. Similarly, in the Northam district, Masters and Milhinch (1974) noted that it had decreased considerably over the previous 20 years because of loss of suitable habitat, particularly jam *Acacia acuminata* woodland.

Fig. 173. Distribution 1900–1937

Yellow-rumped thornbill

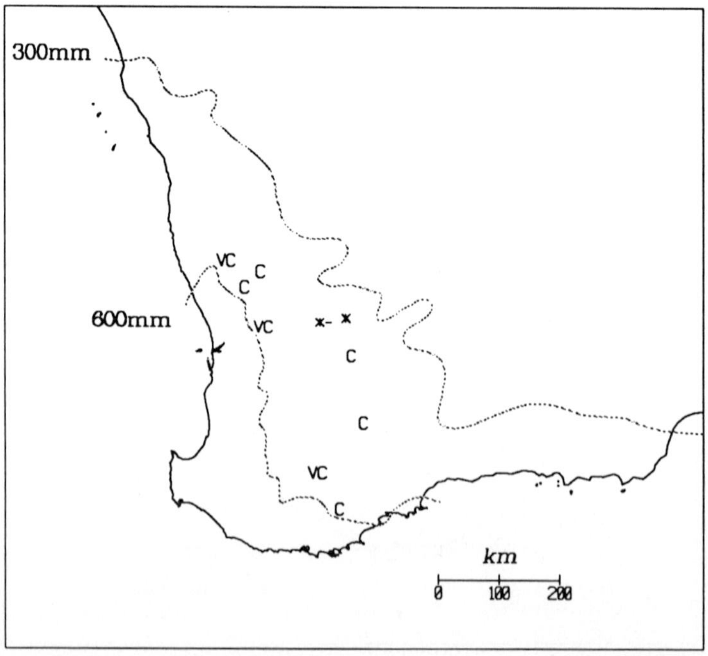

Fig. 173. Distribution 1987–1990

Yellow-rumped thornbill

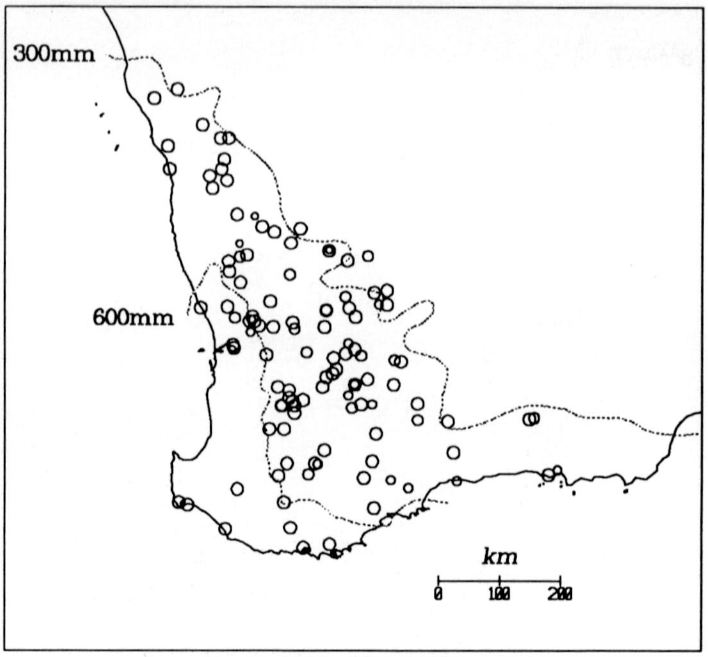

	1900/37	1987	1988	1989	1990
% of observers recording species	100	57	65	66	70
% of recording period species seen	C-VC	34	43	46	50

Fig. 174. Distribution 1900–1937

Western thornbill

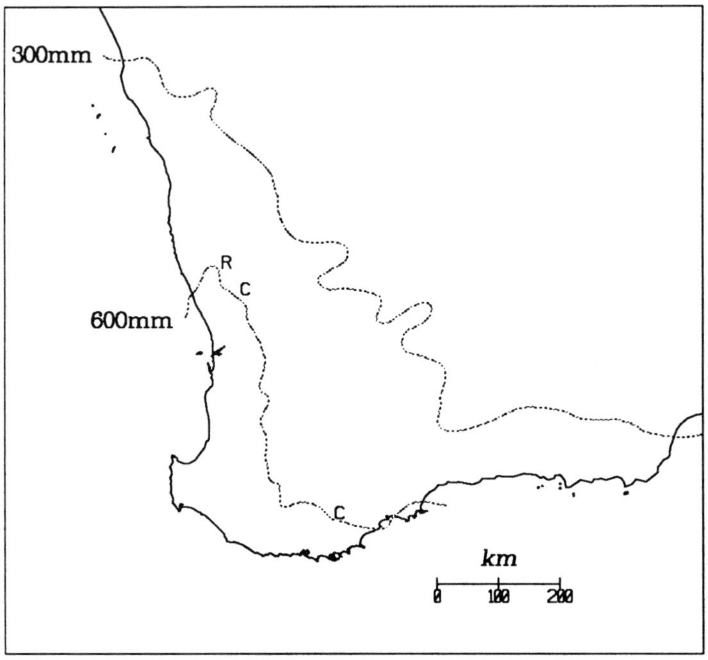

Fig. 174. Distribution 1987–1990

Western thornbill

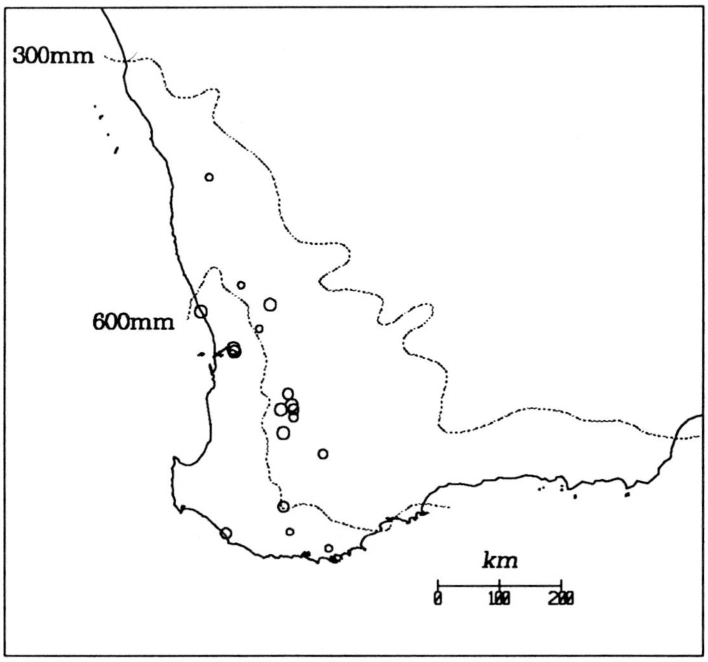

	1900/37	1987	1988	1989	1990
% of observers recording species	30	4.7	7.7	6.8	5.8
% of recording period species seen	R-C	3.3	2.8	3.0	1.8

Western thornbill

Acanthiza inornata

The western thornbill is an uncommon resident of woodland in the higher rainfall districts of the wheatbelt, adjacent to the forests of the south-west (Fig. 174). It forages in trees, shrubs and on the ground for bugs and beetles. It breeds in spring and summer in a dome-shaped nest in a shrub or tree and lays 3–4 eggs.

There is no evidence to suggest that the status of this species has changed over the last 50 years. However, it rapidly declines in remnant patches if the understory of shrubs and ground litter is degraded (J. Masters, in litt.).

Varied sittella

Daphoenositta chrysoptera

OTHER NAME:
Australian sittella

The varied sittella is an uncommon resident of woodland in much of the central and southern wheatbelt (Fig. 175). It forages for grasshoppers, bugs, beetles, crane flies, caterpillars, bees, ants, pseudo-scorpions and spiders on the trunks of trees. It breeds in spring in a cup-shaped nest in a tree and lays 2–3 eggs.

The varied sittella has declined in abundance because of clearing of woodland (Leake 1962).

Fig. 175. Distribution 1900–1937

Varied sittella

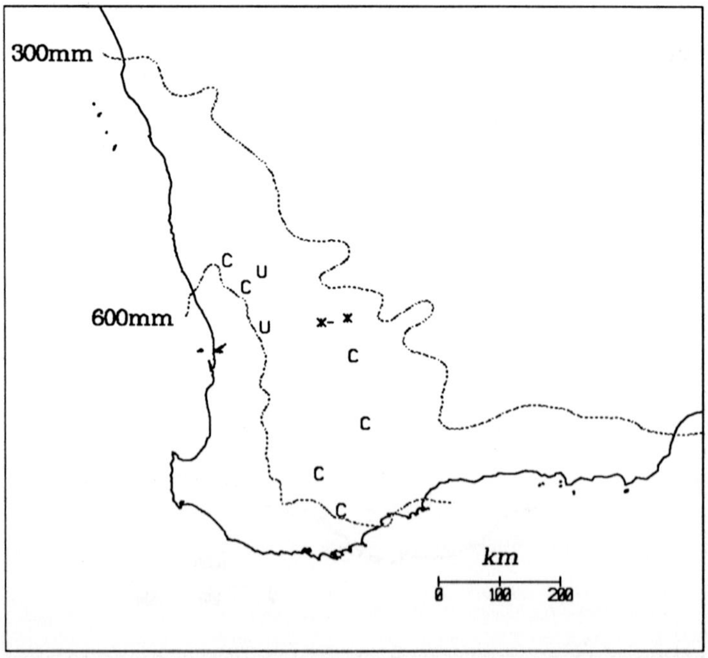

Fig. 175. Distribution 1987–1990

Varied sittella

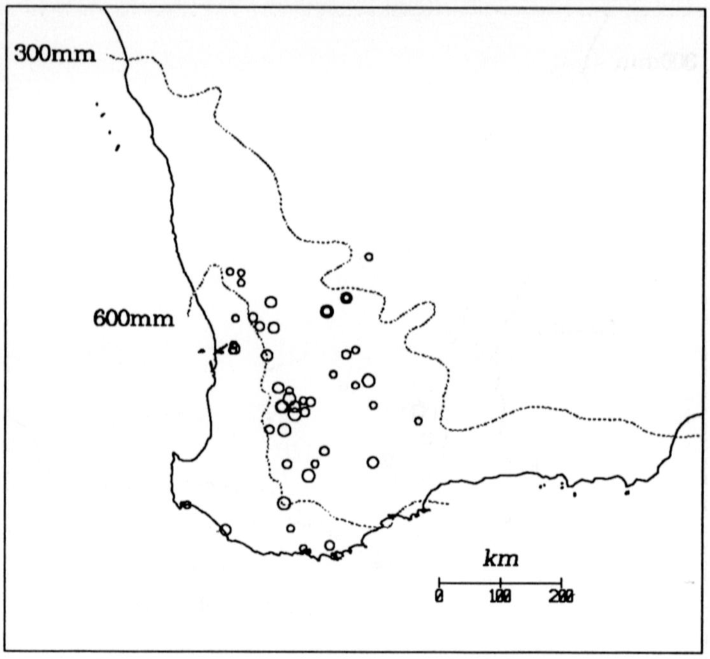

	1900/37	1987	1988	1989	1990
% of observers recording species	100	17	23	26	24
% of recording period species seen	U-C	4.6	5.2	6.1	5.1

Fig. 176. Distribution 1900–1937

Rufous treecreeper

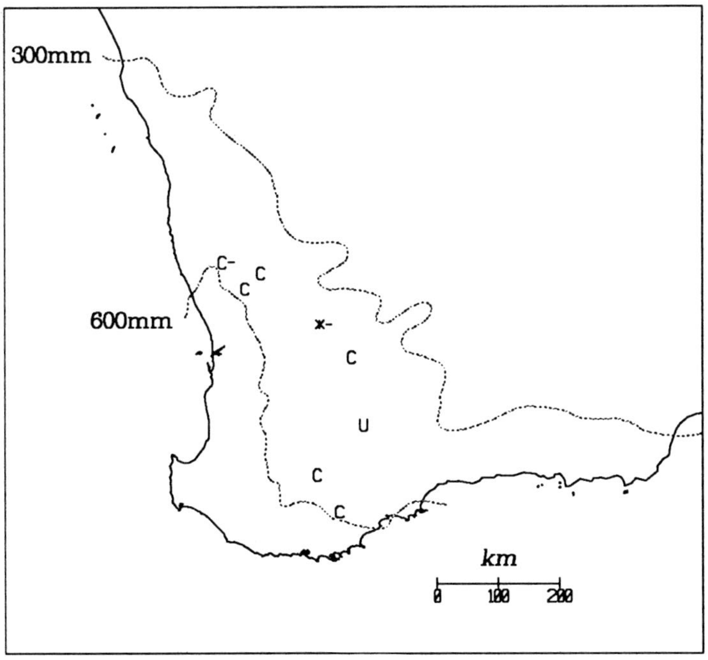

Fig. 176. Distribution 1987–1990

Rufous treecreeper

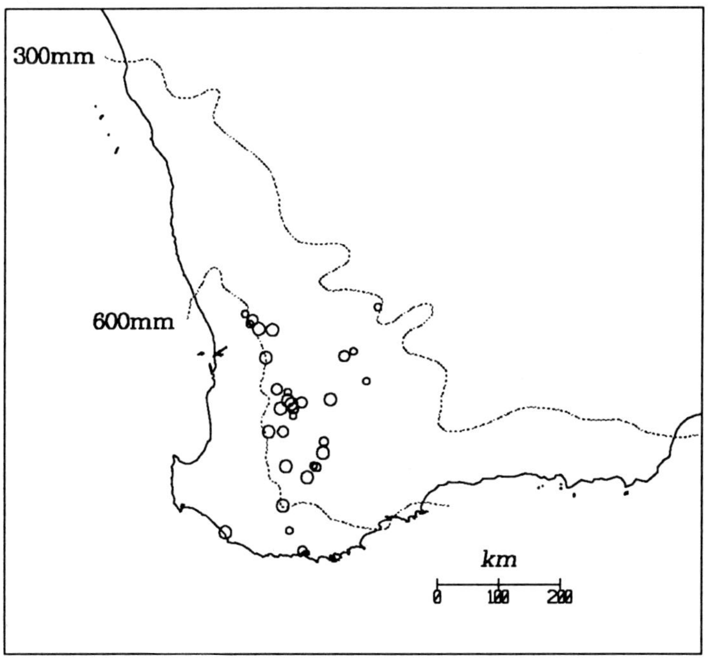

	1900/37	1987	1988	1989	1990
% of observers recording species	80	12	15	14	15
% of recording period species seen	U-C	5.3	7.3	6.0	5.9

Rufous treecreeper

Climacteris rufa

The rufous treecreeper is an uncommon resident in woodland and mallee in the southern half of the wheatbelt (Fig. 176), in areas where there are still considerable patches of remnant woodland. It forages for cockroaches, weevils, caterpillars and ants in trees and on the ground. It breeds in spring in a hollow in a tree, sometimes close to the ground and lays 2–3 eggs.

The rufous treecreeper has declined in range and abundance. Serventy and Whittell (1962) noted that it rapidly disappeared as closer settlement developed. Early this century, it was common everywhere in the Wongan Hills (Milligan 1904), but was not recorded there in the early 1970s (de Rebeira and de Rebeira 1977). It has also disappeared from around Mogumber (Lawson 1905; Loaring and Serventy 1952) and Kellerberrin (Leake 1962; Saunders and Curry 1990). According to Orton and Sandland (1913) it was fairly common around Moora, but was disappearing as the country was being cleared.

Red wattlebird

Anthochaera carunculata

The red wattlebird is a resident of woodland, mallee and shrubland throughout the wheatbelt (Fig. 177). It may also occur in towns and farm gardens. It is common in some areas, but uncommon in others. For example, in the Kellerberrin area it is an uncommon resident that moves widely in its search for food, sometimes travelling over 12 km to suitable food sources and returning the same day. It feeds on nectar, pollen, fruit, mantids, grasshoppers, bugs, cicadas, beetles, weevils, caterpillars, wasps, ants and spiders. The red wattlebird breeds in late winter and spring in a cup-shaped nest in a tree or shrub and lays 2–3 eggs.

Serventy and Whittell (1962) noted that the red wattlebird was resident in the lower southwest, but was a winter visitor to the outer parts of its range. Examining data from this survey for those locations other than those where it was recorded throughout the year, a seasonal movement pattern was demonstrated, with the bird being more commonly recorded in the period from May to November (see Fig. below). Orton and Sandland (1913) remarked that it was plentiful in the Moora district from April to November with nearly all birds absent from the area during the rest of the year.

This species has declined in abundance in the extensively cleared parts of its range in the wheatbelt. For example, early this century it was numerous between Mogumber and the Wongan Hills and in the hills themselves (Milligan 1904). Seventy years later it was widespread but not common in the latter area (de Rebeira and de Rebeira 1977). Similarly, it was plentiful north of Dowerin in the 1930s, but no longer occurs there (G. Evans, in litt.), and it has disappeared from the Merredin area since 1913 (B. Willis, in litt.).

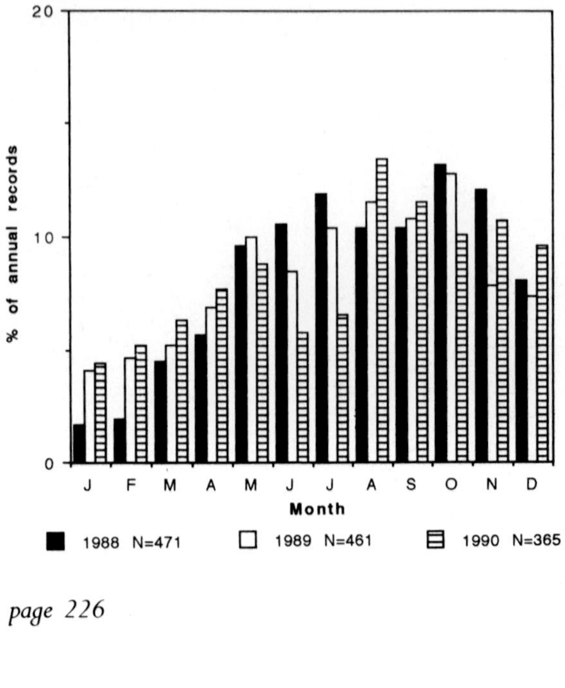

Fig. 177. Distribution 1900–1937

Red wattlebird

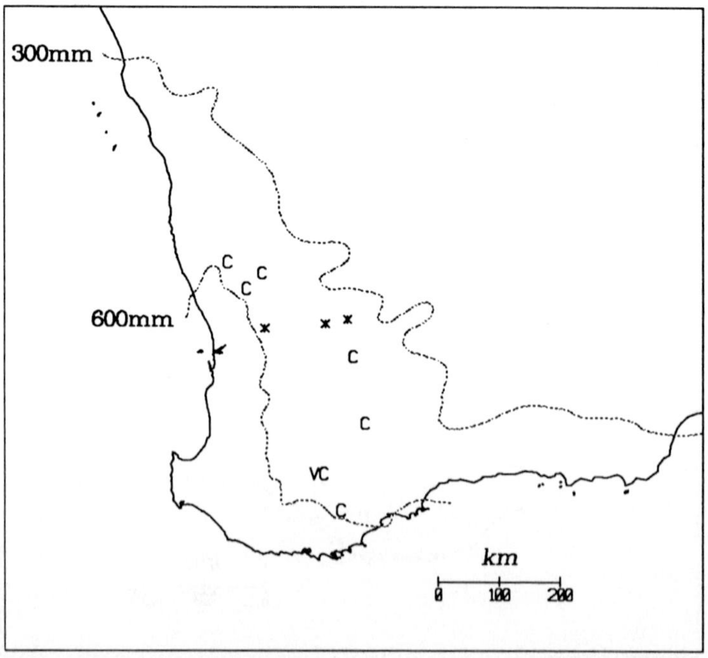

Fig. 177. Distribution 1987–1990

Red wattlebird

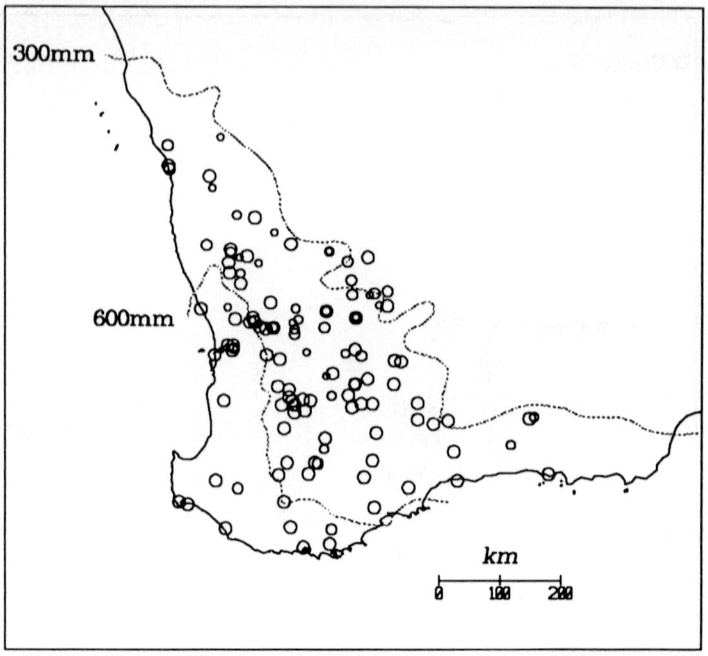

	1900/37	1987	1988	1989	1990
% of observers recording species	100	57	61	61	70
% of recording period species seen	C-VC	30	36	42	41

Fig. 178. Distribution 1900–1937

Little wattlebird

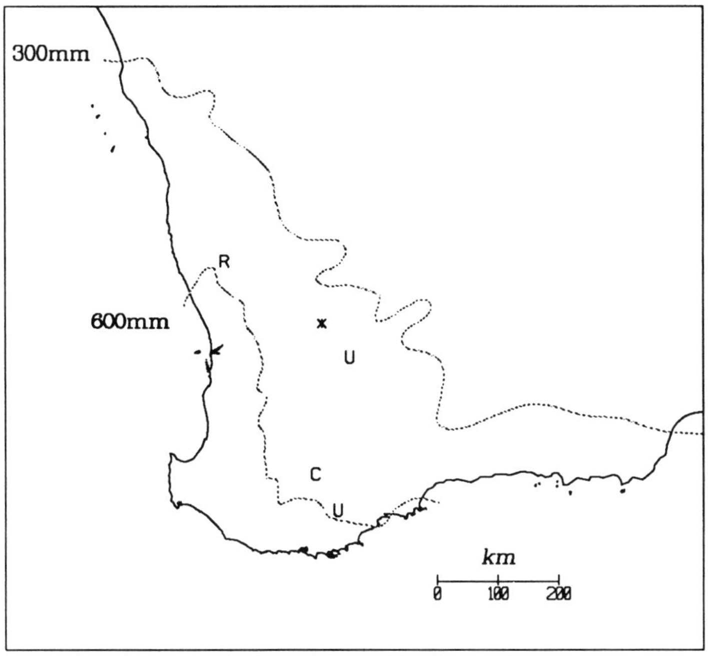

Fig. 178. Distribution 1987–1990

Little wattlebird

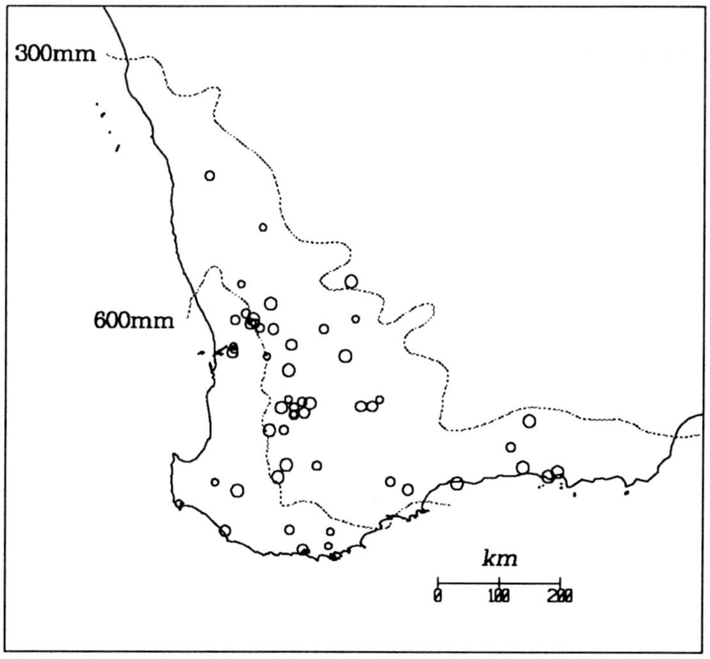

	1900/37	1987	1988	1989	1990
% of observers recording species	50	14	22	24	26
% of recording period species seen	R-C	6.6	6.2	8.4	11

Little wattlebird

Anthochaera chrysoptera

The little wattlebird is an uncommon resident of woodland, shrubland and heath in the higher rainfall districts of the southern half of the wheatbelt (Fig. 178). It feeds on nectar and insects. It breeds in late winter and spring in a cup-shaped nest in a tree or shrub and lays 1–2 eggs.

Leake (1962) recorded the little wattlebird at Kellerberrin, however, it does not occur there now, except as a rare nomad. Over the past 50 years it has declined in abundance and range as a result of clearing of native vegetation.

Spiny-cheeked honeyeater

Acanthagenys rufogularis

The spiny-cheeked honeyeater is an uncommon resident in woodland, mallee and shrubland, and sometimes in towns and farm gardens throughout the northern two-thirds of the wheatbelt (Fig. 179). It eats fruit, seed, nectar, pollen, grasshoppers, bugs, cicadas, beetles, moths, caterpillars, wasps, bees, ants and spiders. It breeds in late winter and spring in a cup-shaped nest in a tree or shrub and lays 2–3 eggs.

This species is predominantly found in native vegetation and has probably declined in abundance as a result of the loss of suitable habitat.

Fig. 179. Distribution 1900–1937

Spiny-cheeked honeyeater

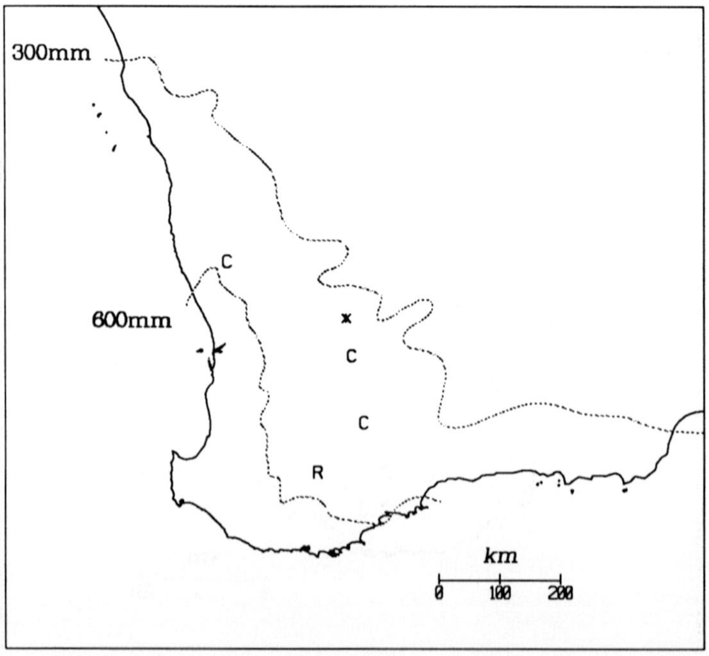

Fig. 179. Distribution 1987–1990

Spiny-cheeked honeyeater

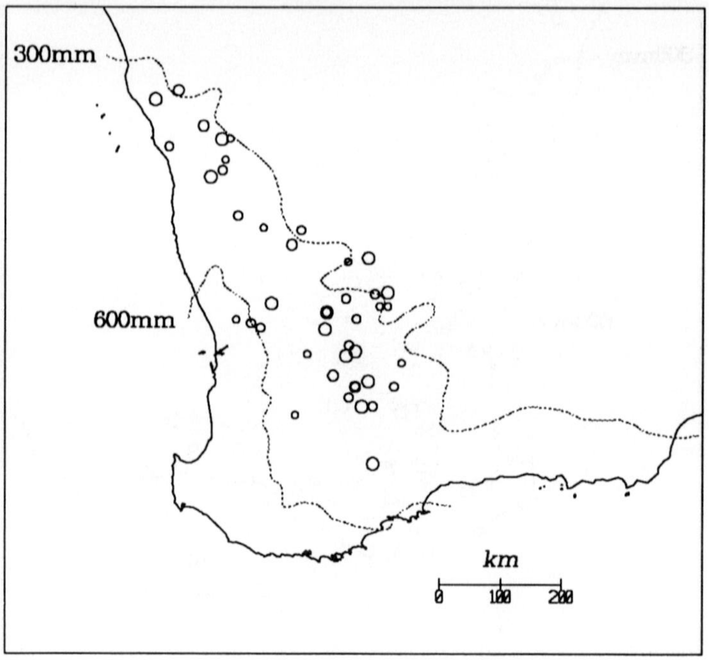

	1900/37	1987	1988	1989	1990
% of observers recording species	50	20	23	24	24
% of recording period species seen	R-C	6.6	7.2	7.5	8.5

Yellow-throated miner

Manorina flavigula

The yellow-throated miner is a common resident in woodland, mallee, shrubland, towns, and in degraded native vegetation on farmland throughout the wheatbelt (Fig. 180). It feeds on seed, nectar, pollen, damselflies, grasshoppers, bugs, lerps, beetles, flies, caterpillars, bees, wasps, ants, pseudo-scorpions and spiders. The yellow-throated miner breeds in late winter and spring in a cup-shaped nest in a tree or shrub and lays 3–4 eggs.

The yellow-throated miner may be one of the most common honeyeaters in the wheatbelt. Although resident, it may wander locally. It is possible that it has increased in range and abundance as a result of changes to the landscape.

Fig. 180. Distribution 1900–1937

Yellow-throated miner

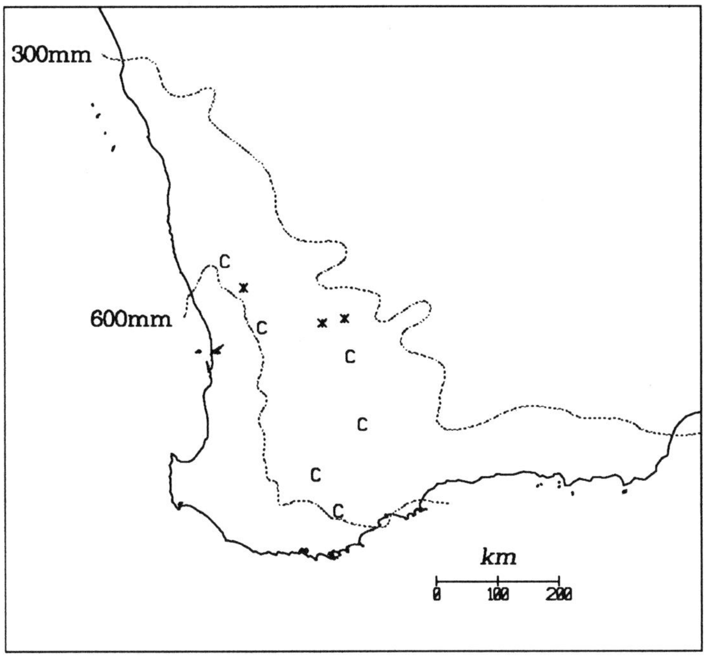

Fig. 180. Distribution 1987–1990

Yellow-throated miner

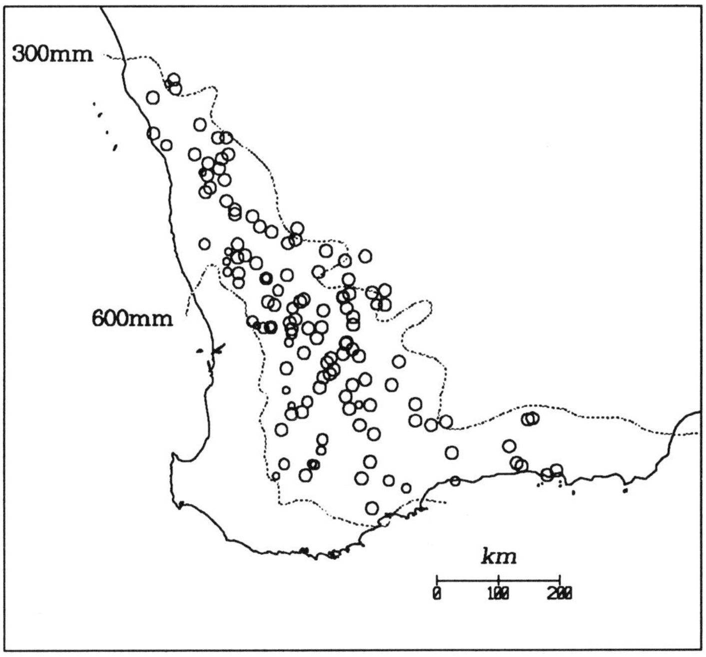

	1900/37	1987	1988	1989	1990
% of observers recording species	90	68	68	71	67
% of recording period species seen	C	54	51	55	55

Singing honeyeater

Lichenostomus virescens

The singing honeyeater is a common resident of woodland, mallee, shrubland, heath, towns and degraded native vegetation on farms throughout the wheatbelt (Fig. 181). It feeds on nectar, berries, seed, pollen, damselflies, mantids, grasshoppers, bugs, lerps, beetles, moths, caterpillars, bees, wasps, ants and spiders. The singing honeyeater breeds in late winter and spring in a cup-shaped nest low in a shrub and lays 2–3 eggs.

Banding studies north of Kellerberrin have shown that the singing honeyeater is sedentary, with 87 per cent of retraps being at the original banding site. However, it does move locally, one individual being caught 10.5 km from the site where it was originally banded (Saunders and de Rebeira 1991). It will fly readily over open agricultural land so the population is not broken up into isolated groups as with species which will not fly over open ground.

It is possible that the singing honeyeater has increased in abundance as a result of changes to the landscape. For example, early this century it was sparsely distributed in the Wongan Hills (Milligan 1904), but by the 1970s it was common and numerous (de Rebeira and de Rebeira 1977).

Fig. 181. Distribution 1900–1937

Singing honeyeater

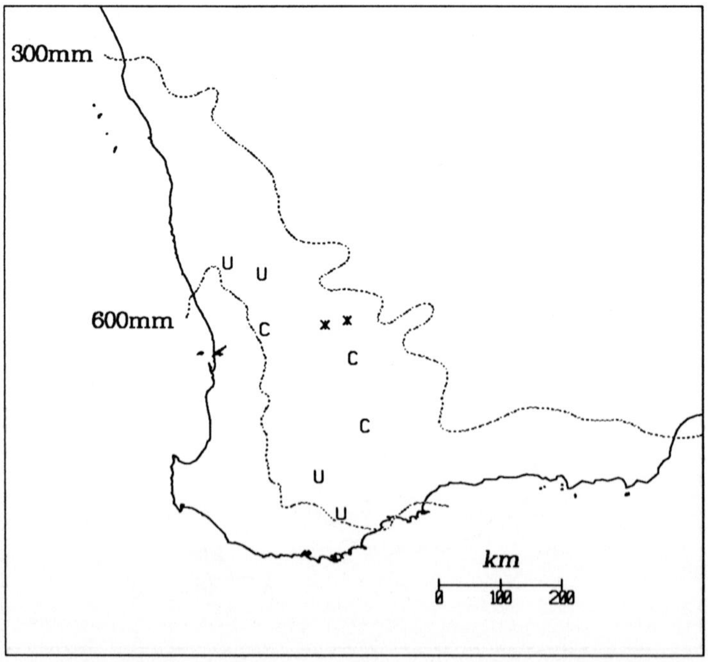

Fig. 181. Distribution 1987–1990

Singing honeyeater

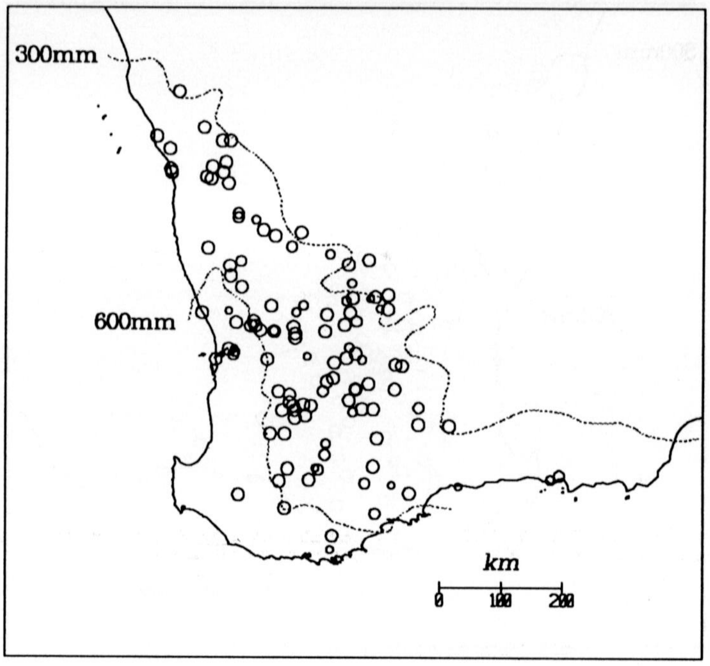

	1900/37	1987	1988	1989	1990
% of observers recording species	90	63	62	63	65
% of recording period species seen	U-C	36	45	46	49

Fig. 182. Distribution 1900–1937

White-eared honeyeater

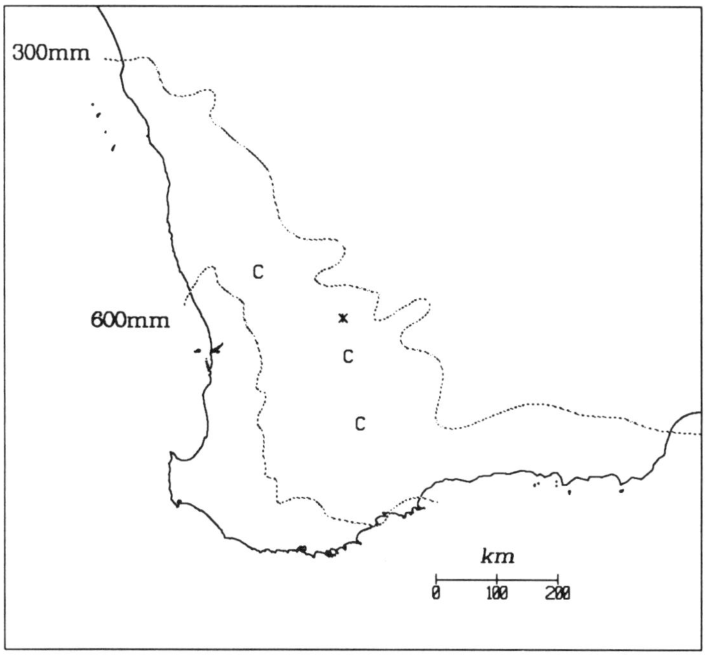

Fig. 182. Distribution 1987–1990

White-eared honeyeater

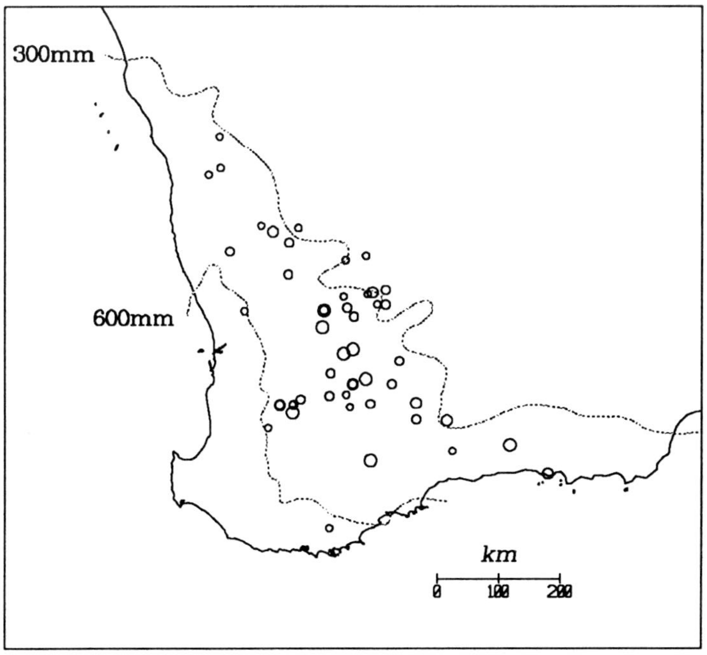

	1900/37	1987	1988	1989	1990
% of observers recording species	40	23	22	23	21
% of recording period species seen	C	6.8	4.1	5.9	5.2

White-eared honeyeater

Lichenostomus leucotis

The white-eared honeyeater is an uncommon resident in woodland, mallee and shrubland throughout the wheatbelt (Fig. 182). It feeds on berries, nectar, manna, flowers, cockroaches, grasshoppers, bugs, lerps, beetles, flies, mosquitoes, caterpillars, bees, wasps, ants and spiders. The white-eared honeyeater breeds in spring in a cup-shaped nest low in a shrub and lays 2–3 eggs.

The white-eared honeyeater moves locally and individuals have been caught 7.5 km from their original banding site (Saunders and de Rebeira 1991). Unlike the singing honeyeater, this species will not readily fly over large open spaces, preferring to move along vegetated links.

This species has probably declined in abundance in the wheatbelt where extensive land clearing has taken place.

Purple-gaped honeyeater

Lichenostomus cracticus

The purple-gaped honeyeater is an uncommon resident in mallee and heath of the southern half of the wheatbelt (Fig. 183). It feeds on seed, nectar, pollen, grasshoppers, bugs, lerps, beetles, fly pupae, wasps, ants and spiders. It breeds in spring in a cup-shaped nest close to the ground and lays 2 eggs.

This species is dependent on native vegetation, particularly mallee, and has declined in abundance because of loss of habitat. For example, it was fairly numerous in the Wongan Hills (Milligan 1904), but had disappeared by the 1970s (de Rebeira and de Rebeira 1977).

Fig. 183. Distribution 1900–1937

Purple-gaped honeyeater

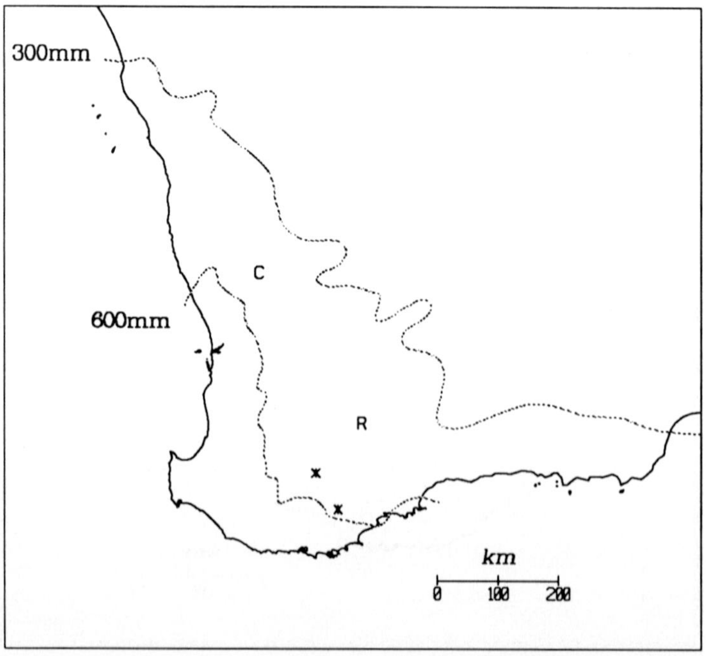

Fig. 183. Distribution 1987–1990

Purple-gaped honeyeater

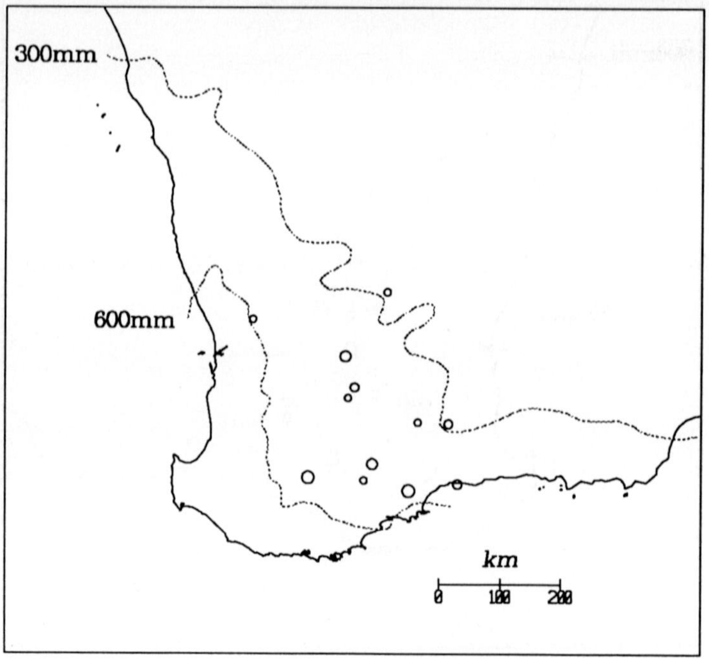

	1900/37	1987	1988	1989	1990
% of observers recording species	40	3.9	4.9	3.4	1.9
% of recording period species seen	R-C	0.4	0.4	0.6	0.7

Fig. 184. Distribution 1900–1937

Yellow-plumed honeyeater

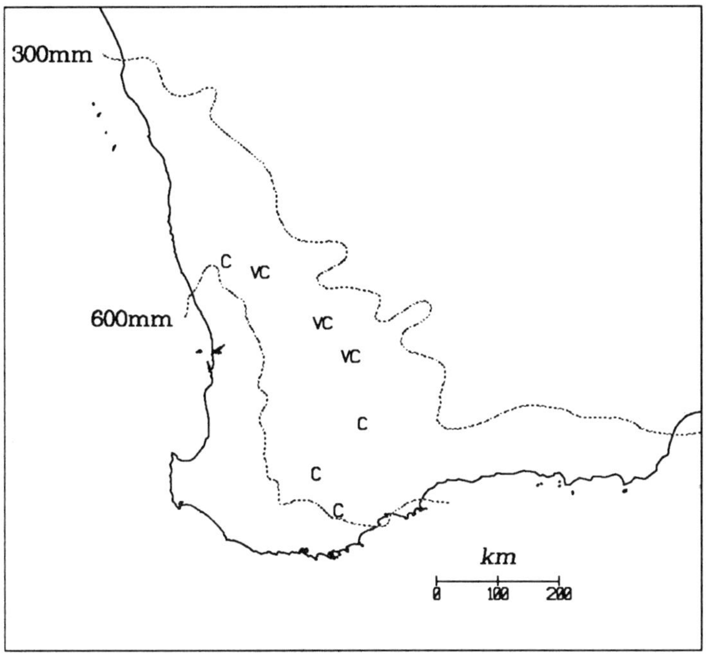

Fig. 184. Distribution 1987–1990

Yellow-plumed honeyeater

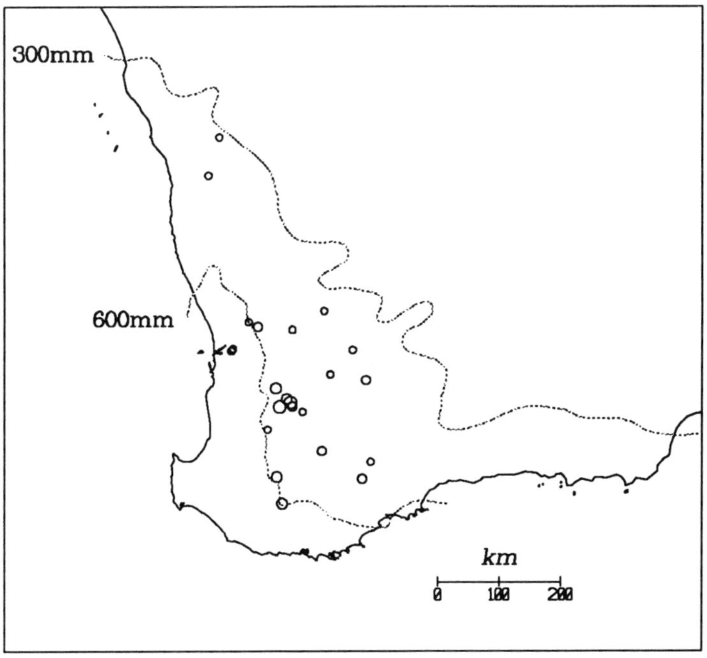

	1900/37	1987	1988	1989	1990
% of observers recording species	70	8.7	7.7	6.0	7.7
% of recording period species seen	C-VC	2.0	1.5	2.1	0.9

Yellow-plumed honeyeater

Lichenostomus ornatus

The yellow-plumed honeyeater is an uncommon resident in woodland and mallee of the higher rainfall districts of the wheatbelt (Fig. 184), where there are still large areas of woodland. It is a rare vagrant in the extensively cleared districts of the central wheatbelt. It feeds on nectar, flowers, cockroaches, bugs, lerps, beetles, flies, caterpillars, ants, spiders and mites. It breeds in spring in a cup-shaped nest in a low tree and lays 2–3 eggs.

Widespread loss and degradation of woodland and mallee has led to the contraction of the range of the yellow-plumed honeyeater. Early this century, it was common to very common in the wheatbelt (Fig. 184). For example, it was common in the "white gum" (wandoo) country in the Stirling Range (Milligan 1903), numerous in eucalypts in the Wongan Hills (Milligan 1904), "the commonest" honeyeater in the area of Cumminin Station, near Bruce Rock (Crossman 1909), common in salmon gum belts in the Moora district (Orton and Sandland 1913), common in open timbered country around Broome Hill (Carter 1923), and the most common honeyeater in the Kellerberrin area (Leake 1962). By the 1950s and 1960s it was uncommon in the Stirling Range (Sedgwick 1964). By the 1970s it had disappeared from the Wongan Hills (de Rebeira and de Rebeira 1977) and was reduced to a small population in wandoo woodland on Kodj Kodjin Nature Reserve north of Kellerberrin (Dell 1978); it no longer occurs there.

Grey-fronted honeyeater

Lichenostomus plumulus

OTHER NAME:

Yellow-fronted honeyeater

The grey-fronted honeyeater is a rare vagrant in woodland, mallee and shrubland of the drier areas of the wheatbelt (Fig. 185). It feeds on berries, seed, nectar, grasshoppers, bugs, cicadas, lerps, beetles, flies, moths, caterpillars, bees, ants and spiders.

There is no evidence to suggest the status of this species has changed.

Fig. 185. Distribution 1987–1990

Grey-fronted honeyeater

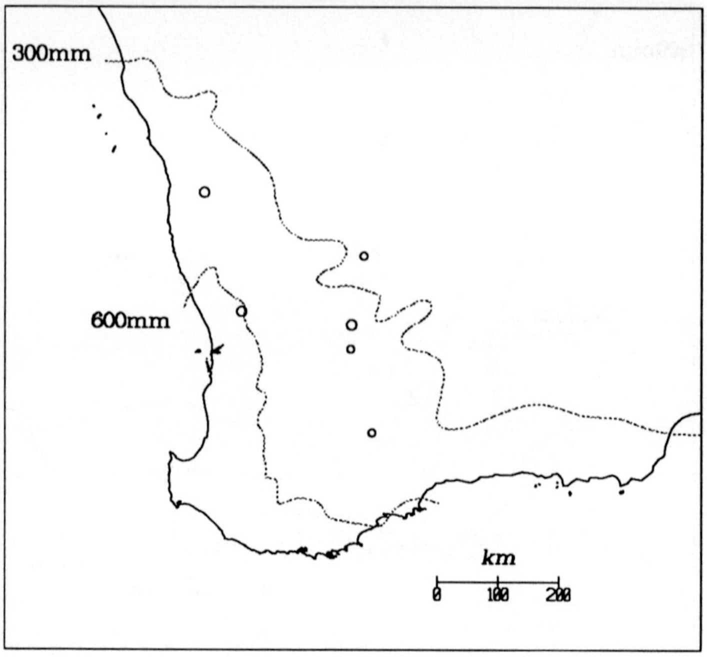

	1900/37	1987	1988	1989	1990
% of observers recording species	0	0	2.1	1.7	2.0
% of recording period species seen	NR	0	0.2	0.2	0.2

White-plumed honeyeater

Lichenostomus penicillatus

The white-plumed honeyeater is an uncommon resident in woodland near water in the northern wheatbelt (Fig. 186). It feeds on seed, nectar, fruit, pollen, manna, grasshoppers, bugs, lerps, lacewings, beetles, flies, moths, caterpillars, bees, wasps, ants, spiders and small lizards. It breeds in winter, spring and summer. It nests in a cup-shaped nest in a tree and lays 2–3 eggs.

There is no evidence to indicate that the status of this species has changed.

Fig. 186. Distribution 1900–1937

White-plumed honeyeater

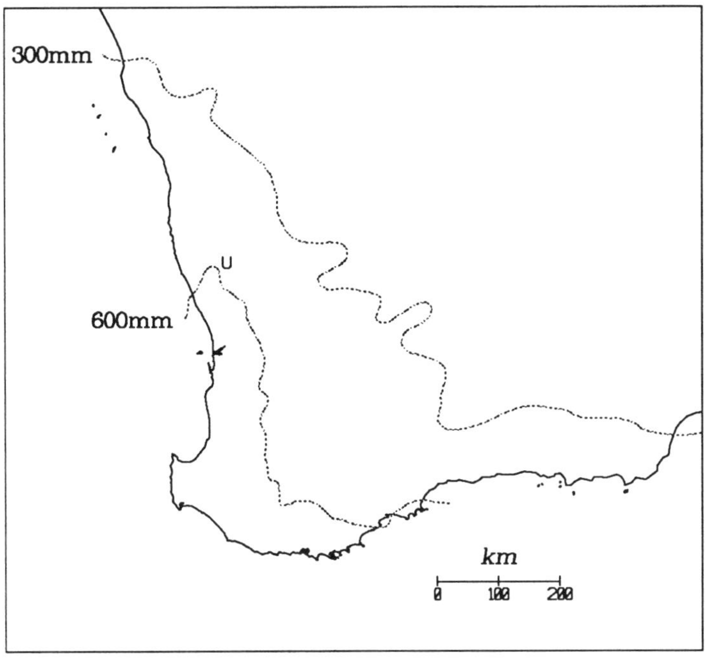

Fig. 186. Distribution 1987–1990

White-plumed honeyeater

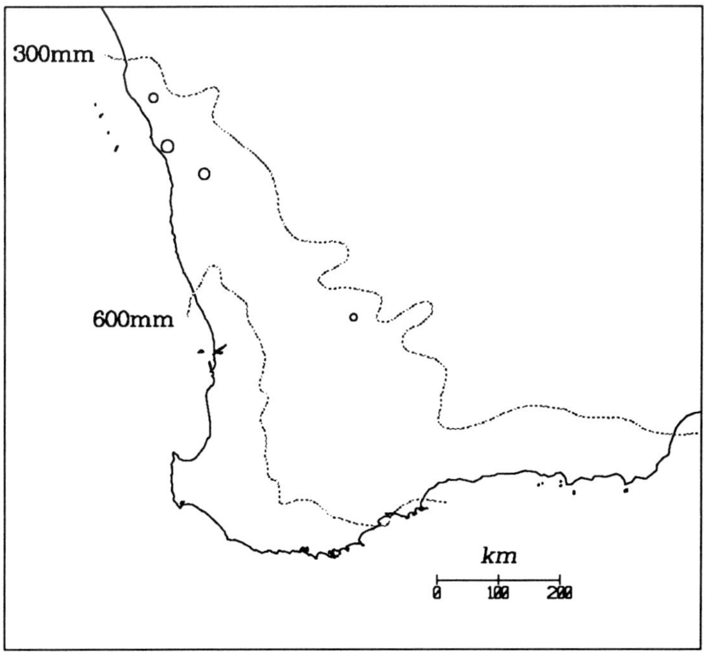

	1900/37	1987	1988	1989	1990
% of observers recording species	10	1.6	0.7	1.7	2.9
% of recording period species seen	U	0.9	0.7	1.1	1.6

Brown-headed honeyeater

Melithreptus brevirostris

The brown-headed honeyeater is an uncommon resident in woodland, mallee and shrubland throughout the wheatbelt (Fig. 187). It feeds on pollen, nectar, manna, lerps, beetles, caterpillars and spiders, although Halse (1978) stated that in the southwest it feeds almost entirely on nectar. It breeds in spring in a cup-shaped nest in a tree and lays 2–3 eggs.

The brown-headed honeyeater forms parties that remain together and wander locally. Banding studies north of Kellerberrin have shown that individuals are retrapped with the same birds they have been caught with previously. Groups have ranged up to 3.1 km along a well-vegetated road verge (Saunders and de Rebeira 1991).

The brown-headed honeyeater is dependent on remnant native vegetation. It will not readily fly over large open spaces and prefers links of native vegetation to move between patches. The brown-headed honeyeater has declined in abundance as a result of loss of suitable habitat.

Fig. 187. Distribution 1900–1937

Brown-headed honeyeater

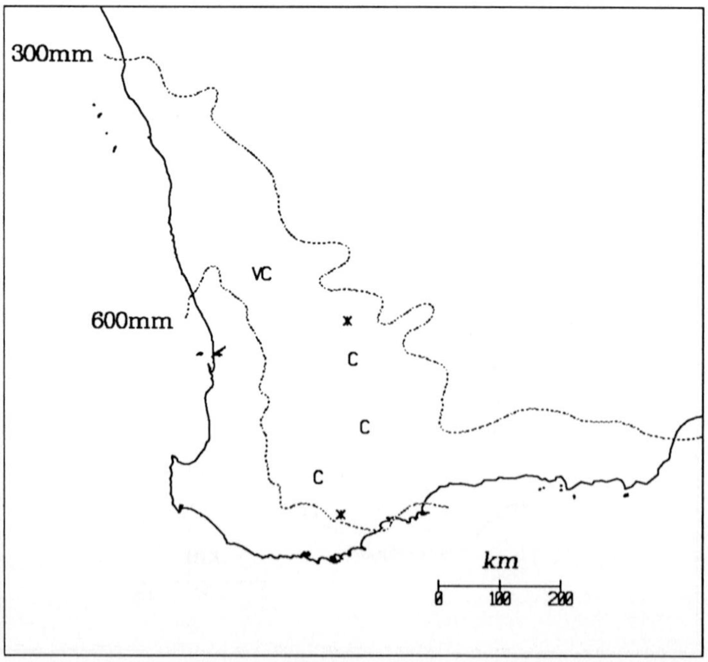

Fig. 187. Distribution 1987–1990

Brown-headed honeyeater

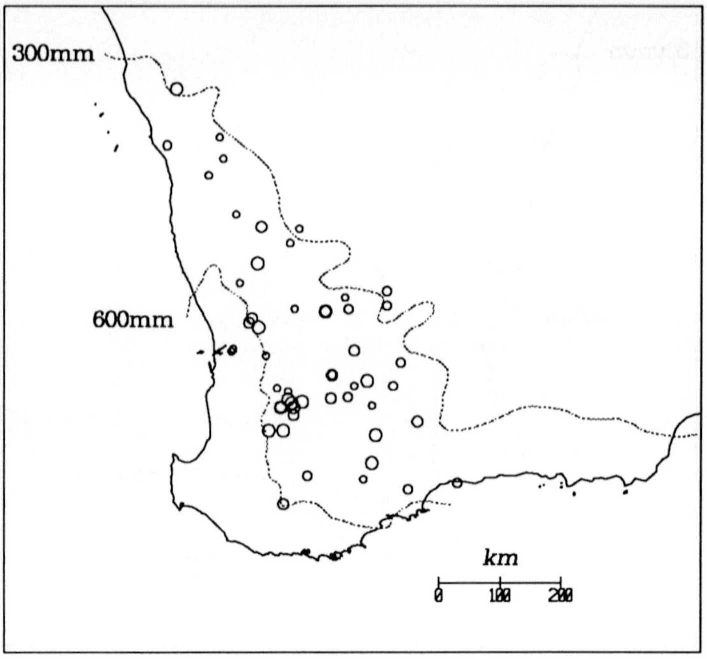

	1900/37	1987	1988	1989	1990
% of observers recording species	60	23	27	25	23
% of recording period species seen	C-VC	6.9	7.3	8.4	6.0

Fig. 188. Distribution 1900–1937

White-naped honeyeater

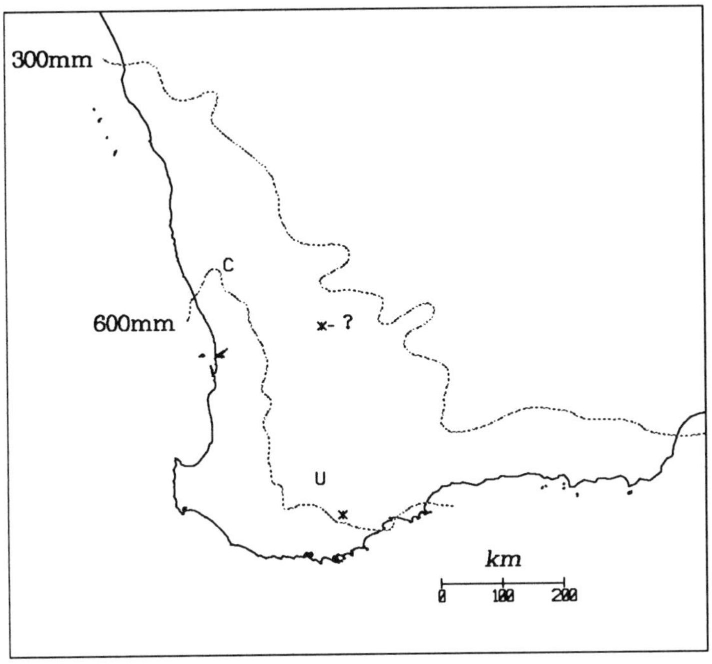

Fig. 188. Distribution 1987–1990

White-naped honeyeater

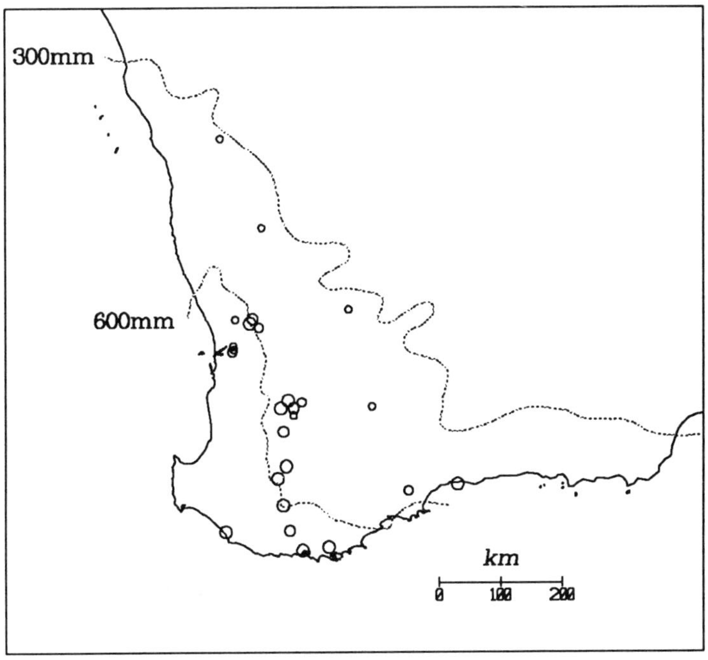

	1900/37	1987	1988	1989	1990
% of observers recording species	50	10	15	13	13
% of recording period species seen	U-C	2.3	7.5	6.4	7.1

White-naped honeyeater

Melithreptus lunatus

The white-naped honeyeater is an uncommon resident in woodland of the higher rainfall districts of the wheatbelt (Fig. 188). It is a rare nomad in other areas. It feeds on pollen, nectar, flowers, manna, bugs, lerps, aphids, thrips, beetles, flies, caterpillars, wasps, ants and spiders. The white-naped honeyeater breeds in late winter through early summer in a cup-shaped nest in a tree and lays 2–3 eggs.

This species has disappeared from parts of its former range in the wheatbelt which have lost most of their woodlands. For example, Leake (1962) recorded it at Kellerberrin, noting it was declining due to habitat destruction. It no longer occurs there.

Brown honeyeater

Lichmera indistincta

The brown honeyeater is a common resident in woodland, mallee, shrubland, heath, and in exotic vegetation in towns and farm gardens throughout the wheatbelt (Fig. 189). It feeds on nectar, flowers, bugs, beetles, flies, moths, wasps, bees, ants and spiders. In southwestern Australia it feeds almost entirely on nectar (Halse 1978). The brown honeyeater breeds in late winter through early summer in a cup-shaped nest low in a bush. It lays 2–3 eggs.

The brown honeyeater may wander widely while foraging, concentrating on areas where native vegetation is in flower, hence the term "blossom nomad" (Storr 1991). Results of a banding study, carried out north of Kellerberrin, revealed that 75 per cent of individuals re-trapped were caught at the site where they were originally banded (Saunders and de Rebeira 1991). Some individuals were caught up to 12.5 km from their original banding site. Some birds were caught at a series of different sites when foraging in flowering bushes, indicating the birds move between patches and revisit them in subsequent years.

In the Northam district, Masters and Milhinch (1974) noted that some birds were always present, but numbers increased during late winter and spring. Analyses of the data gathered during this survey revealed that there were slightly more recordings of the brown honeyeater in late winter and spring, but this was not a strong seasonal pattern.

The brown honeyeater is dependent on remnant native vegetation throughout the wheatbelt, and movements of birds are fostered by vegetated links such as remnant vegetation along roads. The loss of extensive areas of native vegetation has resulted in the decline of this species.

Fig. 189. Distribution 1900–1937

Brown honeyeater

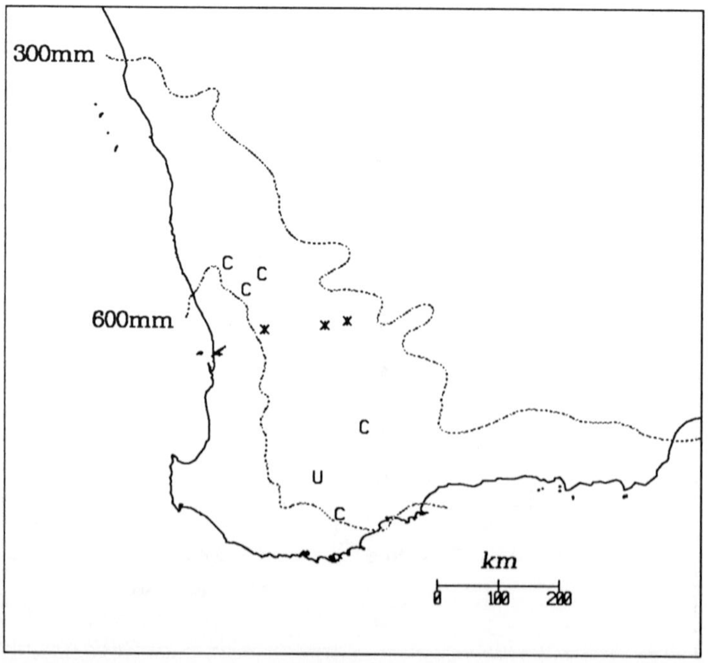

Fig. 189. Distribution 1987–1990

Brown honeyeater

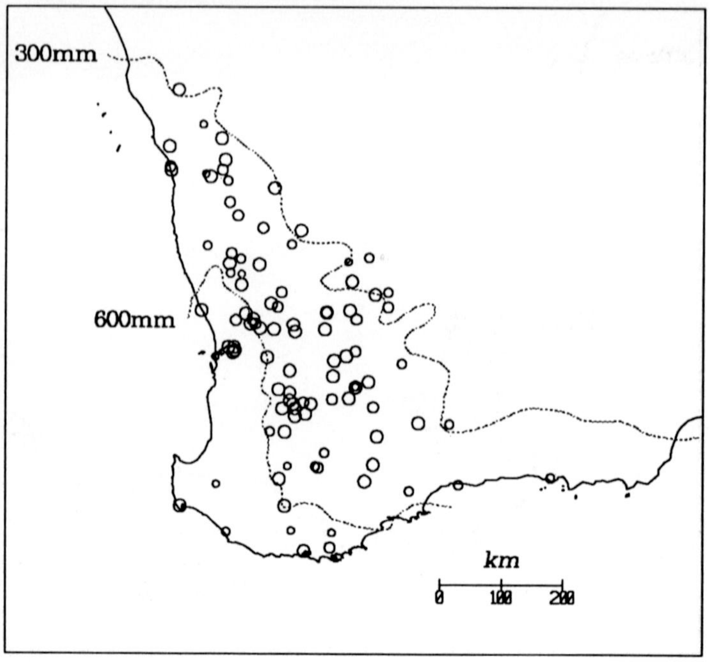

	1900/37	1987	1988	1989	1990
% of observers recording species	90	39	54	56	60
% of recording period species seen	U-C	25	29	32	32

Fig. 190. Distribution 1900–1937

New Holland honeyeater

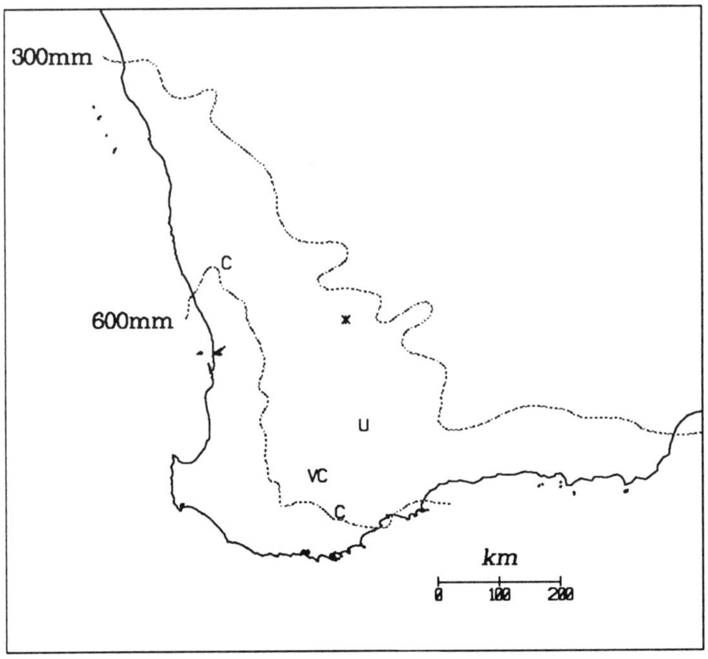

Fig. 190. Distribution 1987–1990

New Holland honeyeater

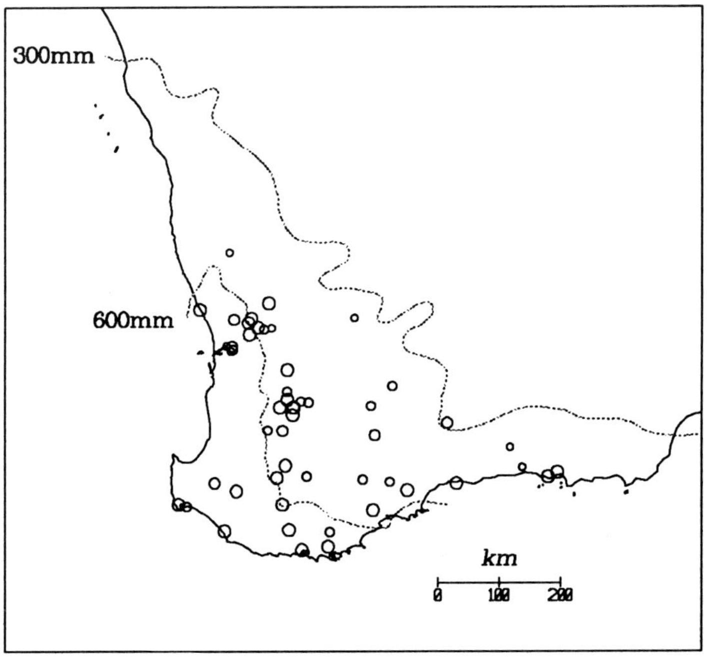

	1900/37	1987	1988	1989	1990
% of observers recording species	50	15	27	30	30
% of recording period species seen	U-VC	6.7	14	18	18

New Holland honeyeater

Phylidonyris novaehollandiae

The New Holland honeyeater is an uncommon nomad in woodland, shrubland and heath of the higher rainfall districts of the wheatbelt (Fig. 190). It feeds on nectar, pollen, fruit, termites, crickets, grasshoppers, bugs, lerps, beetles, flies, mosquitoes, midges, moths, caterpillars, wasps, bees, ants and spiders. It breeds in winter and spring in a cup-shaped nest low in a tree or shrub and lays 2–3 eggs.

There is no evidence to suggest that the status of this species has changed.

White-cheeked honeyeater

Phylidonyris nigra

The white-cheeked honeyeater is an uncommon nomad in woodland, shrubland and heath of the middle and higher rainfall districts of the wheat-belt (Fig. 191). It feeds on nectar, flowers, pollen, lerps, beetles, flies, moths and ants. In southwestern Australia it spends equal time feeding on insects and nectar (Halse 1978). It breeds in winter and spring in a cup-shaped nest close to the ground and lays 2–3 eggs.

There is no evidence to indicate that the status of this species has changed. However, the white-cheeked honeyeater is dependent on native vegetation and the loss of suitable habitat may have resulted in its decline.

Fig. 191. Distribution 1900–1937

White-cheeked honeyeater

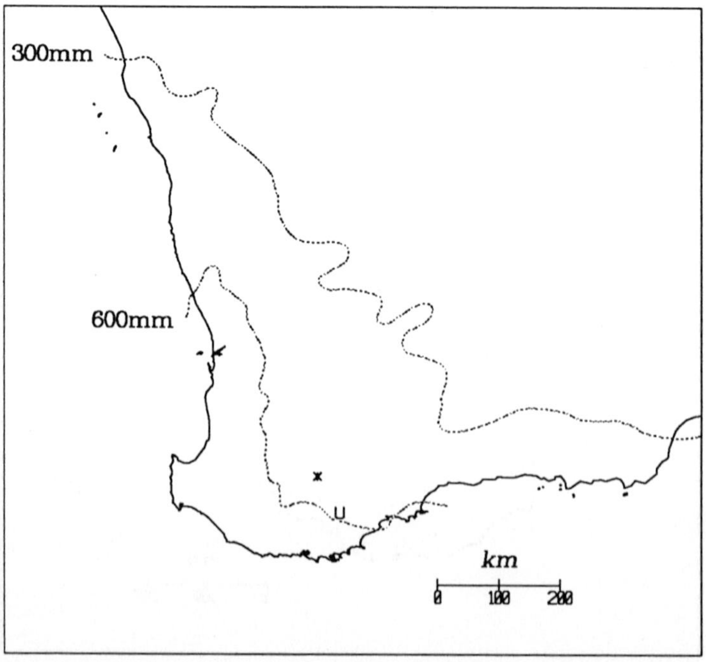

Fig. 191. Distribution 1987–1990

White-cheeked honeyeater

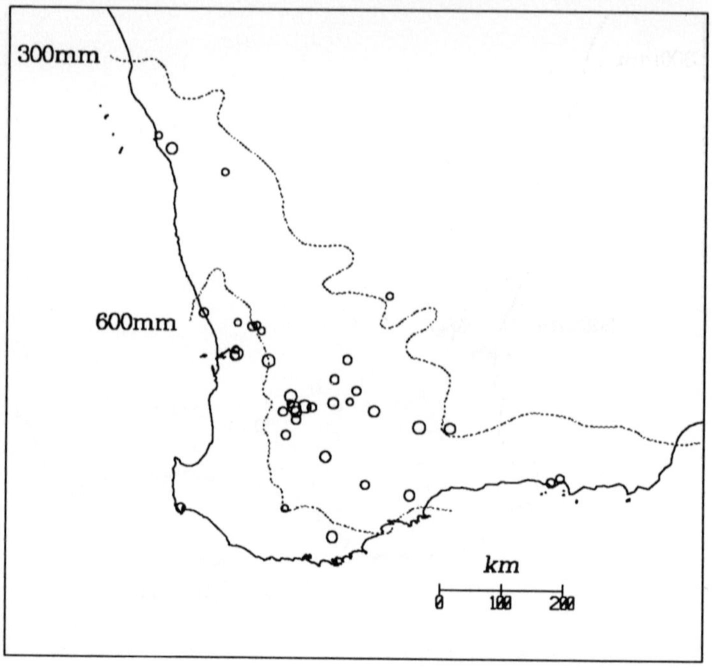

	1900/37	1987	1988	1989	1990
% of observers recording species	20	14	15	15	17
% of recording period species seen	U	3.2	4.4	4.4	5.5

Fig. 192. Distribution 1900–1937

White-fronted honeyeater

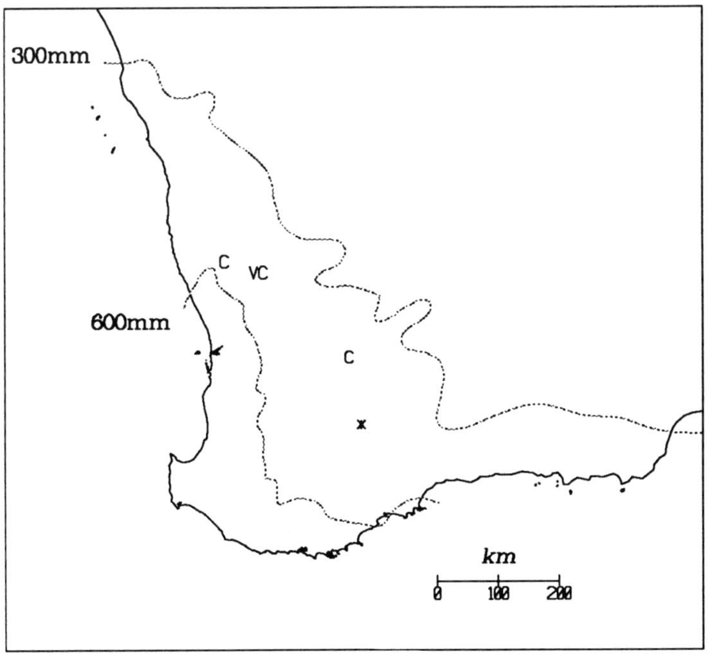

Fig. 192. Distribution 1987–1990

White-fronted honeyeater

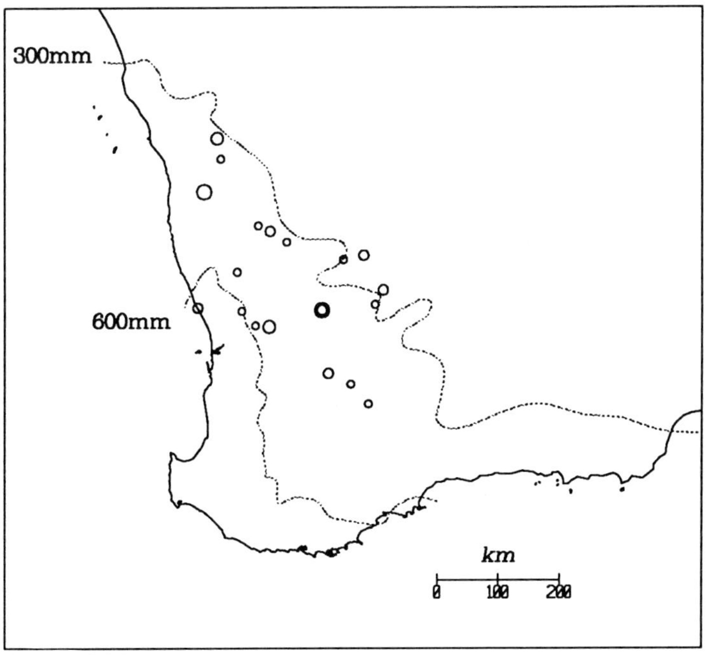

	1900/37	1987	1988	1989	1990
% of observers recording species	40	4.7	8.5	6.8	7.7
% of recording period species seen	C-VC	0.8	0.9	0.4	1.7

White-fronted honeyeater

Phylidonyris albifrons

The white-fronted honeyeater is a species of the arid zone and an uncommon nomad in shrubland and heath of the northern half of the wheatbelt (Fig. 192). It feeds on nectar, flowers, seed, bugs, beetles, flies, mosquitoes, moths, wasps, bees, ants and spiders. It breeds in spring and summer in a cup-shaped nest in a low tree or shrub and lays 2–3 eggs.

During this survey, recording rates for this species varied, while the percentage of observers recording the species did not vary markedly. In 1990 the recording rate was twice that of 1988 and four times that of 1989, indicating the birds were more abundant.

Early this century, several observers in the wheatbelt recorded that this species was common to very common. Milligan (1904) visited the Wongan Hills in late Sepember and early October 1903 and noted that it was very common and breeding. During a four year survey of the same area over 70 years later, de Rebeira and de Rebeira (1977) only recorded the species once in the early part of the survey, but it became plentiful between February and April 1975. Orton and Sandland (1913) recorded that it visited the Moora district during March and April in large numbers in odd years to feed on white gum blossom. The removal of native vegetation would have resulted in its decline.

Tawny-crowned honeyeater

Phylidonyris melanops

The tawny-crowned honeyeater is an uncommon resident of mallee, shrubland and heath in the southern wheatbelt and along the coast of the northern wheatbelt (Fig. 193). It is a rare nomad in other areas. It feeds on nectar, pollen, bugs, flies, bees, ants and spiders. In south-western Australia it spends equal time feeding on insects and nectar (Halse 1978). It breeds in spring and summer in a cup-shaped nest on or near the ground and lays 2–3 eggs.

During the survey, recording rates were higher in autumn, winter and spring than in summer (see Fig. below).

This species has decreased in range and abundance over the past half century, as a direct result of loss of native vegetation. For example, Milligan (1904) recorded that the tawny-crowned honeyeater was common at many places along the route from Mogumber to the Wongan Hills. Over seventy years later it was recorded during only one of four years of surveys around the Wongan Hills (de Rebeira and de Rebeira 1977).

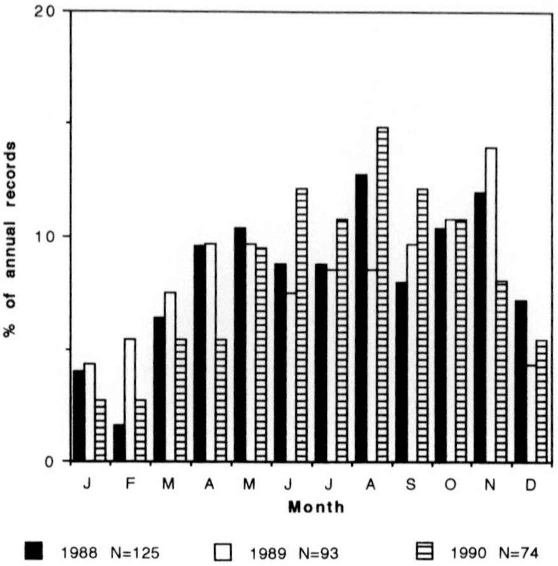

Fig. 193. Distribution 1900–1937

Tawny-crowned honeyeater

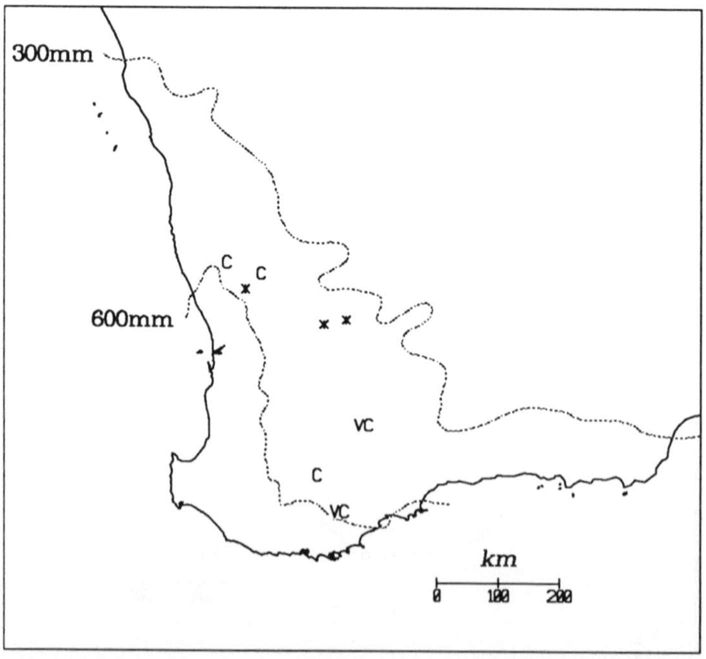

Fig. 193. Distribution 1987–1990

Tawny-crowned honeyeater

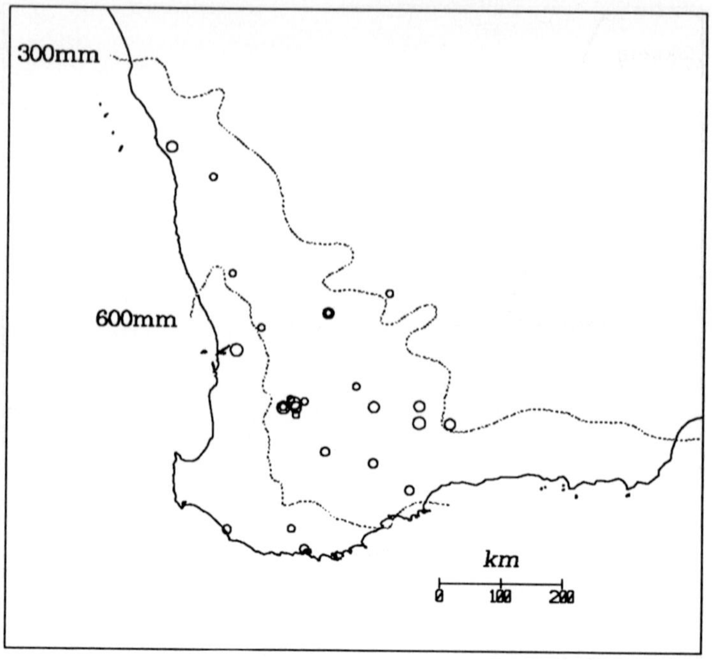

	1900/37	1987	1988	1989	1990
% of observers recording species	80	8.7	13	7.7	11
% of recording period species seen	C-VC	2.5	2.5	2.1	1.8

Fig. 194. Distribution 1900–1937

Western spinebill

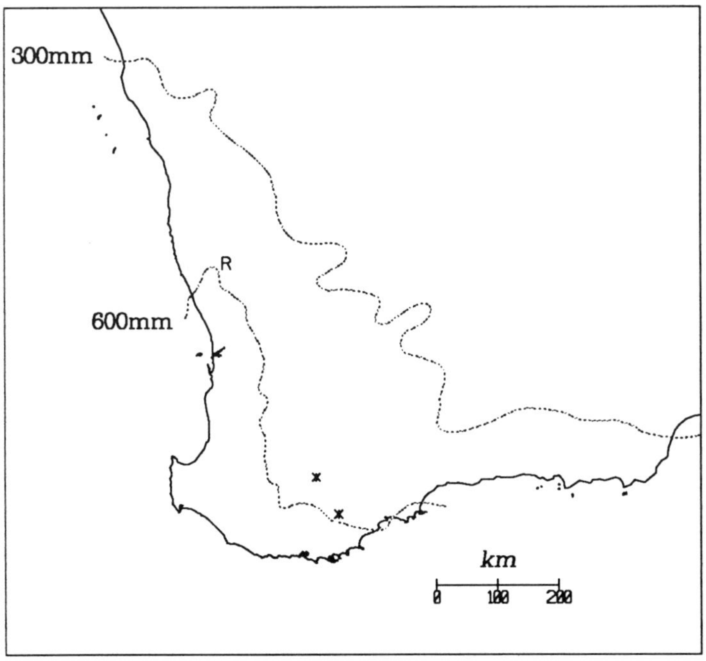

Fig. 194. Distribution 1987–1990

Western spinebill

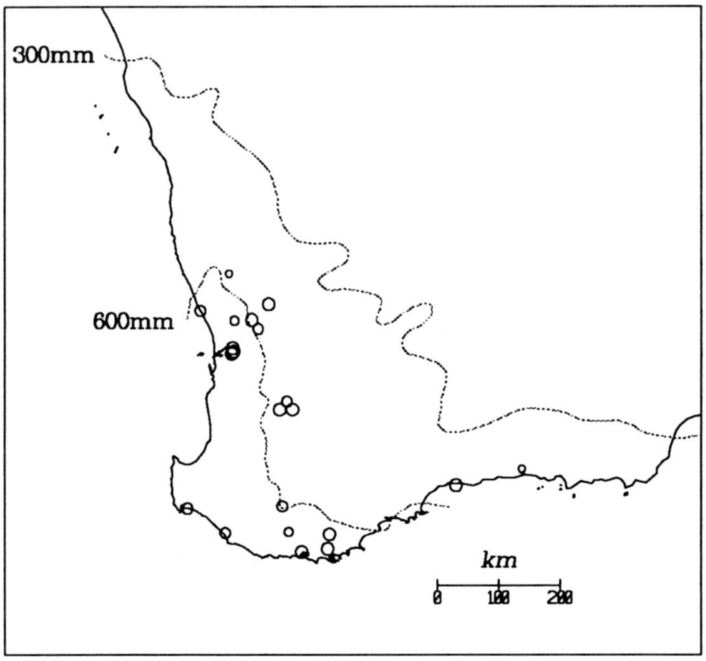

	1900/37	1987	1988	1989	1990
% of observers recording species	30	5.5	13	12	11
% of recording period species seen	R	3.7	6.1	5.1	4.5

Western spinebill

Acanthorhynchus superciliosus

The western spinebill is endemic to south-western Autralia. It is found in woodland and heath of the wheatbelt adjacent to the forests of the southwest and along the south coast (Fig. 194). It feeds on nectar, flowers and insects. It spends equal time feeding on insects and nectar (Halse 1978). The western spinebill breeds in spring and summer in a cup-shaped nest in a shrub and lays 1–2 eggs.

The increase in percentage of observers recording the western spinebill and the increase in recording rates between 1987 and the subsequent three years of the survey are due to the fact that more observers from within the range of the species joined the scheme and collected data in the latter three years.

This species is dependent on native vegetation and has possibly declined in those areas of its former range in the wheatbelt where extensive clearing has taken place.

Black honeyeater

Certhionyx niger

The black honeyeater is an arid zone species which is a rare nomad in shrubland of the northern half of the wheatbelt (Fig. 195). It feeds on seed, nectar, flowers, termites, caterpillars and ants. It breeds in spring in a cup-shaped nest close to the ground and lays 2–3 eggs.

There is no evidence to suggest that the status of the black honeyeater has changed.

Fig. 195. Distribution 1900–1937

Black honeyeater

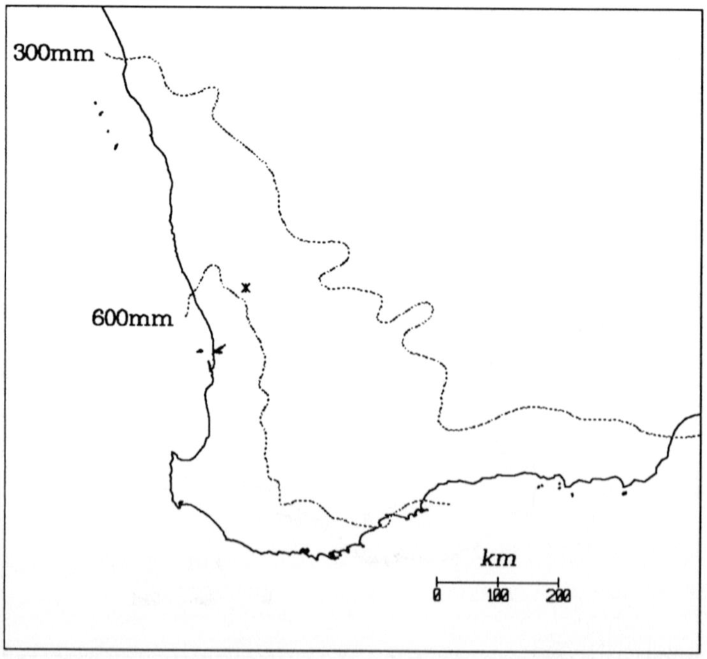

Fig. 195. Distribution 1987–1990

Black honeyeater

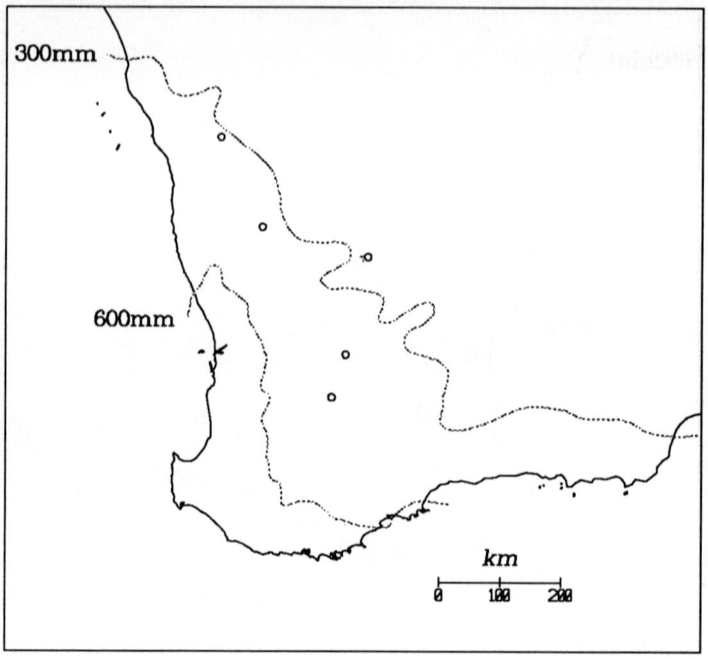

	1900/37	1987	1988	1989	1990
% of observers recording species	10	1.6	1.4	0	1.0
% of recording period species seen	—	0.1	0.1	0	0.07

Fig. 196. Distribution 1900–1937

Pied honeyeater

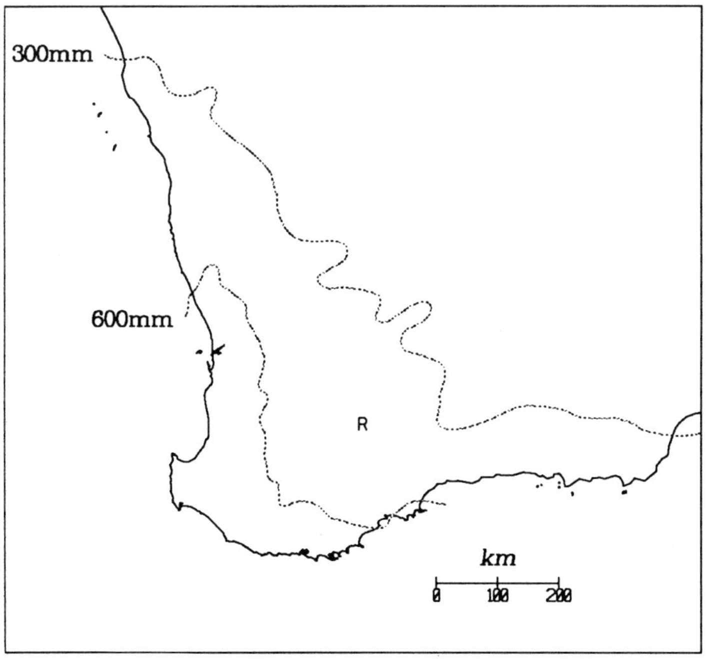

Fig. 196. Distribution 1990

Pied honeyeater

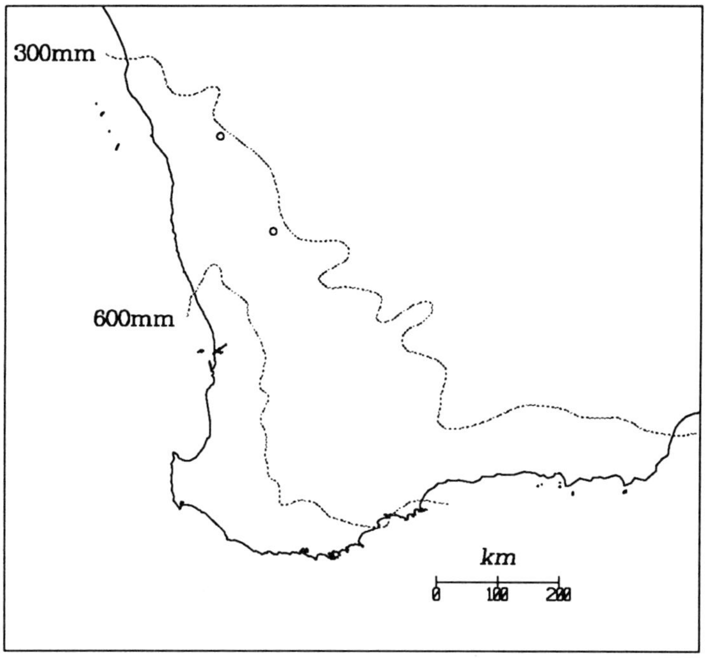

	1900/37	1987	1988	1989	1990
% of observers recording species	10	0	0	0	1.9
% of recording period species seen	R	0	0	0	0.07

Pied honeyeater

Certhionyx variegatus

The pied honeyeater is an arid zone bird which is a rare nomad in the northern wheatbelt (Fig. 196). It was recorded by only 2 observers in 1990. It feeds on seed, berries, nectar, beetles and caterpillars. It breeds in spring and summer in a cup-shaped nest in a low bush. It lays 2–4 eggs.

Crimson chat

Ephthianura tricolor

The crimson chat is an arid zone species which is an occasional, seasonal (May–December) nomad in shrubland and heath of the drier rainfall districts of the wheatbelt (Fig. 197). It feeds on seed, grasshoppers, bugs, beetles, moths, caterpillars, wasps, ants and spiders. It breeds in spring in a cup-shaped nest close to the ground and lays 3–4 eggs.

The distribution of the crimson chat varies from year to year (Fig. 197), because of its nomadic nature. Irruptions have seen it occur round Perth (Serventy and Whittell 1962) and Yanchep (pers. obs.). It was more widely distributed and more commonly seen in 1990, than in the period 1988–1989 (Fig. 197).

There is no evidence to suggest that the status of this species has changed.

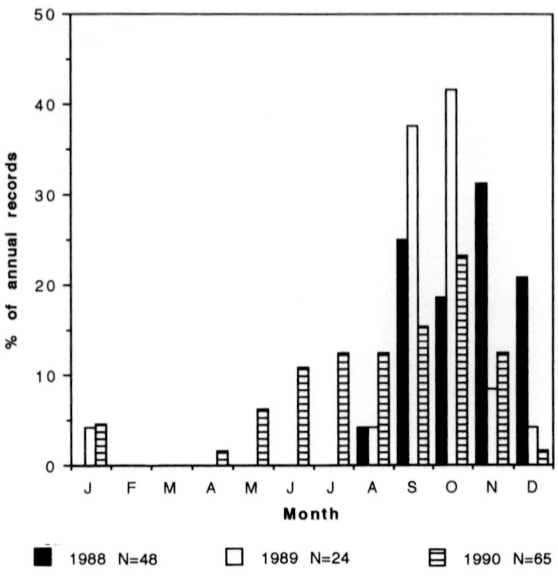

Fig. 197. Distribution 1987

Crimson chat

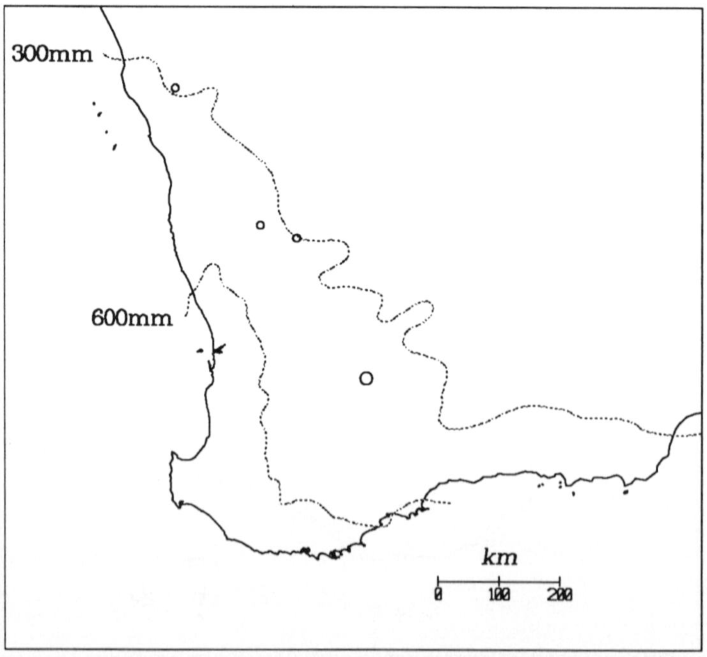

Fig. 197. Distribution 1988

Crimson chat

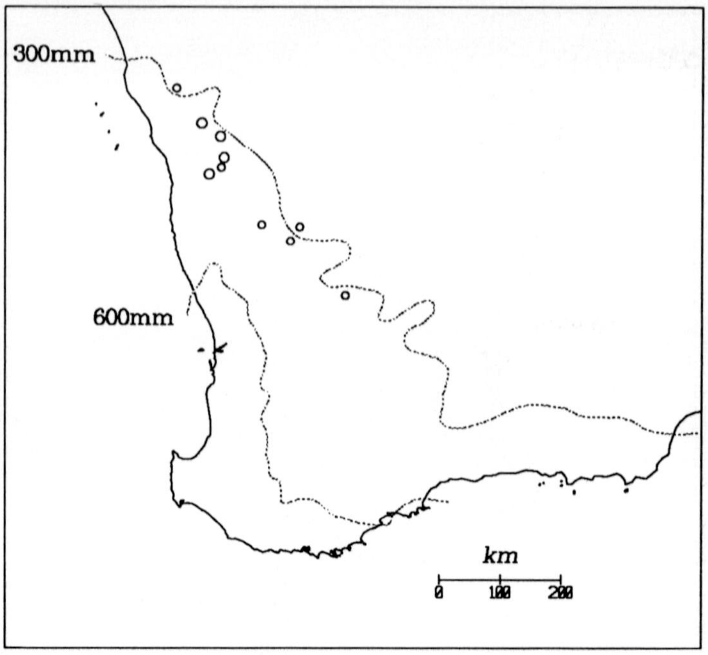

	1900/37	1987	1988	1989	1990
% of observers recording species	0	3.1	7.0	5.1	9.6
% of recording period species seen	NR	0.4	0.9	0.5	1.5

Fig. 197. Distribution 1989

Crimson chat

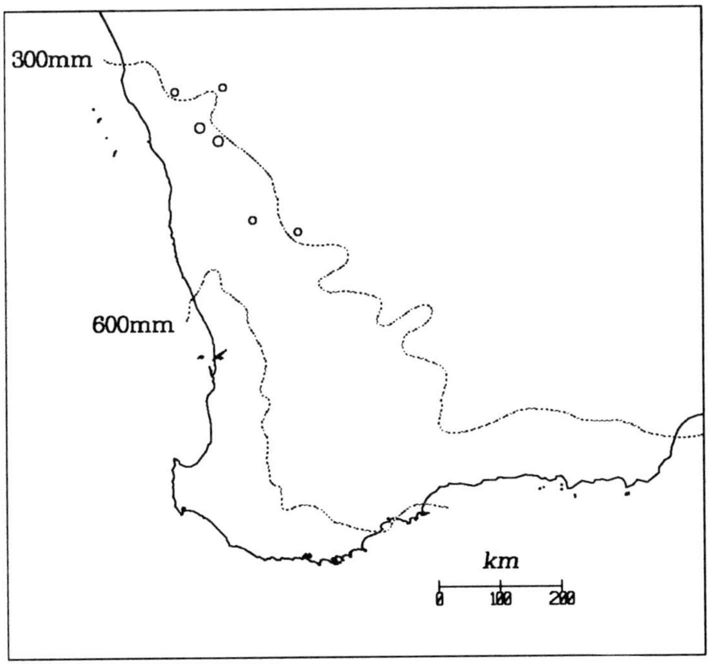

Fig. 197. Distribution 1990

Crimson chat

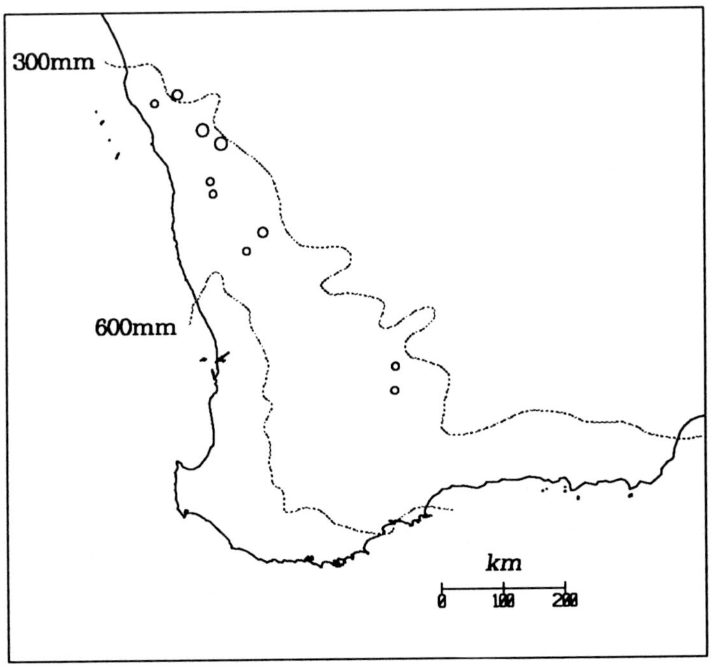

Orange chat

Ephthianura aurifrons

The orange chat is a bird of the arid zone that is a rare vagrant in samphire around salt lakes in the northern wheatbelt (Fig. 198). It feeds on seed, grasshoppers, bugs, beetles, caterpillars, wasps and spiders. It breeds in spring in a cup-shaped nest in a low shrub and lays 3–4 eggs.

There is no evidence to indicate that the status of the orange chat has changed.

Fig. 198. Distribution 1987–1990

Orange chat

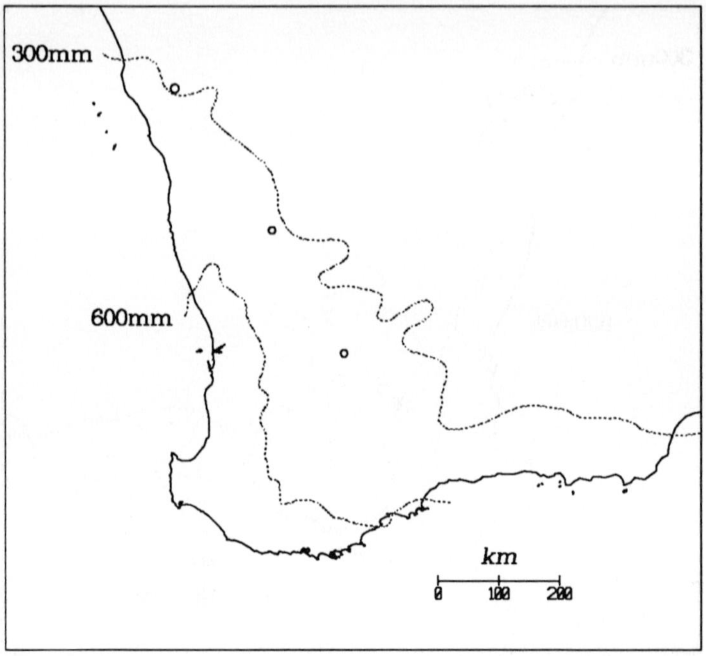

	1900/37	1987	1988	1989	1990
% of observers recording species	0	0	2.1	0	1.0
% of recording period species seen	NR	0	0.3	0	0.02

Fig. 199. Distribution 1900–1937

White-fronted chat

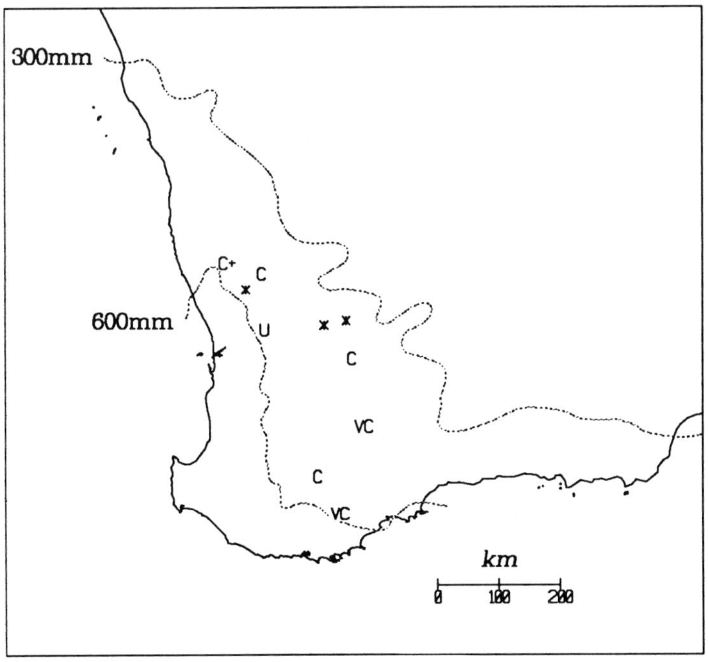

Fig. 199. Distribution 1987–1990

White-fronted chat

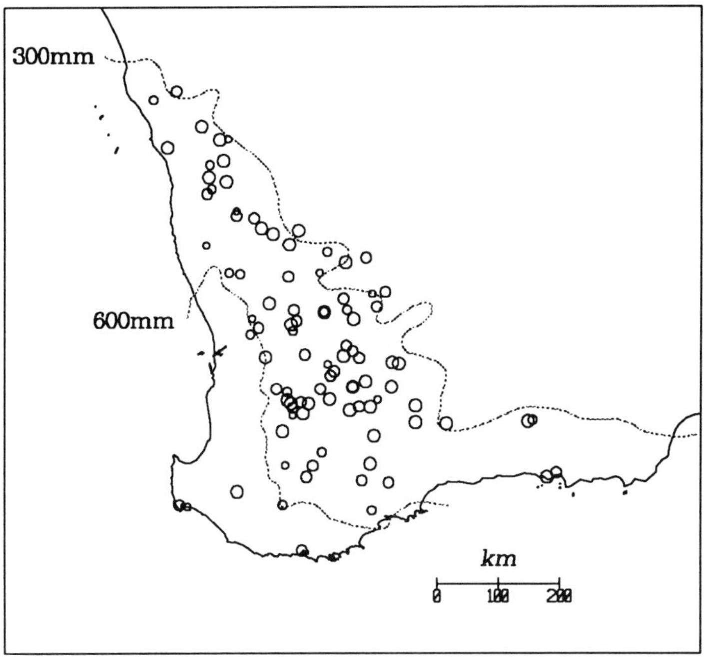

	1900/37	1987	1988	1989	1990
% of observers recording species	100	47	50	55	54
% of recording period species seen	U-VC	20	19	22	24

White-fronted chat

Ephthianura albifrons

The white-fronted chat is an uncommon nomad in heath, around the margins of salt lakes and occasionally on farmland throughout the wheatbelt (Fig. 199). It feeds on seed, cockroaches, earwigs, crickets, bugs, beetles, caterpillars, bees, ants and spiders. It breeds in late winter to early summer in a cup-shaped nest in a low bush and lays 2–4 eggs.

During this survey, recording rates for the white-fronted chat showed some seasonality in occurrence with recording rates being lowest in the period January to April (see Fig. below).

The white-fronted chat may have benefited from development for agriculture. Orton and Sandland (1913) noted that it had become more plentiful as clearing advanced. Sedgwick (1974) recorded that it appeared on newly-cleared farmland in the eastern wheatbelt as soon as cultivation began.

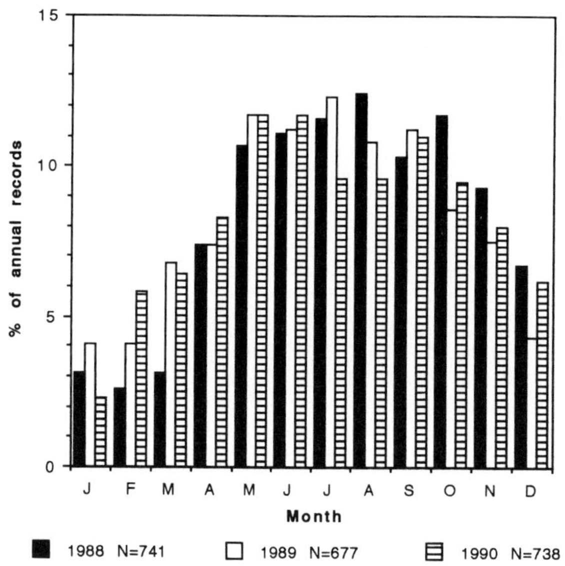

Mistletoe bird

Dicaeum hirundinaceum

The mistletoe bird is an uncommon nomad in woodland and shrubland of the higher rainfall districts of the wheatbelt (Fig. 200). It feeds predominantly on the berries of mistletoe, and also eats other berries, fruit, insects and spiders. It breeds in spring and summer in a suspended pear-shaped nest in a tree or shrub and lays 3 eggs.

This species has declined in abundance in the wheatbelt as a result of the loss of suitable habitat. For example, the mistletoe bird was recorded in the Wongan Hills early this century (Milligan 1904), but was not seen there in the 1970s (de Rebeira and de Rebeira 1977). In the mid-1970s, it was seen in Durokoppin and Kodj Kodjin Nature Reserves north of Kellerberrin (Dell 1978); it had disappeared by the mid-1980s.

Fig. 200. Distribution 1900–1937

Mistletoe bird

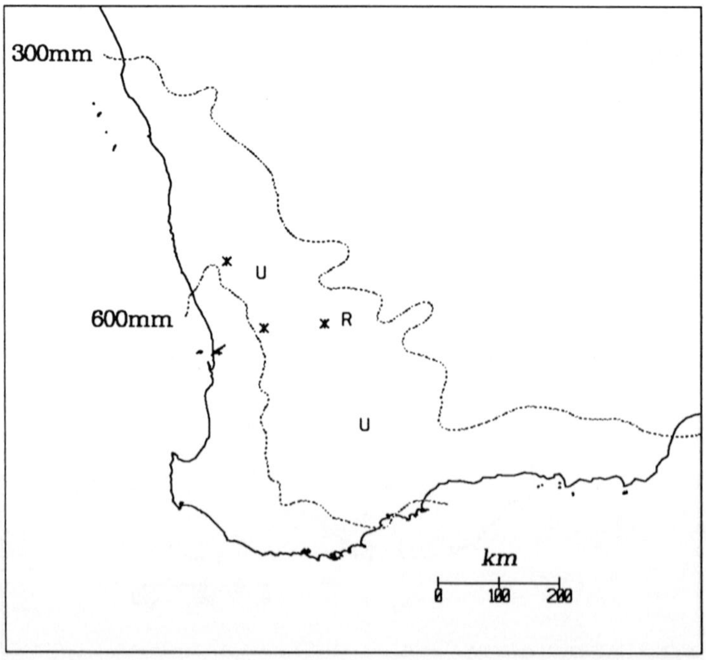

Fig. 200. Distribution 1987–1990

Mistletoe bird

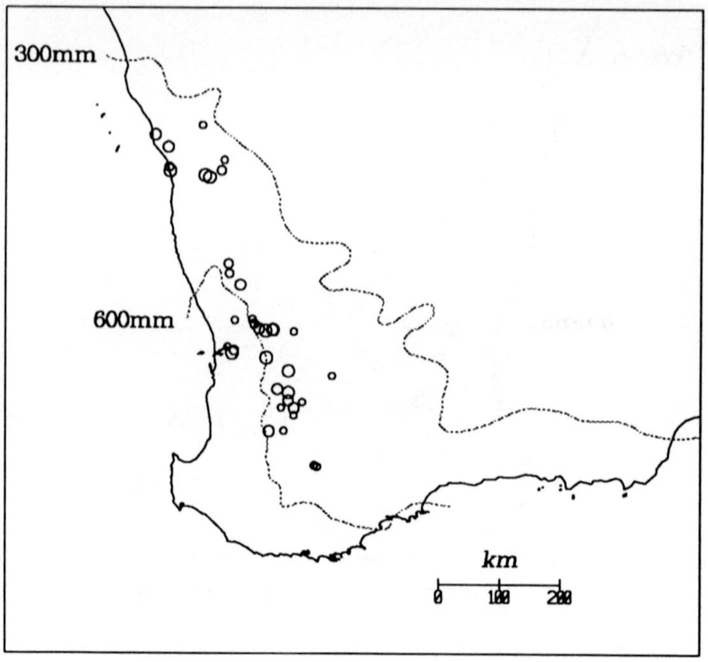

	1900/37	1987	1988	1989	1990
% of observers recording species	60	15	18	15	20
% of recording period species seen	R-U	5.6	5.1	5.3	5.7

Fig. 201. Distribution 1900–1937

Spotted pardalote

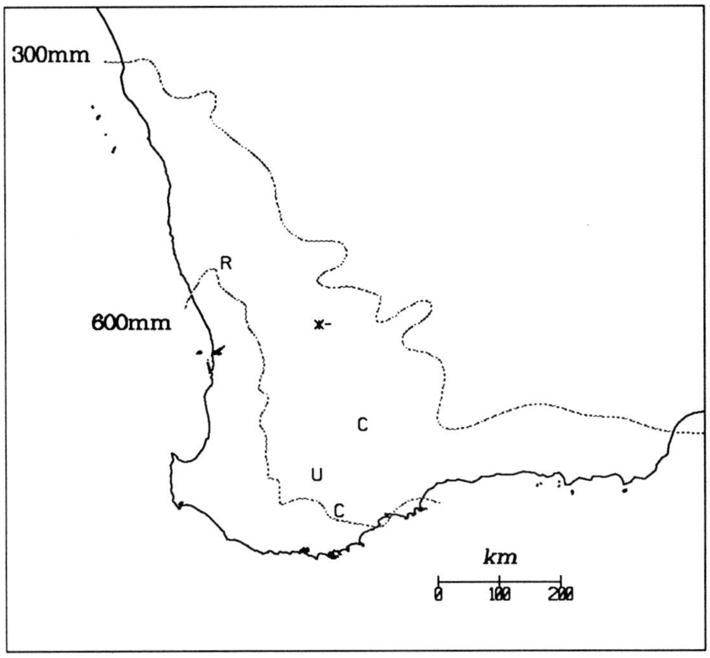

Fig. 201. Distribution 1987–1990

Spotted pardalote

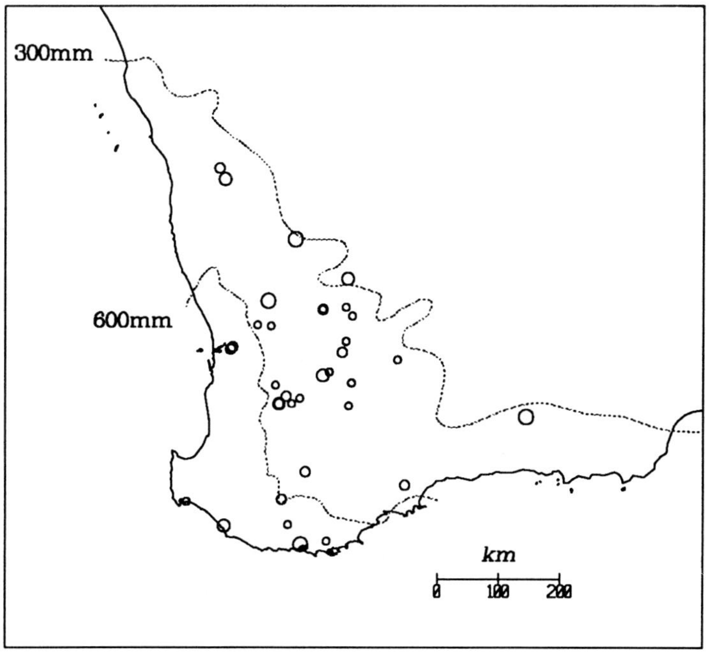

	1900/37	1987	1988	1989	1990
% of observers recording species	50	13	13	6.0	15
% of recording period species seen	R-C	3.3	1.8	1.6	3.4

Spotted pardalote

Pardalotus punctatus

The spotted pardalote is an uncommon resident of woodland in the higher rainfall districts of the wheatbelt (Fig. 201), and a rare nomad in woodland and mallee of the drier districts. It feeds on plant material, cockroaches, mantids, cicadas, leafhoppers, thrips, beetles, crane flies, moths, caterpillars, ants and spiders. The spotted pardalote breeds in spring and summer in a burrow and lays 3–4 eggs.

The spotted pardalote was a common breeding species in the Lake Grace area (Carnaby 1933), however, it is not common now. It may have declined in abundance and range.

Yellow-rumped pardalote

Pardalotus xanthopygus

The yellow-rumped pardalote is an uncommon resident in mallee of the southeastern wheatbelt and a rare vagrant in the central wheatbelt (Fig. 202). It feeds on plant material, earwigs, beetles, caterpillars, ants and spiders. It breeds in late winter and spring in a burrow and lay 3–4 eggs.

Its status as a vagrant in the central wheatbelt was confirmed in April 1992 when one yellow-rumped pardalote was caught and banded north of Kellerberrin. This was the first time this species had been recorded in the area.

There is no evidence to suggest that the status of this species has changed.

Fig. 202. Distribution 1987–1990

Yellow-rumped pardalote

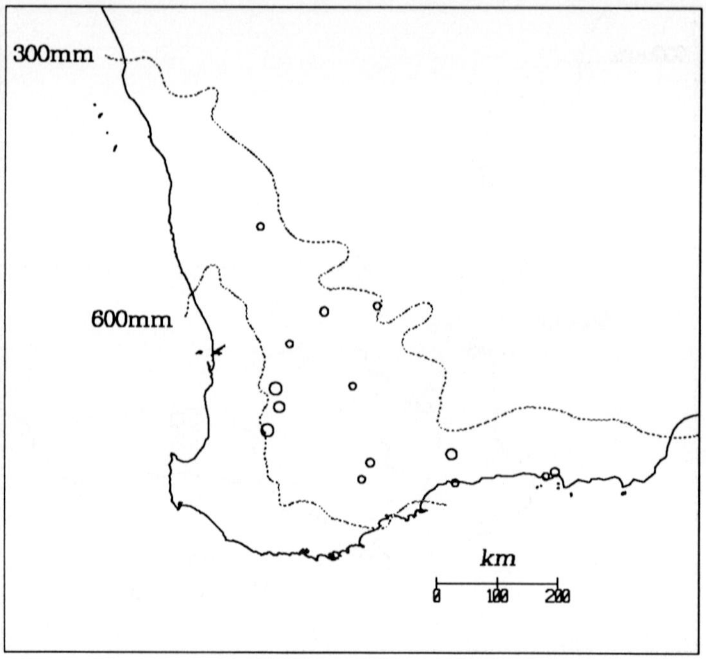

	1900/37	1987	1988	1989	1990
% of observers recording species	0	5.5	4.2	3.4	4.8
% of recording period species seen	NR	1.6	1.3	1.3	1.1

Fig. 203. Distribution 1900–1937

Striated pardalote

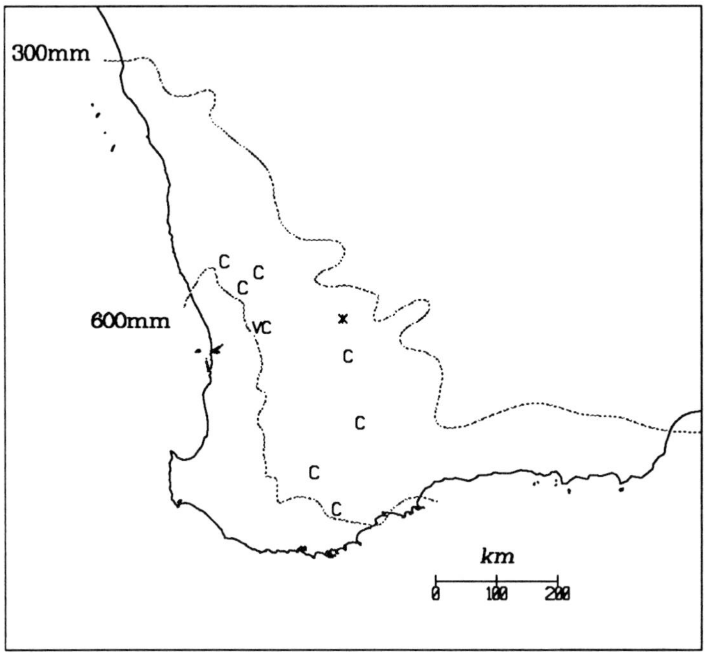

Fig. 203. Distribution 1987–1990

Striated pardalote

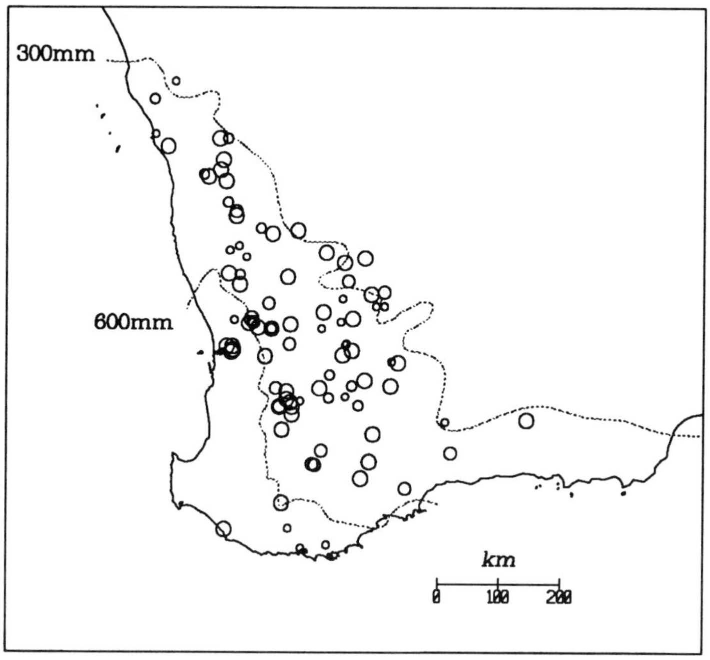

	1900/37	1987	1988	1989	1990
% of observers recording species	90	43	44	48	49
% of recording period species seen	C-VC	18	21	4	24

Striated pardalote

Pardalotus striatus

The striated pardalote is a common resident in woodland and mallee in parts of the wheatbelt (Fig. 203) and a breeding migrant in others. It feeds on cockroaches, bugs, beetles, caterpillars and spiders. It breeds in winter and spring in a hollow in a tree or creek bank and lays 3–4 eggs.

The status and occurrence of the striated pardalote varies in different parts of the wheatbelt (see Table below where arrival and departure times are listed). This seasonal pattern is also shown by the percentage of annual recordings for each month of the three complete years of this survey (see Fig. below).

	1988	1989	1990
Tardun	Apr. Wk 4 Aug. Wk 3	Apr. Wk 2 Oct. Wk 4	Apr. Wk 1 Dec. Wk 4
Arrino	Aug. Wk 2 May Wk 1	Aug. Wk 1 Apr. Wk 3	Aug. Wk 1 *
Avon River between Northam and Toodyay	Resident		
Bruce Rock	Apr. Wk 3 Dec. Wk 2	Apr. Wk 3 Nov. Wk 2	Jul. Wk 1 Oct. Wk 3
Pingrup	Resident		

*The departure time was not noted as the survey ended on 31 December, 1990.

The striated pardalote is dependent on woodland and mallee in the wheatbelt and must have declined in abundance as a result of loss of habitat.

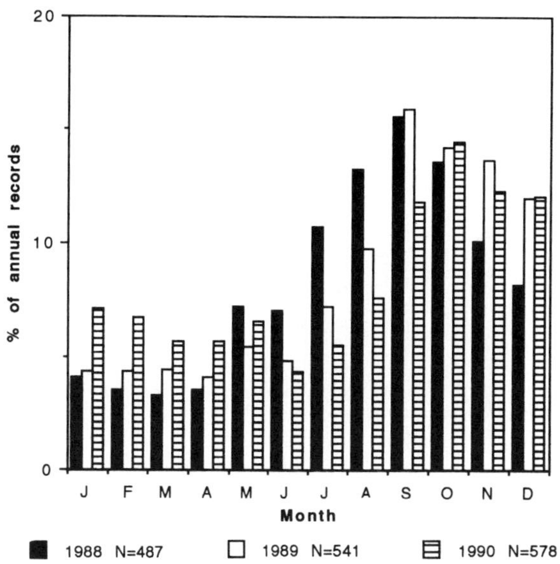

Silvereye

Zosterops lateralis

OTHER NAME:

Grey-breasted white-eye

The silvereye is an uncommon resident in woodland, mallee, shrubland and heath, and farm and town gardens in the higher rainfall districts of the wheatbelt (Fig. 204), and close to the coast. It is a seasonal migrant in adjacent areas and a rare nomad in the drier areas of the wheatbelt. It feeds on flowers, nectar, fruit, berries, seed, scale insects, thrips, lacewings, beetles, flies, midges, caterpillars, ants, spiders and mites, and is occasionally a nuisance in gardens. It breeds in spring and summer in a suspended cup-shaped nest in a shrub and lays 2–4 eggs.

The seasonal pattern of occurence of the silvereye in those localities where it is not resident, is shown by the percentage of annual recordings for each month of the three complete years of this survey (see Fig. below). Recording rates were lowest in the period June to October. The status of this species may vary between years at some locations as they may remain all year during some years, but become seasonal migrants in other years.

Common in many parts of the wheatbelt early this century, and regarded by Serventy and Whittell (1962) as the most common small bird over much of the southwest, the silvereye has declined in abundance and possibly range in the drier districts of the wheatbelt. Early this century, the silvereye was numerous everywhere around the Wongan Hills (Milligan 1904), but in the 1970s de Rebeira and de Rebeira (1977) recorded that "small groups and flocks occur most times of the year." Around Mogumber the silvereye was common (Lawson 1905), but had apparently decreased in abundance by the 1950s (Loaring and Serventy 1952). Extensive loss of habitat is the most likely reason for this decline.

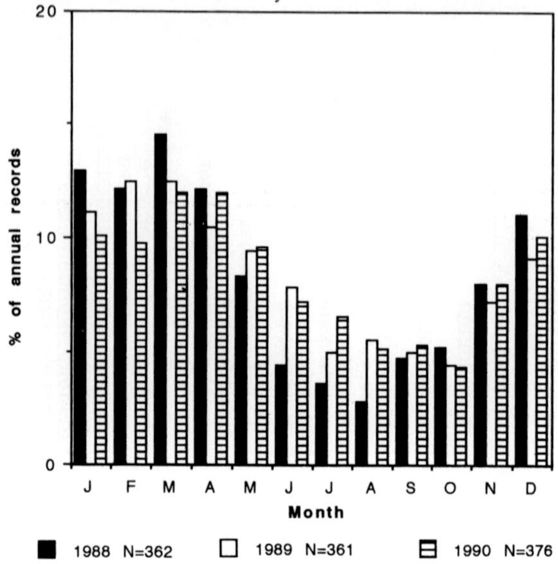

Fig. 204. Distribution 1900–1937

Silvereye

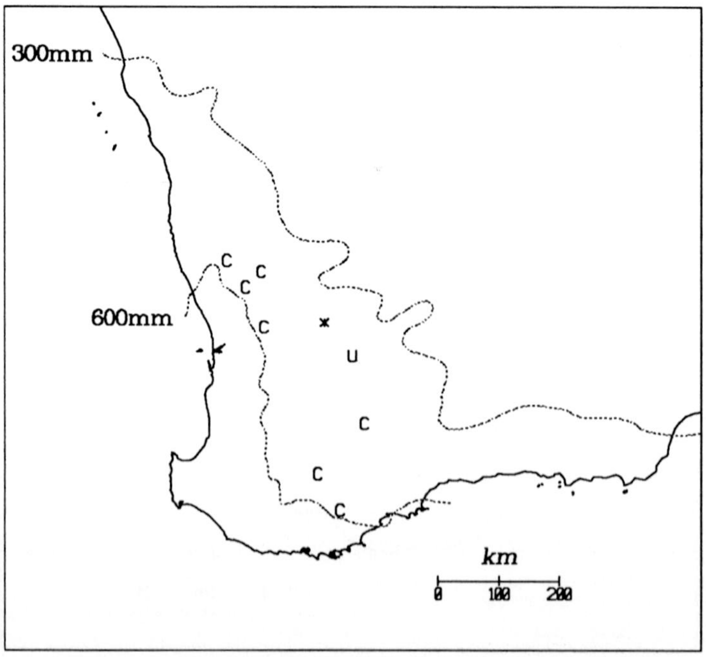

Fig. 204. Distribution 1987–1990

Silvereye

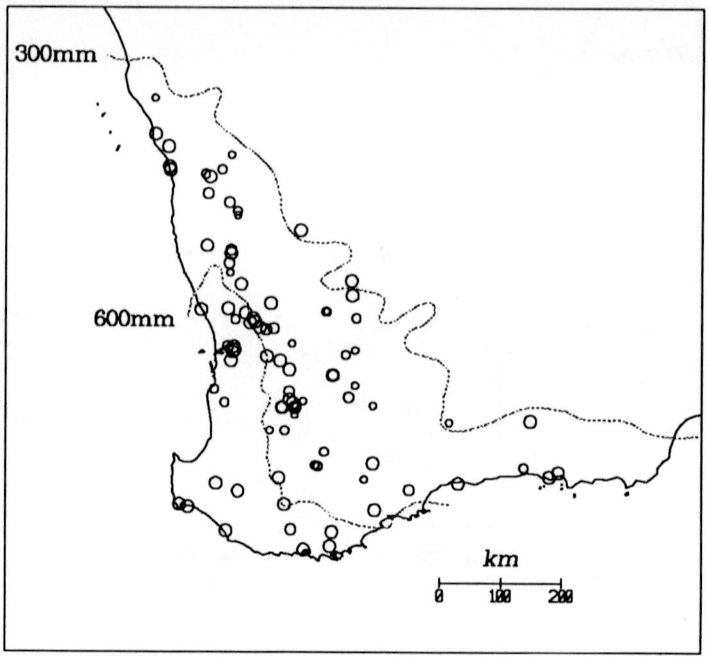

	1900/37	1987	1988	1989	1990
% of observers recording species	90	31	44	39	44
% of recording period species seen	U-C	16	20	22	24

Red-eared firetail

Emblema oculata

The red-eared firetail is an uncommon resident in woodland and heath around the south coast (Fig. 205), and in streamside vegetation in jarrah and karri forests (Serventy and Whittell 1962; Blakers *et al.* 1984). It feeds on seed and insects. It breeds in spring and early summer in a bottle-shaped structure in a tree or shrub and lays 4–6 eggs.

Serventy and Whittell (1962) noted that the red-eared firetail had disappeared from parts of its former range because of extensive clearing of riverside and swamp vegetation.

Fig. 205. Distribution 1987–1990

Red-eared firetail

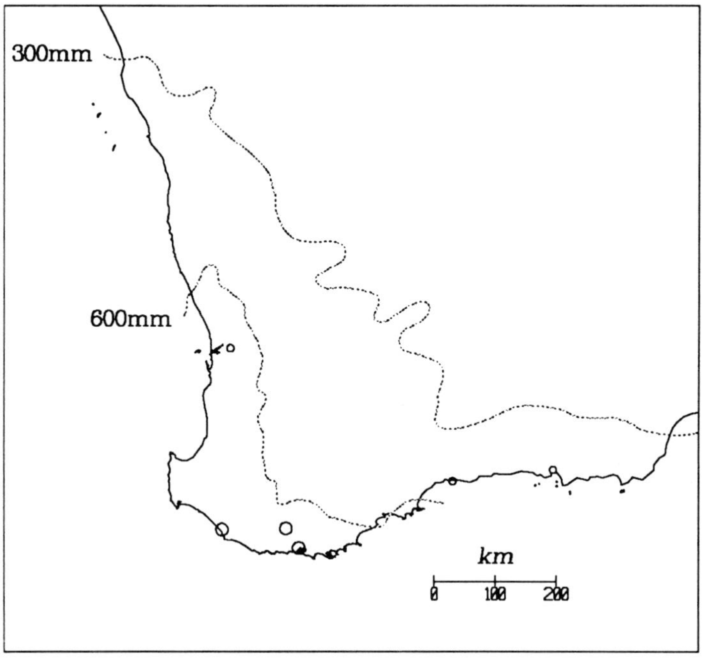

	1900/37	1987	1988	1989	1990
% of observers recording species	0	1.6	2.8	3.4	4.8
% of recording period species seen	NR	0.1	1.7	2.3	2.1

Zebra finch

Poephila guttata

The zebra finch is an uncommon resident in woodland, mallee, shrubland, and on some farmland through much of the wheatbelt (Fig. 206). It does not occur in the southwest forested corner of the state. It feeds mainly on seed, but does eat insects. It breeds in spring and early summer in a dome-shaped nest in a low tree or bush and lays 4–5 eggs.

Being a seed eater, the zebra finch needs to drink fresh water often. The provision of water and increase in grass seed throughout the wheatbelt as part of the development for farming may have favoured it (Blakers *et al.* 1984) and allowed it to increase its population.

Fig. 206. Distribution 1900–1937

Zebra finch

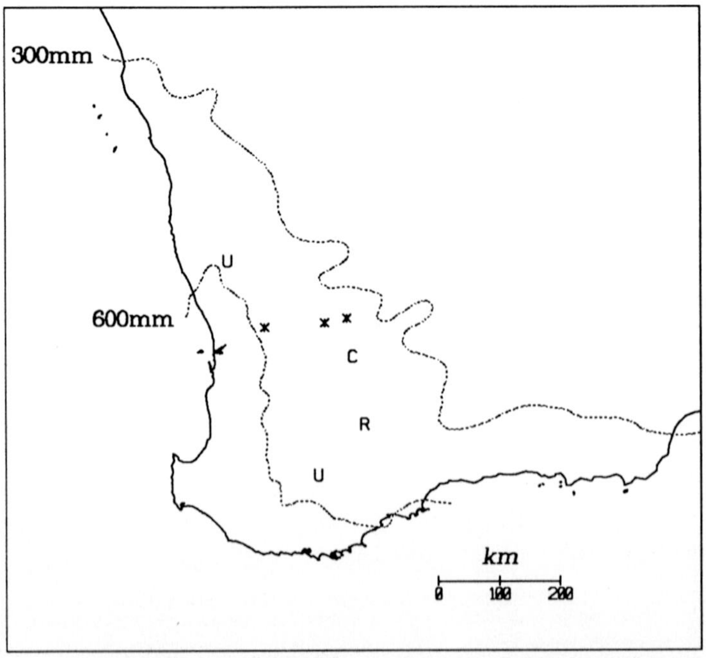

Fig. 206. Distribution 1987–1990

Zebra finch

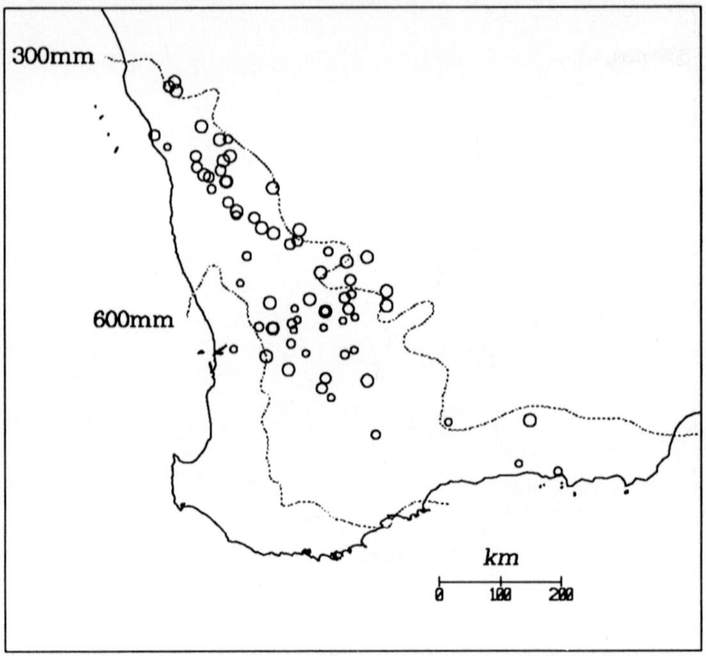

	1900/37	1987	1988	1989	1990
% of observers recording species	70	32	30	37	33
% of recording period species seen	R-C	13	12	16	13

Fig. 207. Distribution 1900–1937

Australian magpie lark

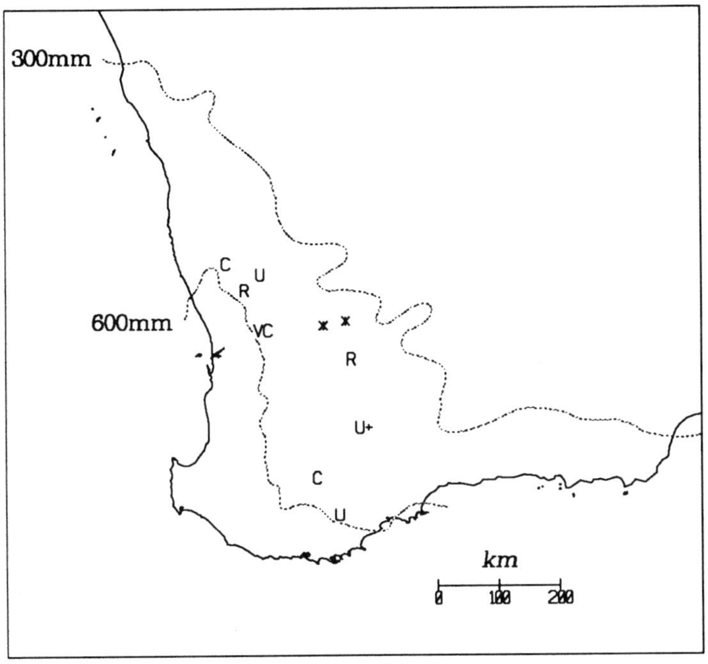

Fig. 207. Distribution 1987–1990

Australian magpie lark

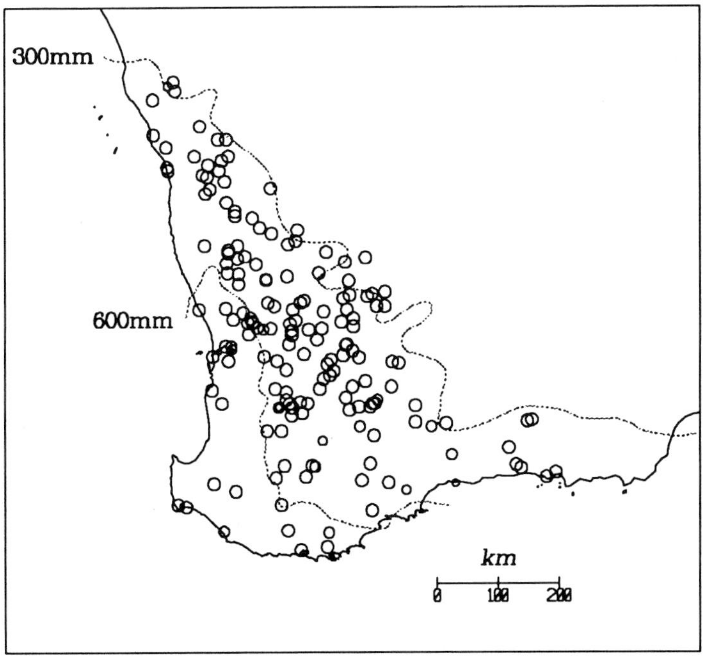

	1900/37	1987	1988	1989	1990
% of observers recording species	100	91	98	97	95
% of recording period species seen	R-VC	76	86	85	85

Australian magpie lark

Grallina cyanoleuca

OTHER NAMES:

Magpie-lark, Mudlark

The Australian magpie lark is a common resident in woodland, farmland and towns throughout the wheatbelt (Fig. 207). It feeds on seed, slugs, snails, millipedes, cockroaches, termites, earwigs, grasshoppers, locusts, bugs, cicadas, beetles, flies, moths, caterpillars, wasps, bees, ants, spiders, ticks and frogs. It breeds in winter and spring in a bowl-shaped nest in a tree, telegraph pole or old shed. It lays 3–5 eggs.

The Australian magpie lark is a bird of open areas and clearing for agriculture has favoured it. The provision of farm dams and other surface water has also been beneficial as it needs mud to build its nest (Serventy and Whittell 1962). Early this century it was uncommon (Milligan 1903). For example, in the Wongan Hills area, it was not numerous and only seen near settlements (Milligan 1904). Over seventy years later it was common in the area, particularly near water (de Rebeira and de Rebeira 1977). Similarly, it was scarce around Lake Grace, but increasing in numbers (Carnaby 1933), as it was around Mogumber (Loaring and Serventy 1952).

Masked woodswallow

Artamus personatus

The masked woodswallow is an uncommon, seasonal nomad in open woodland and farmland in the northern and central wheatbelt (Fig. 208). The masked woodswallow feeds on nectar, pollen, termites, grasshoppers, bugs, cicadas, beetles, moths, caterpillars, wasps, bees and ants. It breeds in spring in a bowl-shaped nest in a low shrub and lays 2–3 eggs. Flocks may appear for some time and disappear. This apparently happened in 1990 when recording rates doubled, as did the percentage of observers recording the species. The seasonal pattern of occurrance of the masked woodswallow is shown by the percentage of annual records each month for the three complete years of the survey (see Fig. below). It was more commonly recorded in the period September to December.

Serventy and Whittell (1962) commented on the propensity of this species to flock into the southwest and breed there if conditions were suitable. Orton and Sandland (1913) noted that it visited the Moora district at odd intervals. It was fairly common around Broome Hill in the summer and occasionally in great numbers (Carter 1923). In 1950 and 1951 large flocks arrived in October in the Kellerberrin/Kwolyin districts, bred and left in January (Ford and Stone 1957); it did not nest in the area every year.

There is no evidence to indicate any change in status of this species in the wheatbelt.

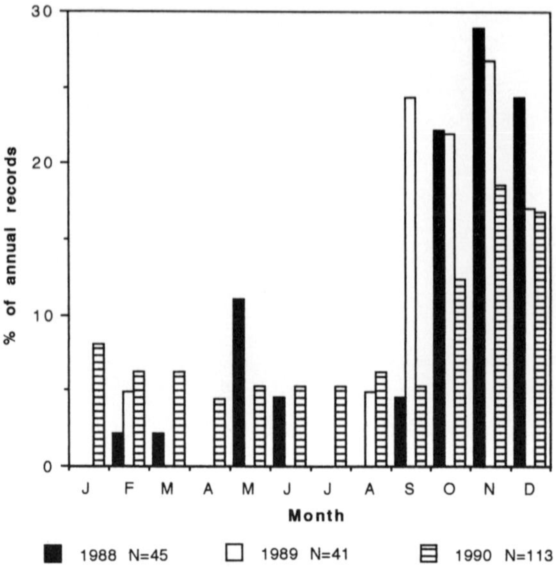

Fig. 208. Distribution 1900–1937

Masked woodswallow

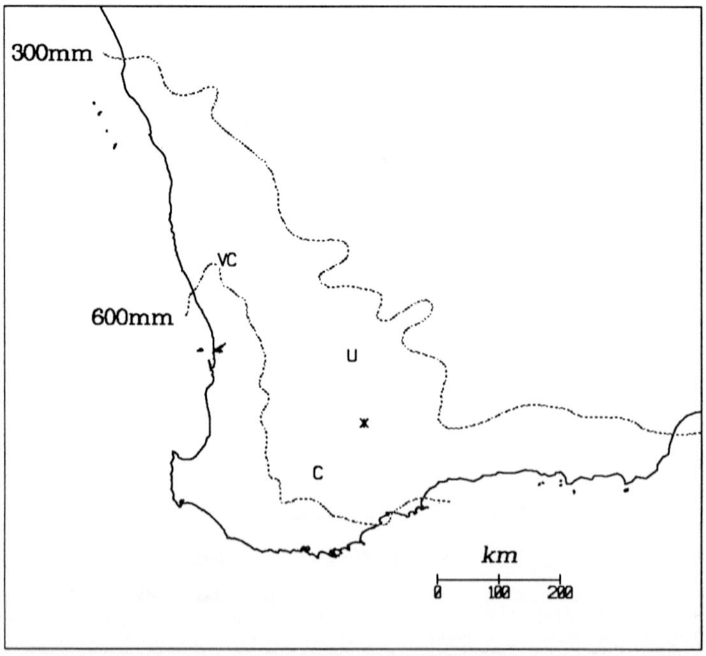

Fig. 208. Distribution 1987–1990

Masked woodswallow

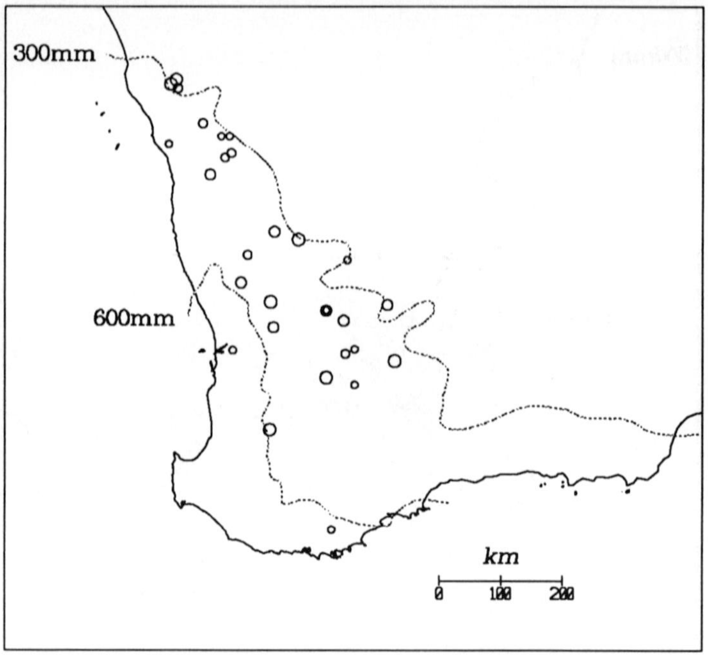

	1900/37	1987	1988	1989	1990
% of observers recording species	40	9.4	7.7	6.8	18
% of recording period species seen	U-VC	4.2	1.7	1.8	3.3

Fig. 209. Distribution 1900–1937

Black-faced woodswallow

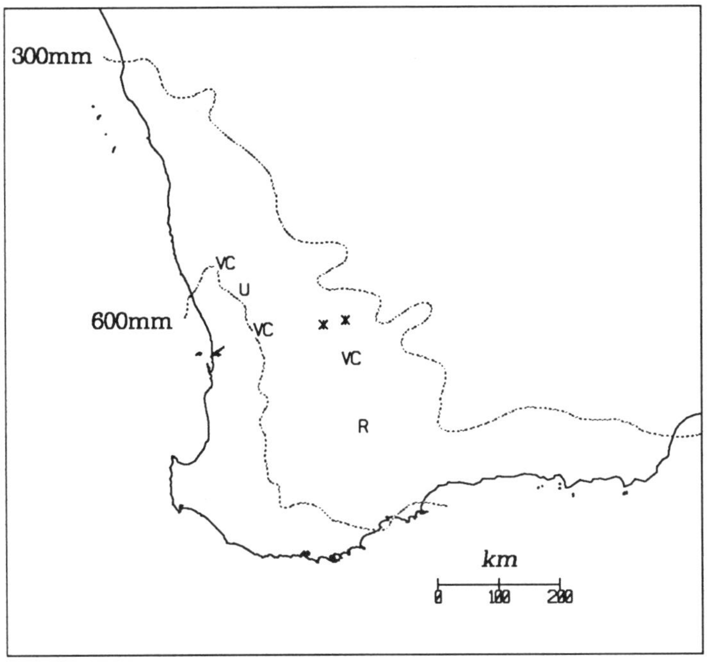

Fig. 209. Distribution 1987–1990

Black-faced woodswallow

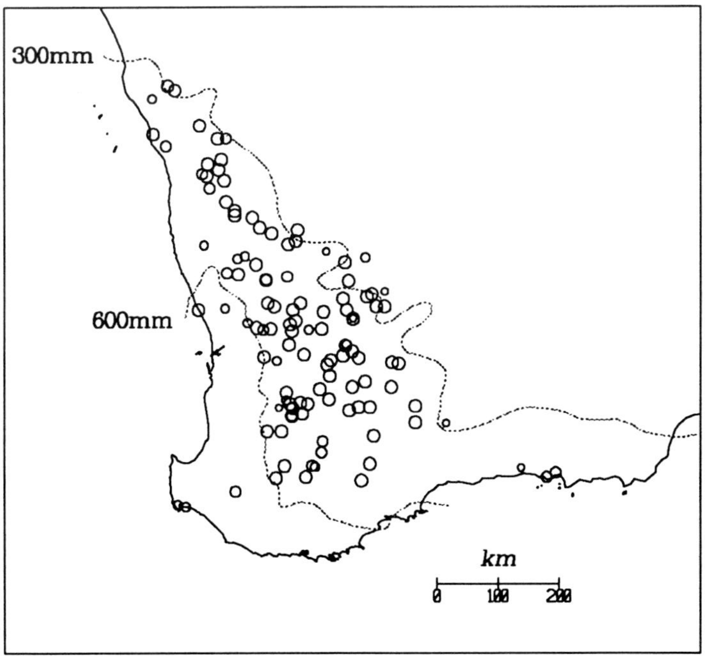

	1900/37	1987	1988	1989	1990
% of observers recording species	70	57	59	59	59
% of recording period species seen	R-VC	40	39	42	44

Black-faced woodswallow

Artamus cinereus

The black-faced woodswallow is the most common woodswallow in the wheatbelt (Fig. 209) occuring in woodland, mallee, shrubland, farmland and towns. It feeds on seed, nectar, centipedes, damselflies, cockroaches, termites, grasshoppers, crickets, cicadas, beetles, flies, moths, caterpillars, wasps, bees, ants, scorpions and spiders. The black-faced woodswallow breeds in spring and summer in a saucer-shaped nest in a tree or shrub and lays 3–4 eggs.

Clearing for agriculture has created open country favoured by this species. Serventy and Whittell (1962) noted that in the past the black-faced woodswallow was much more restricted, and that it did not range far south of a line through Northam to Merredin. Now it occurs through most of the southern wheatbelt.

Little woodswallow

Artamus minor

The little woodswallow is a bird of the arid zone which is a rare vagrant in the extreme north of the wheatbelt (Fig. 210), being seen by one observer once in 1987 and 1988. It feeds on beetles, flies, mosquitoes, wasps, bees and ants.

Fig. 210. Distribution 1987–1990

Little woodswallow

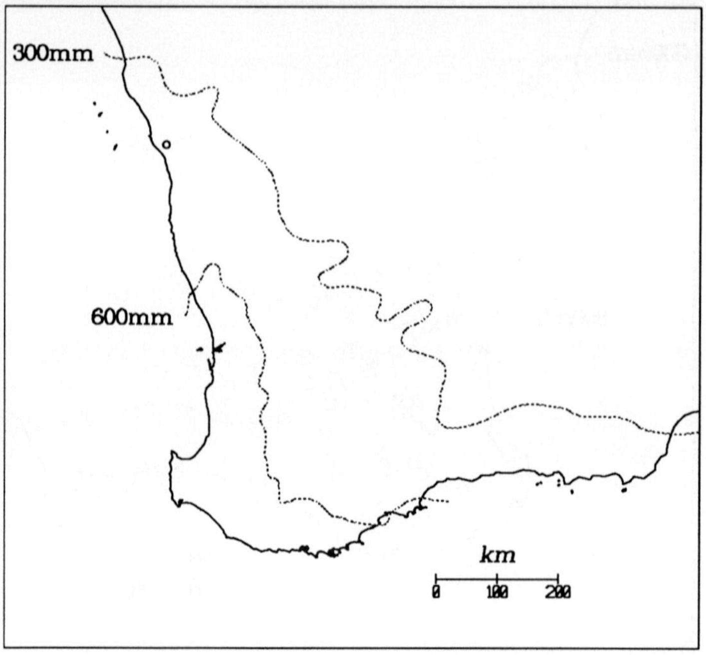

	1900/37	1987	1988	1989	1990
% of observers recording species	0	0.8	0.7	0	0
% of recording period species seen	NR	0.03	0.02	0	0

Fig. 211. Distribution 1900–1937

Dusky woodswallow

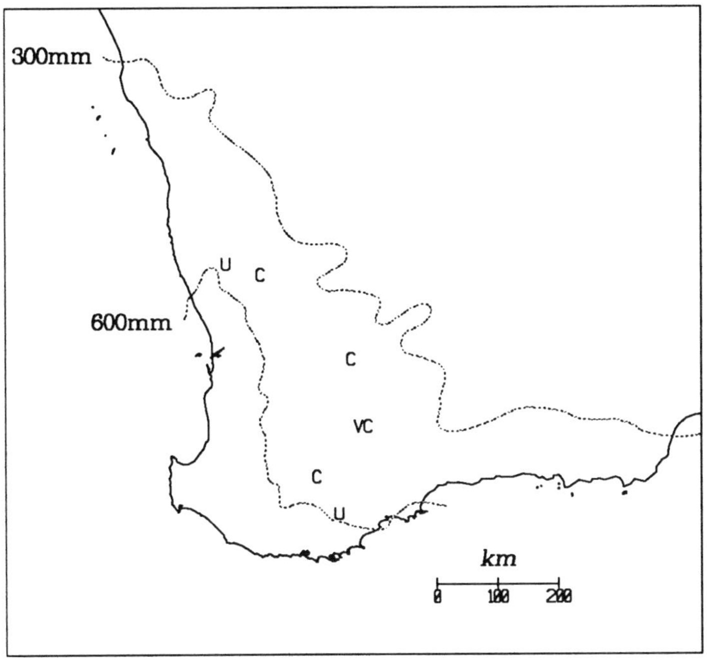

Fig. 211. Distribution 1987–1990

Dusky woodswallow

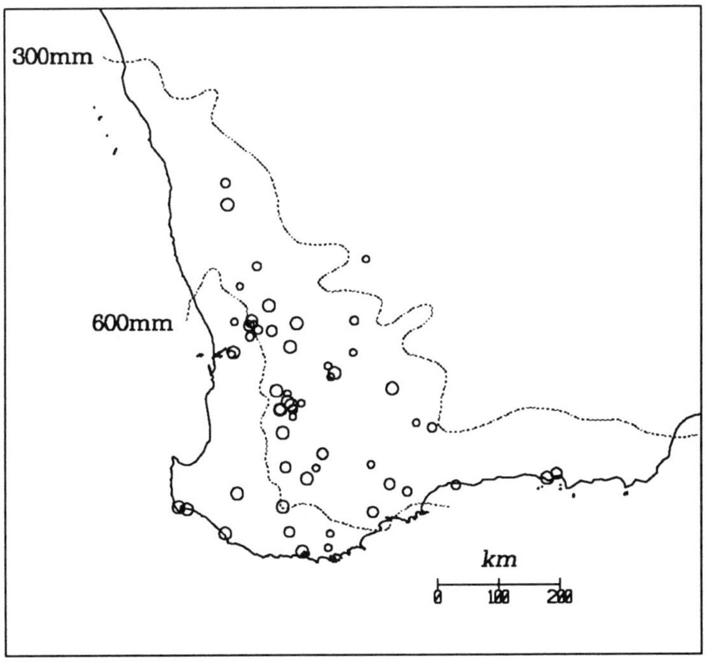

	1900/37	1987	1988	1989	1990
% of observers recording species	60	17	25	27	30
% of recording period species seen	U-VC	9.9	14	12	12

Dusky woodswallow

Artamus cyanopterus

The dusky woodswallow is a common resident of woodland in the higher rainfall districts of the wheatbelt (Fig. 211), and a nomad on the drier edges of this range. It feeds on seed, berries, grasshoppers, locusts, crickets, beetles, flies, caterpillars, wasps, bees, ants and spiders. It breeds in spring and summer in a basket-shaped nest in a tree or stump and lays 3–4 eggs.

Serventy and Whittell (1962) noted that in the past, this was the common woodswallow of the southwest and that the range had been greatly reduced. Formerly it occurred in Geraldton, Perenjori and Kalgoorlie. The removal of native vegetation has probably disadvantaged this species, while favouring the black-faced woodswallow that has replaced it (Serventy and Whittell 1962).

Grey butcherbird

Cracticus torquatus

The grey butcherbird is a common resident of woodland and mallee throughout the wheatbelt (Fig. 212). It feeds on berries, seed, fruit, cockroaches, mantids, earwigs, grasshoppers, bugs, cicadas, beetles, flies, moths, caterpillars, wasps, bees, ants, spiders, as well as small lizards, birds and mammals. The grey butcherbird breeds in late winter and spring in a cup-shaped nest in a tree and lays 3–4 eggs.

The grey butcherbird occurs mainly in remnant vegetation, including well-wooded areas on farmland. It is possible that this species has declined in abundance in the more extensively cleared parts of its range as a result of loss of habitat. For example, Milligan (1904) listed this species as fairly numerous during his return trip from Mogumber to the Wongan Hills, but de Rebeira and de Rebeira (1977) recorded the grey butcherbird as widespread, but uncommon. Similarly, Masters and Milhinch (1974) reported that in Northam Shire the grey butcherbird had decreased considerably over the previous 25 years.

Fig. 212. Distribution 1900–1937

Grey butcherbird

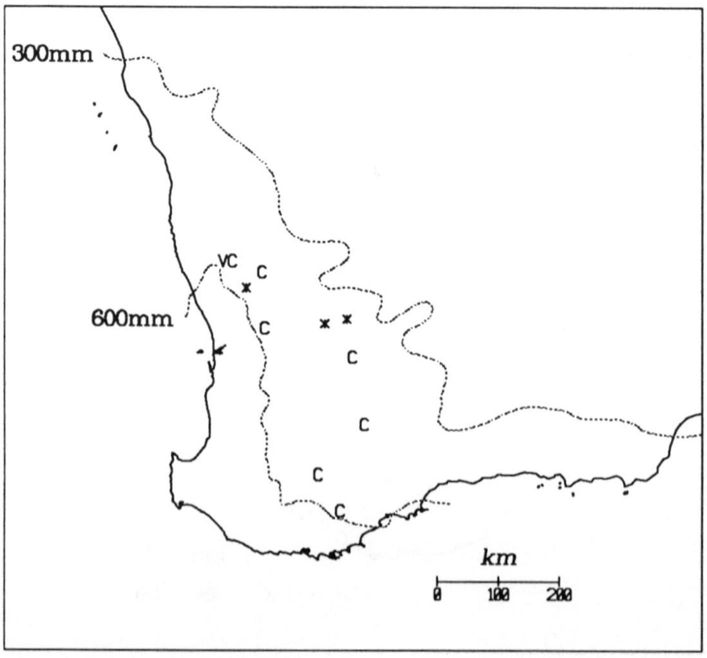

Fig. 212. Distribution 1987–1990

Grey butcherbird

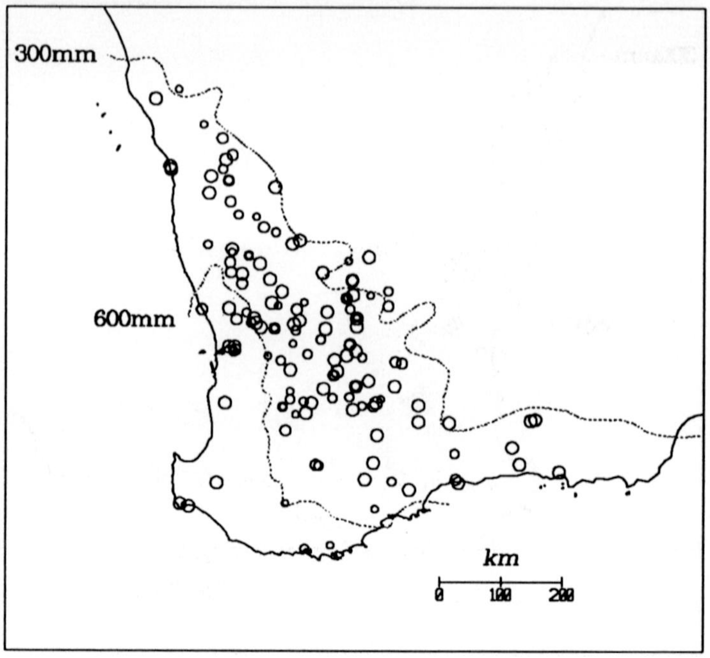

	1900/37	1987	1988	1989	1990
% of observers recording species	100	59	70	64	61
% of recording period species seen	C-VC	25	30	32	33

Fig. 213. Distribution 1900–1937

Pied butcherbird

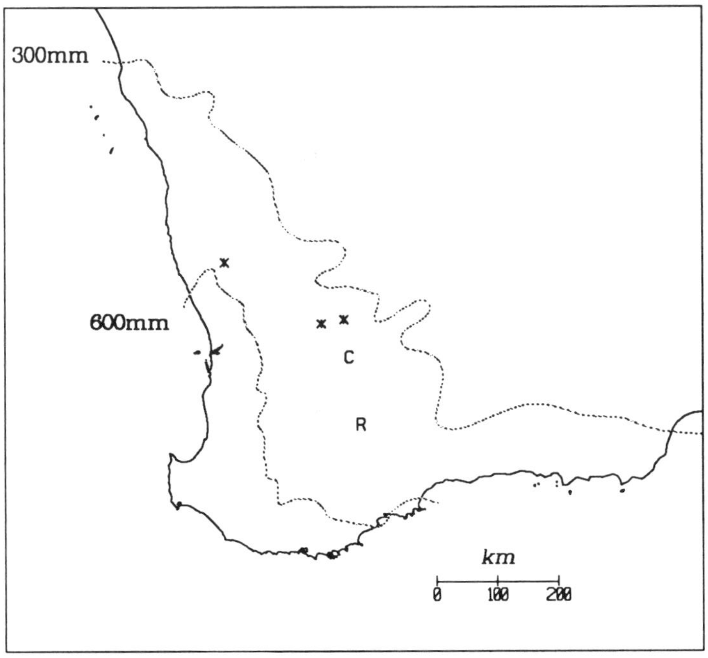

Fig. 213. Distribution 1987–1990

Pied butcherbird

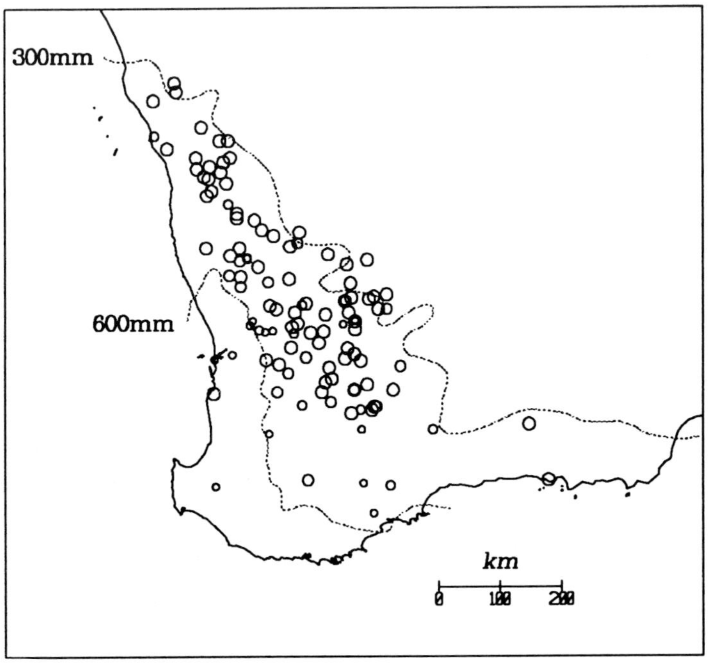

	1900/37	1987	1988	1989	1990
% of observers recording species	50	61	58	57	57
% of recording period species seen	R-C	40	39	35	35

Pied butcherbird

Cracticus nigrogularis

The pied butcherbird is a common resident of open woodland and farmland throughout the northern, central and parts of the southern wheatbelt (Fig. 213). It feeds on worms, mayflies, dragonflies, termites, grasshoppers, cicadas, beetles, caterpillars, ants, ostracods, spiders, frogs, and small lizards, birds and mammals. The pied butcherbird breeds in spring in a cup-shaped nest in a tree and lays 3–4 eggs.

The pied butcherbird prefers open woodland, and has been slowly extending its range into the southwest as a result of clearing for agriculture. It invaded the Kellerberrin district in 1899 (Leake 1962) and colonized the Moora district in 1912 (Orton and Sandland 1913). Serventy and Whittell (1962) stated that the southern extent of its distribution was Gingin, near Toodyay and Lake Grace. It is well south of those locations and will continue to extend its range until it colonizes all suitable open woodland habitat.

Australian magpie

Gymnorhina tibicen

The Australian magpie is one of the best recognized birds in the wheatbelt. It is a common resident in open woodland and on farmland throughout the wheatbelt (Fig. 214). It feeds on seed and other plant material, earthworms, snails, millipedes, cockroaches, mantids, grasshoppers, crickets, phasmids, cicadas, beetles, moths, caterpillars, bees, ants, crustacea, spiders, frogs, and small lizards, birds and mammals. The Australian magpie breeds in spring in a basket-shaped stick nest in a tree and lays 2–4 eggs.

There is evidence to suggest that the status of the Australian magpie in the wheatbelt has changed. Clearing for agriculture may have benefited this species by providing more food and suitable open habitat. For example, early this century it was numerous around the mission station at New Norcia, but less common inland on the route to the Wongan Hills (Milligan 1904). Seventy years later, de Rebeira and de Rebeira (1977) noted it was present around the Wongan Hills in groups up to 12 birds. Sedgwick (1974) recorded that the Australian magpie colonized newly-cleared farmland in the eastern wheatbelt within the first year after clearing commenced.

Fig. 214. Distribution 1900–1937

Australian magpie

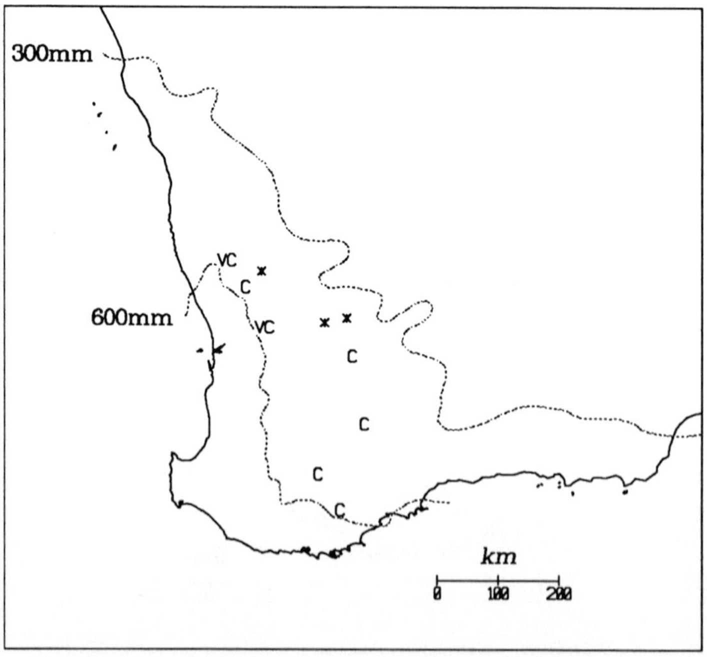

Fig. 214. Distribution 1987–1990

Australian magpie

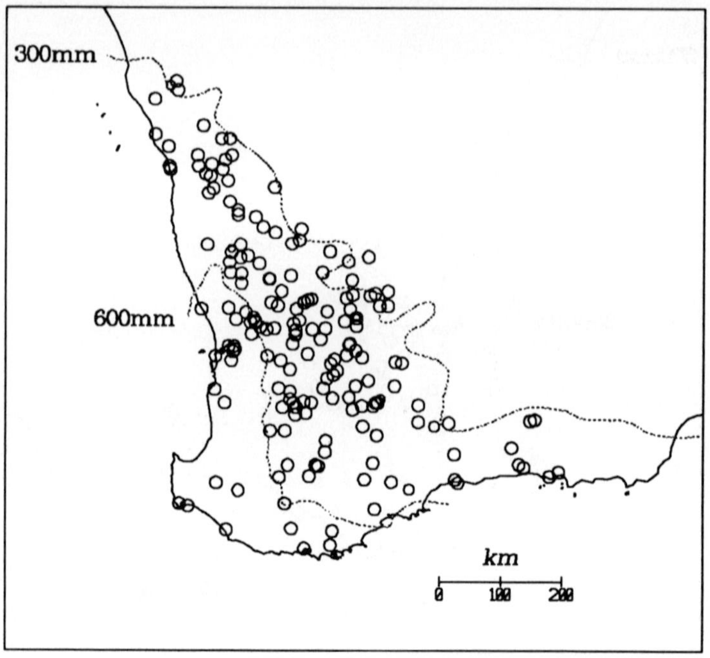

	1900/37	1987	1988	1989	1990
% of observers recording species	100	94	96	99	97
% of recording period species seen	C-VC	84	91	93	93

Fig. 215. Distribution 1900–1937

Grey currawong

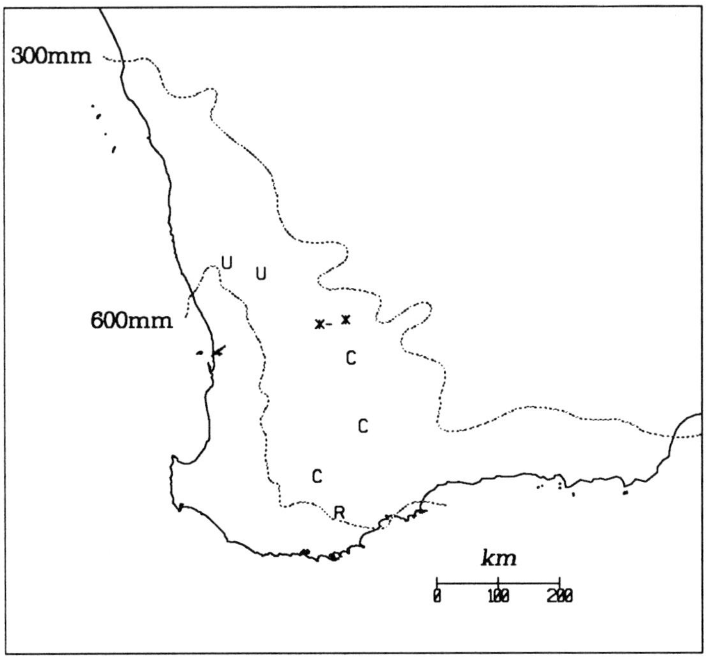

Fig. 215. Distribution 1987–1990

Grey currawong

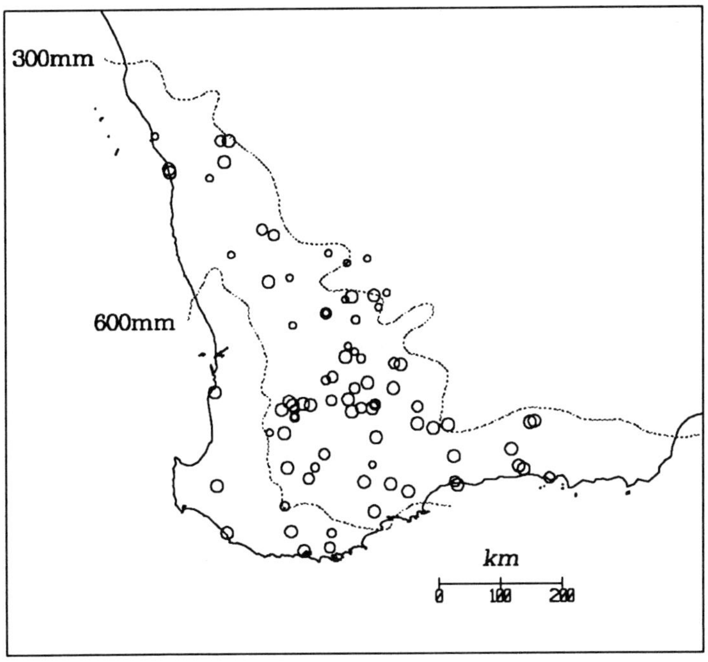

	1900/37	1987	1988	1989	1990
% of observers recording species	80	36	42	33	35
% of recording period species seen	R-C	16	19	17	17

Grey currawong

Strepera versicolor

OTHER NAME:

Squeaker

The grey currawong is an uncommon resident of woodland and mallee in the higher rainfall districts of the wheatbelt (Fig. 215), and a rare nomad in the drier more extensively cleared districts. It feeds on fruit, seed, berries, cockroaches, earwigs, grasshoppers, beetles, ants, wood lice, spiders, frogs, and small lizards, birds and mammals. It breeds in late winter and spring in a cup-shaped nest in a tree and lays 2–4 eggs.

The grey currawong has declined in range and abundance in the wheatbelt. For example, Leake (1962) noted that it was declining at Kellerberrin due to clearing for agriculture; it has disappeared from there. It has also disappeared from Merredin (B. Willis, in litt.), Ogilvie (K. Chick, in litt.) and is declining around Northam (J. Masters, in litt.).

Australian raven

Corvus coronoides

The Australian raven is a common resident of open woodland and farmland throughout the wheatbelt (Fig. 216). It feeds on fruit, seed, grain, flowers, berries, snails, millipedes, centipedes, cockroaches, termites, mantids, earwigs, grasshoppers, crickets, stick insects, lice, bugs, cicadas, beetles, moths, caterpillars, cutworms, ants, slaters, spiders, ticks, fish, frogs, lizards, birds and their eggs, mammals and carrion. It breeds in late winter and early spring in a cup-shaped stick nest in a tree and lays 4–5 eggs.

The Australian raven has benefited greatly from development for agriculture. Milligan (1903) saw no corvids during a 14-day return journey from Tenterden to the Stirling Range in the spring of 1902. It is common there now (Fig. 216). The raven was rarely seen around Broome Hill in the early 1900s. By 1922 it was numerous and settlers were complaining about raven numbers and damage (Carter 1923). The raven was rare around Lake Grace until sheep raising increased in the area (Carnaby 1933).

Fig. 216. Distribution 1900–1937

Australian raven

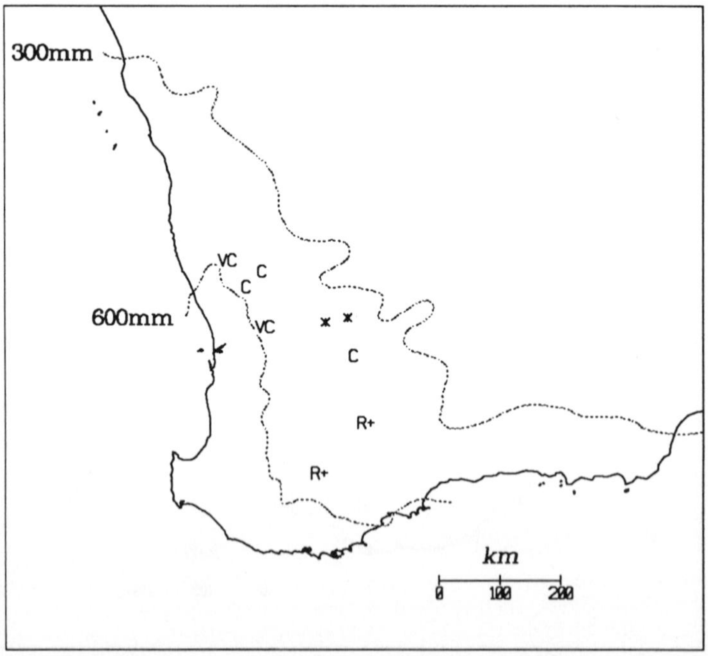

Fig. 216. Distribution 1987–1990

Australian raven

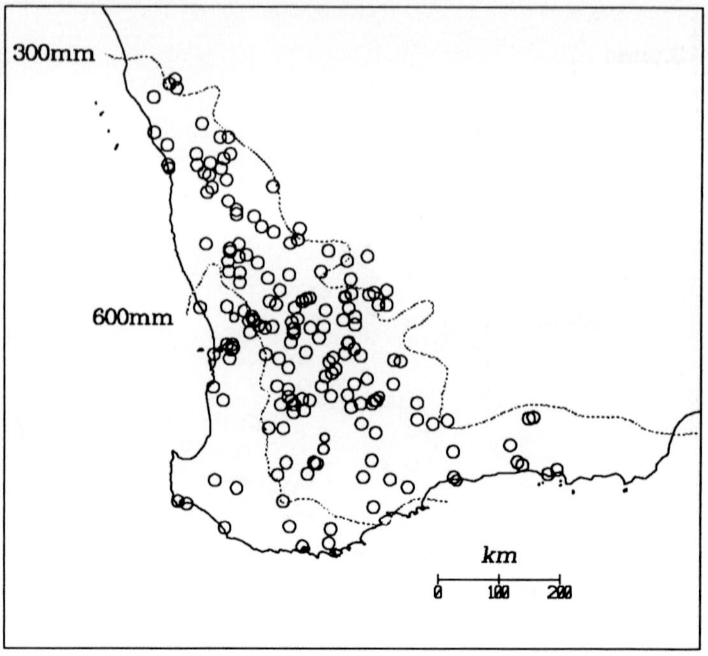

	1900/37	1987	1988	1989	1990
% of observers recording species	90	97	97	98	94
% of recording period species seen	R-VC	88	92	92	91

Fig. 217. Distribution 1900–1937

Little crow

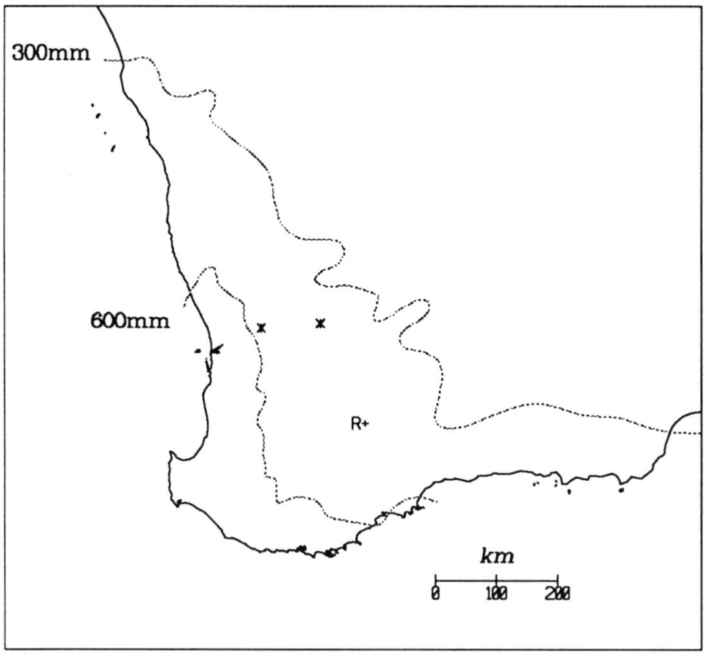

Fig. 217. Distribution 1987–1990

Little crow

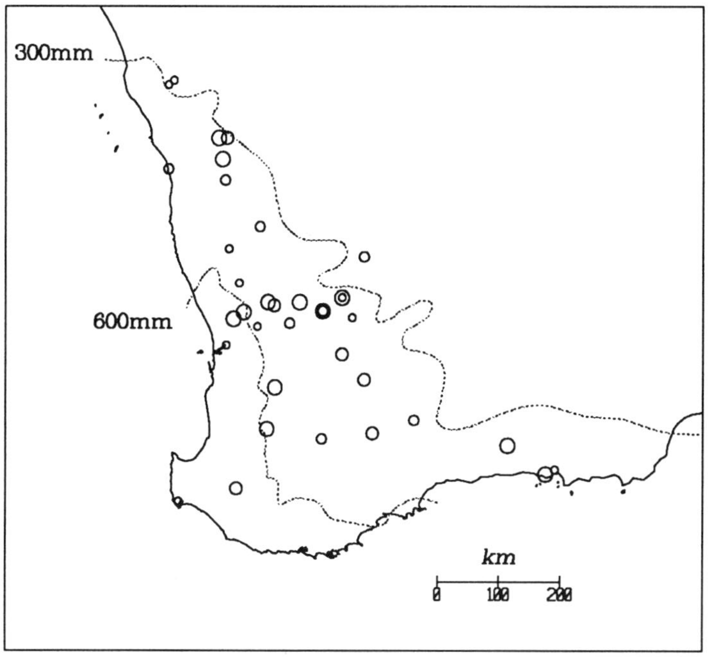

	1900/37	1987	1988	1989	1990
% of observers recording species	30	17	15	16	16
% of recording period species seen	R	6.9	4.0	3.8	3.5

Little crow

Corvus bennetti

The little crow and Australian raven are confused by those not well acquainted with their differences. As a result, the little crow may have been under-sampled during this survey. It is an uncommon seasonal migrant in open woodland and farmland of the wheatbelt (Fig. 217), although this seasonality is variable. This is illustrated by the percentage of annual records each month for the period 1988 to 1990 (see Fig. below). During 1988 and 1990 the little crow was more common in the wheatbelt from October to May, however, in 1989 it was apparently equally common throughout the year. Analyses of its distribution at different times of year did not reveal a clear pattern of movement.

It feeds on seed, fruit, berries, tubers, earthworms, millipedes, centipedes, crickets, cockroaches, grasshoppers, cicadas, beetles, moths, caterpillars, ants, spiders, frogs, and small lizards, birds and their eggs as well as carrion. The little crow breeds in spring in a basket-shaped stick nest in a tree and lays 3–4 eggs.

There is little evidence to suggest that the status of the little crow has changed, although numbers of the little crow on migration in the Avon Valley have decreased markedly in the last 20 years (J. Masters, in litt.).

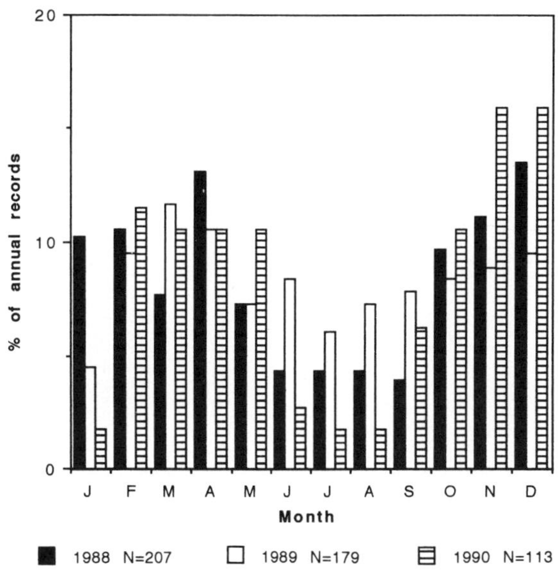

(Top photo) Mist nets are used to catch birds to study their ecology and particularly their movements. It is necessary to understand what habitat characteristics are required to allow particular species to survive and move around fragmented agricultural landscapes so that habitat reconstruction plans can be developed for species that are threatened in such landscapes (photo by D. A. Saunders).

(Bottom photo) Striated pardalote caught in a mist net prior to being banded with an aluminium leg band (photo by D. A. Saunders).

chapter five

Changes in distribution and/or abundance of birds of the wheatbelt

IT is not possible to remove over 90 per cent of the original area of native vegetation from any region without having major effects on the flora and fauna. The data presented in this book make it clear that the avifauna of the wheatbelt has not escaped these effects. A summary of the changes in range and/or abundance, together with diet, habitat prefences, nest site and status of all species recorded from the wheatbelt is presented in Table 1. An examination of these data shows that there has been a massive reduction in the range and/or abundance of a large proportion of wheatbelt birds. Excluding 14 species which are vagrants, 195 species of bird have been recorded from the region since settlement by Europeans. Ninety-five (49%) of these have declined in range and/or abundance since the turn of the century, 34 (17%) have increased, and no change could be demonstrated for the remaining 66 (34%) species. When nearly half of the avifauna has been adversely affected and only 17 per cent have benefited, there is an urgent need to take stock of the situation from a conservation point of view, identify the species and habitats on which we should concentrate our resources, and develop guidelines for reversing the situation. We need to establish how to manage our agricultural landscapes so that we take the principles of ecologically sustainable development into account and integrate management for nature conservation into management for agricultural production across the wheatbelt.

Twenty-one species of non-passerine and 13 species of passerine (perching or songbirds) have increased in range and/or abundance over the past 90 years. Three (9% of the species which have increased) are exotic. The domestic pigeon and laughing turtledove were brought in from overseas and probably established via escapes from aviaries (domestic pigeon), or from deliberate releases from Perth or country towns (laughing turtledove) (Long 1981). The laughing kookaburra was brought in from southeastern Australia and released in Perth and a number of country centres (Long 1981). The pigeon and turtledove are commensals and do not occur far from towns or farm buildings, although the former does nest in hollows in trees on patches of remnant vegetation. The extent to which these species compete with native species is unknown, although the pigeon and dove are preyed upon by raptors, and the kookaburra preys on native vertebrates including birds.

Seven species (21%) which have increased are birds of the arid zone which have invaded the wheatbelt because of the creation of suitable habitat. Widespread clearing of shrublands and woodlands to create open plains with associated patches of remnant vegetation, scattered trees, provision of water (stock troughs, dams), and food (cereal crops, pasture grasses, weed seed, insects in agricultural land, and mice) have suited species like the black-shouldered kite (the only raptor which has increased in range), crested pigeon, red-tailed black cockatoo, galah, little corella, mulga parrot, and pied butcherbird. Some of these species are still expanding their ranges.

The creation of this open "riverine plain" has allowed 14 species (41%) which favour open habitats, and which formerly occurred in the wheatbelt, or parts of it, to increase in numbers and/or range. These are the mountain duck, wood duck, Australian dotterel, long-billed corella, Port Lincoln parrot, elegant parrot, Richard's pipit, willie wagtail, white-fronted chat, zebra finch, Australian magpie lark, black-faced woodswallow, Australian magpie, and Australian raven.

The changes in status of the mountain duck and wood duck are of interest because they raise a point of conservation concern. A combination of creation of farm dams and an ability to exist on water bodies that have become saline as a result of rising water tables, has allowed the mountain duck to benefit from changing land use. The wood duck has also benefited from the provision of farm dams and is also widespread and common. Both species nest in hollows in trees which are still available throughout the wheatbelt, although as many old trees die and blow over, fall down or are deliberately destroyed, there may be a shortage of nest hollows. It will be a long time before there are sufficient hollows in trees being planted to replace them, as it may take a salmon gum somewhere over 130 years to grow large enough to accommodate a hollow of sufficient size for a duck or a cockatoo (Saunders *et al.* 1982). Unless this issue is addressed specifically in revegetation plans at the landscape scale, there will be a critical shortage of nest sites for species which are obligate hole nesters. This will have more impact on non-passerines (particularly parrots and owls) as they account for 25 of the 27 species known to nest only in tree hollows.

Table 1. Summary of change in range and/or abundance, diet, habitat preferences, nest site and status of species recorded during this and earlier surveys of birds of the wheatbelt of Western Australia. Change: 0 = no change in range and/or abundance or not enough information to assess change; — = decrease in range and/or abundance; + = increase in range and/or abundance. Diet: Pl = plant material (leaves, stems, roots, etc.); Fr = fruit; Ne = nectar; Se = seed; Po = pollen; In = invertebrates; Fi = fishes; Fro = frogs; Re = reptiles; Bi = birds; Ma = mammals; and Ca = carrion. Habitat: W = woodland; M = mallee; Sh = shrubland; He = heath; Aq = aquatic; Fa = farmland; and To = town. Status: BM = breeding migrant; M = migrant; N = nomad; R = resident; and V = vagrant.

Species	Change	Diet	Habitat	Nest	Status
Non-passerines					
Emu	—	Pl, Fr, Se, In	W, M, Sh, He, Fa	Ground	R
Hoary-headed grebe	—	Se, In, Fi	Aq	Raft on water	N/R
Australasian little grebe	0	Pl, Se, In, Fi	Aq	Raft on water	N/R
Great crested grebe	?	Pl, Se, In, Fi	Aq	Raft on water	V
Australian pelican	0	In, Fi, Bi	Aq	Ground	N
Darter	0	Pl, In, Fi	Aq	Tree-stick nest	N
Little black cormorant	0	Pl, In, Fi, Fro	Aq	Tree/shrub-stick nest	N
Black cormorant	0	In, Fi, Fro, Bi	Aq	Tree/shrub/ground-stick nest	N
Little pied cormorant	—	In, Fi, Fro	Aq	Tree-stick nest	R/N
White-necked heron	0	In, Ma	Aq, Fa	Tree-stick nest	N
White-faced heron	—	In, Fi, Fro, Re, Ma	Aq, Fa	Tree-stick nest	R
Nankeen night heron	0	Pl, Se, In, Fi, Fro, Bi, Ma	Aq	Tree/shrub/ground-stick nest	N
Cattle egret	+	In	Fa	Tree-stick nest	N
Large egret	+	In, Fi, Fro, Bi	Aq	Tree-stick nest	N
Glossy ibis	0	In, Fro	Aq	Not recorded in area	V
White ibis	+	In, Fi, Fro, Re, Ma	Aq	Tree/shrub/ground-stick nest	N
Straw-necked ibis	?	Se, In, Fi, Fro, Re, Ma	Aq, Fa	Tree/shrub-stick nest	N
Australian brown bittern	—	Pl, In, Fro	Aq	Reeds	N
Royal spoonbill	?	Pl, Se, In, Fi	Aq	Not recorded in area	V
Yellow-billed spoonbill	+	Pl, In, Fi	Aq	Tree/shrub/rushes-stick nest	N
Black swan	—?	Pl, Se, In	Aq	Ground	N
Freckled duck	0	Pl, Se, In, Fi	Aq	Ground	N
Mountain duck	+	Pl, Se, In	Aq, Fa	Tree hollow	R/BM
Black duck	—	Pl, Se, In, Fi	Aq	Ground/tree hollow	R/N
Grey teal	0	Pl, Se, In, Fi	Aq	Ground/tree hollow	R/N
Chestnut teal	0	Pl, Se, In	Aq	Ground/tree hollow	N
Shoveller	0	Pl, Se, In	Aq	Ground	N
Pink-eared duck	0	Pl, Se, In	Aq	Ground/shrub	N
White-eyed duck	0	Pl, Se, In	Aq	Tree/shrub	N
Wood duck	+	Pl, Se, In	Aq, Fa	Tree hollow	R/BM
Blue-billed duck	0	Pl, Se, In	Aq	Ground	N
Musk duck	0	Pl, Se, In. Fi, Fro, Bi	Aq	Ground	N
Black-shouldered kite	+	In, Re, Bi, Ma	W, Fa	Tree-stick nest	R
Square-tailed kite	0	In, Re, Bi	W, M, He	Tree-stick nest	N
Black kite	0	In, Fi, Fro, Re, Bi, Ma, Ca	W, He, Fa	Not recorded in area	V
Whistling kite	—	In, Fi, Re, Bi, Ma, Ca	W, He, Aq, Fa	Tree-stick nest	R
Brown goshawk	—	In, Fro, Re, Bi, Ma	W, Sh, Fa	Tree-stick nest	R
Collared sparrowhawk	—	In, Re, Bi, Ma	W, Sh, Fa	Tree-stick nest	R
Wedge-tailed eagle	0	Re, Bi, Ma, Ca	W, Sh, He, Fa	Tree-stick nest	R
Little eagle	—	In, Re, Bi, Ma	W, Sh, He, Fa	Tree-stick nest	R
Spotted harrier	0	In, Re, Bi, Ma	Sh, He, Fa	Tree-stick nest	R
Swamp harrier	—	Se, In, Fi, Re, Bi, Ma, Ca	Aq	Ground	R
Black falcon	0	In, Bi, Ma	W	Not recorded in area	V
Peregrine falcon	0	Bi, Ma	W, Sh, He, Fa	Tree hollow/cliff	R
Little falcon	—	In, Bi, Ma	W, Sh, He, Fa	Tree-stick nest	R
Grey falcon	0	Se, In, Bi, Ma	?	Not recorded in area	V
Brown falcon	—	In, Fi, Re, Bi, Ma, Ca	W, Sh, He, Fa	Tree-stick nest	R
Nankeen kestrel	—	In, Fro, Re, Bi, Ma	W, Sh, He, Fa	Tree hollow/tree-stick nest	R
Mallee fowl	—	Pl, Se, Fr, In	M, Sh	Ground	R

Table 1 — continued

Species	Change	Diet	Habitat	Nest	Status
Stubble quail	+	Se, In, Fro	Sh, He, Fa	Ground	R
Brown quail	0	Pl, Se, In, Re	Edge of creeks, etc.	Ground	N
Painted button quail	—	Se, Fr, In	W, Sh, He	Ground	R
Little button quail	0	Se, In	W, Sh, He, Fa	Ground	M/N
Spotted crake	?	Se, In	Aq	Not recorded in area	?
Marsh crake	?	Se, In	Aq	Reeds	?
Black-tailed native hen	0?	Pl, Se, In	Aq	Ground	N
Dusky moorhen	0	Pl, Se, In	Aq	Ground	R/N
Swamp hen	—?	Pl, Se, In, Fi, Bi	Aq	Ground	N
Coot	0	Pl	Aq	Ground	R
Bustard	—	Pl, Se, In, Re, Bi, Ma	W, He, Fa	Ground	R/N
Bush stone curlew	—	Pl, Se, In, Fro, Re	W	Ground	R
Banded plover	0	Pl, Se, In	Fa	Ground	R
Red-kneed dotterel	0	Se, In	Aq	Ground	N
Hooded dotterel	0	In	Salt lake	Ground	N
Red-capped dotterel	0	Pl, In	Aq	Ground	R
Black-fronted dotterel	—	Se, In	Aq	Ground	R
Australian dotterel	+	Pl, Se, In	Fa	Ground	R
Pied stilt	+?	Se, In	Aq	Ground	R/N
Banded stilt	0	In	Salt lake	Ground	N
Red-necked avocet	0	Pl, Se, In	Salt lake	Ground	N
Wood sandpiper	0	In	Aq	Northern Hemisphere	M
Greenshank	0	In, Fi, Fro	Aq	Northern Hemisphere	M
Marsh sandpiper	0	In	Aq	Northern Hemisphere	M
Common sandpiper	0	In	Aq	Northern Hemisphere	M
Sharp-tailed sandpiper	0	Se, In	Aq	Northern Hemisphere	M
Red-necked stint	0	Se, In	Aq	Northern Hemisphere	M
Long-toed stint	0	In	Aq	Northern Hemisphere	M
Curlew sandpiper	0	Pl, Se, In	Aq	Northern Hemisphere	M
Silver gull	+	Omnivore	Aq	Ground	V
Whiskered tern	0	In, Fi, Fro, Bi, Ma	Aq	Raft on water	N
Domestic pigeon	+(exotic)	Se	W, Fa, To	Tree hollow/tree	R
Laughing turtledove	+(exotic)	Se, Fr	W, Fa, To	Tree-stick nest	R
Diamond dove	0	Se	W	Tree-stick nest	N
Brush bronzewing	—	Se	Sh, He	Tree/shrub-stick nest	R
Common bronzewing	—	Se, Fr, In	W, M, Sh	Tree/shrub-stick nest	R
Crested pigeon	+	Se, Pl, In	W, M, Sh, He, Fa	Tree	R
Red-tailed black cockatoo	+	Se	W, Fa	Tree hollow	R
Carnaby's cockatoo	—	Se, In	W, Sh, Fa	Tree hollow	R/BM
Galah	+	Se, Fr, In	W, M, Fa	Tree hollow	R
Little corella	+	Se, In	W, Fa	Tree hollow	R
Long-billed corella	—/+	Pl, Se, In	W, Fa	Tree hollow	R
Pink cockatoo	—	Se, Fr, In	W, Fa	Tree hollow	R
Sulphur-crested cockatoo	Aviary esc	Pl, Se, In	W, Fa	Tree hollow	R
Purple-crowned lorikeet	—	Po, Ne, Fr, In	W, M	Tree hollow	R/N
Regent parrot	—	Se, Fr	W, M, Sh	Tree hollow	R/N
Cockatiel	0	Se, Fr	W, Fa	Tree hollow	BN
Budgerigar	0	Se	W, Sh, Fa	Tree hollow	N
Red-capped parrot	—	Se, Fr	W	Tree hollow	R/N
Western rosella	—	Se, Fr	W	Tree hollow	R/N
Port Lincoln parrot	+	Se, Fr, In	W, M, Sh, Fa	Tree hollow	R
Mulga parrot	+	Se, Fr, In	W, Sh, Fa	Tree hollow	R
Elegant parrot	+	Se, Fr	W, Sh, Fa	Tree hollow	R
Bourke's parrot	0	Se	?	Tree hollow	V
Pallid cuckoo	0	Fr, In	W, M, Sh, Fa	Nest parasite	BM
Fan-tailed cuckoo	—	In	W, M, Fa	Nest parasite	BM

Table 1 — continued

Species	Change	Diet	Habitat	Nest	Status
Black-eared cuckoo	—	In	W, M, Sh	Nest parasite	N
Horsfield's bronze cuckoo	—	Se, Fr, In	W, M, Sh, He	Nest parasite	BM
Golden bronze cuckoo	—	In	W	Nest parasite	BM
Boobook owl	—	In, Fro, Re, Bi, Ma	W, M, Sh	Tree hollow	R
Barking owl	0	In, Re, Bi, Ma	W, M	Tree hollow	R
Barn owl	0	In, Fro, Re, Bi, Ma	W, Fa	Tree hollow	R
Masked owl	?	In, Ma	W	Tree hollow	R
Tawny frogmouth	—	Se, In, Fro, Re, Bi, Ma	W, M, Sh	Tree-stick nest	R
Owlet-nightjar	—	In	W, M	Tree hollow	R
Spotted nightjar	0	In	W, M	Ground	R
Fork-tailed swift	0	In	Aerial	Northern Hemisphere	V
Laughing kookaburra	+(exotic)	In, Fi, Re, Bi, Ma	W	Tree hollow	R
Red-backed kingfisher	0	In, Fro, Re, Ma	W, M	Burrow	BM
Sacred kingfisher	—	In, Fi, Fro, Re, Ma	W, Sh	Tree hollow/burrow	BM
Rainbow bee-eater	0	In	W, M, Sh, He, Fa	Burrow	BM
Passerines					
White-backed swallow	+	In	He, Fa	Burrow	R
Welcome swallow	+	In	Fa, To	On buildings, etc.–cup-shaped nest	R
Tree martin	—	In	W, M, Fa, To	Tree hollow	R
Fairy martin	—	In	W, Fa	Mud nest on buildings, etc.	BM?
Richard's pipit	+	Se, In	Fa	Ground	R
Black-faced cuckoo-shrike	0	Se, Fr, In	W, M, Sh, He, Fa, To	Tree–saucer-shaped nest	R
Ground cuckoo-shrike	—	In, Bi	W, Sh	Tree–bowl-shaped nest	N
White-winged triller	—	Se, Ne, In	W, M, Sh, He	Tree/shrub–saucer-shaped nest	BM
Southern scrub robin	—	Se, In	M, Sh	Ground/low shrub–cup-shaped nest	R
Scarlet robin	—	In	W	Tree–cup-shaped nest	R
Red-capped robin	—	In	W, Sh	Tree/shrub–cup-shaped nest	R
Hooded robin	—	In	W, M, Sh	Tree/shrub–cup-shaped nest	R
Western yellow robin	—	In	W, M, Sh	Tree–cup-shaped nest	R
Jacky winter	—	In	W, M	Tree–saucer-shaped nest	R
Crested shrike-tit	—	Se, Fr, In	W	Tree–goblet-shaped nest	R
Gilbert's whistler	—	Pl, Se, In	W, M	Shrub–cup-shaped nest	R/N
Golden whistler	—	Se, Fr, In	W, M, Sh	Tree/shrub–cup-shaped nest	R
Rufous whistler	—	Se, In	W, M, Sh	Tree/shrub–cup-shaped nest	R
Grey shrike-thrush	—	Pl, Se, In, Fro, Re, Bi, Ma	W, M, Sh	Tree/shrub–cup-shaped nest	R
Crested bellbird	—	Pl, In	W, M, Sh	Tree/shrub–cup-shaped nest	R
Restless flycatcher	—	In	W, M	Tree–cup-shaped nest	R/N
Grey fantail	0	In	W, M, Sh	Tree–cup-shaped nest	R/BM
Willie wagtail	+	Se, In	W, M, Sh, He, Fa, To	Tree/shrub/verandah–cup-shaped nest	R
Chestnut quail-thrush	—	Se, Fr, In	Sh	Ground	N
White-browed babbler	—	Se, Fr, Bi, In	W, M, Sh, He	Shrub–enclosed nest	R
Clamorous reed warbler	0	Se, In	Aq	Shrub/reeds–cup-shaped nest	BM
Little grassbird	0	In	Aq	Shrub/rushes–cup-shaped nest	N
Rufous song lark	—	Se, In	W, Fa	Ground	BM
Brown song lark	0/+	Se, In	Fa	Ground	BM
Splendid wren	—	Se, In	W, M, Sh	Shrub–ball-shaped nest	R
Variegated wren	—	Se, In	Sh, He	Shrub–oval-shaped nest	R
Red-winged wren	—	In	Riperian	Shrub–globular-shaped nest	R
Blue-breasted wren	—	In	M, Sh, He	Shrub–oval-shaped nest	R
White-winged wren	0	In	Sh, He	Shrub–oval-shaped nest	R
Southern emu wren	—	Se, In	Sh, He	Shrub–oval-shaped nest	R
Thick-billed grass wren	—	Se, In	Sh	Extinct in area	

Table 1 — continued

Species	Change	Diet	Habitat	Nest	Status
White-browed scrub wren	—	Se, In	M, Sh, He	Ground–dome-shaped nest	R
Shy hylacola	—	Se, In	M, Sh, He	Ground–dome-shaped nest	R
Redthroat	—	Se, In	W, M, Sh	Shrub–globe-shaped nest	R
Field wren	—	Se, In	Sh, He	Shrub/ground–globe-shaped nest	R
Southern whiteface	0	Se, In	W	Tree/shrub–dome-shaped nest	R
Weebill	—	Se, In	W, M	Tree/shrub–pear-shaped nest	R
Western warbler	—	In	W, M	Tree/shrub–oval-shaped nest	R/M
Broad-tailed thornbill	—	Se, Fr, In	W, M, Sh	Shrub–dome-shaped nest	R
Chestnut-rumped thornbill	—	Se, In	W, M, Sh	Tree hollow–dome-shaped nest	R
Yellow-rumped thornbill	—	Se, In	W, M, Sh, Fa	Tree/shrub–dome-shaped nest	R
Western thornbill	0	In	W	Tree/shrub/hollow–dome-shaped nest	R
Varied sittella	—	In	W	Tree–cup-shaped nest	R
Rufous treecreeper	—	In	W, M	Tree hollow/near ground	R
Red wattlebird	—	Ne, Po, Fr, In	W, M, Sh	Tree/shrub–cup-shaped nest	R
Little wattlebird	—	Ne, In	W, Sh, He	Tree/shrub–cup-shaped nest	R
Spiny-cheeked honeyeater	—	Se, Fr, Ne, In	W, M, Sh, To	Tree/shrub–cup-shaped nest	R
Yellow-throated miner	+?	Se, Ne, Po, In	W, M, Sh, Fa, To	Tree/shrub–cup-shaped nest	R
Singing honeyeater	+	Se, Fr, Ne, Po, In	W, M, Sh, He, To	Shrub–cup-shaped nest	R
White-eared honeyeater	—	Fr, Ne, In	W, M, Sh	Shrub–cup-shaped nest	R
Purple-gaped honeyeater	—	Se, Ne, Po, In	M, He	Near ground–cup-shaped nest	R
Yellow-plumed honeyeater	—	Ne, In	W, M	Tree–cup-shaped nest	R
Grey-fronted honeyeater	0	Fr, Se, Ne, In	W, M, Sh	Not recorded in area	V
White-plumed honeyeater	0	Se, Ne, Po, In	W	Tree–cup-shaped nest	R
Brown-headed honeyeater	—	Ne	W, M, Sh	Tree–cup-shaped nest	R
White-naped honeyeater	—	Ne, Po, In	W	Tree–cup-shaped nest	R/N
Brown honeyeater	—	Ne	W, M, Sh, He	Shrub–cup-shaped nest	R
New Holland honeyeater	0	Ne, Po, Fr, In	W, Sh, He	Tree/shrub–cup-shaped nest	N
White-cheeked honeyeater	—	Ne, In	W, Sh, He	Near ground–cup-shaped nest	N
White-fronted honeyeater	—	Ne, Se, In	Sh, He	Shrub–cup-shaped nest	N
Tawny-crowned honeyeater	—	Ne, In	M, Sh, He	Near ground–cup-shaped nest	R
Western spinebill	—	Ne, In	W, He	Shrub–cup-shaped nest	R
Black honeyeater	0	Se, Ne, In	Sh	Near ground–cup-shaped nest	N
Pied honeyeater	0	Se, Fr, Ne, In	Sh	Shrub–cup-shaped nest	N
Crimson chat	0	Se, In	Sh, He	Near ground–cup-shaped nest	N
Orange chat	0	Se, In	Edge of salt lake	Near ground–cup-shaped nest	V
White-fronted chat	+?	Se, In	He, Fa	Near ground–cup-shaped nest	N
Mistletoe bird	—	Fr, In	W, Sh	Tree–pear-shaped nest	N
Spotted pardalote	—	Pl, In	W	Burrow	R/N
Yellow-rumped pardalote	0	Pl, In	M	Burrow	R/V
Striated pardalote	—	In	W, M	Tree hollow	R/BM
Silvereye	—	Ne, Fr, In	W, M, Sh, He, To	Shrub–cup-shaped nest	R/M
Red-eared firetail	—	Se, In	W, He	Shrub–bottle-shaped nest	R
Zebra finch	+	Se, In	W, M, Sh, Fa	Shrub–dome-shaped nest	R
Australian magpie lark	+	Se, In, Fro	W, Fa, To	Tree–bowl-shaped nest	R
Masked woodswallow	0	Ne, Po, In	W, Fa	Shrub–bowl-shaped nest	N
Black-faced woodswallow	+	Se, Ne, In	W, M, Sh, Fa, To	Tree/shrub–saucer-shaped nest	R
Little woodswallow	0	In	W	Not recorded in area	V
Dusky woodswallow	—	Se, Fr, In	W	Tree–basket-shaped nest	R
Grey butcherbird	—	Se, Fr, In, Re, Bi, Ma	W, M	Tree–cup-shaped nest	R
Pied butcherbird	+	In, Fro, Re, Bi, Ma	W, Fa	Tree–cup-shaped nest	R
Australian magpie	+	Se, In, Fro, Re, Bi, Ma	W, Fa	Tree–basket-shaped nest	R
Grey currawong	—	Se, Fr, In, Fro, Re, Bi, Ma	W, M	Tree–cup-shaped nest	R/N
Australian raven	+	Se, Fr, In, Fi, Fro, Re, Bi, Ma, Ca	W, Fa	Tree–cup shaped nest	R
Little crow	0	Se, Fr, In, Fro, Re, Bi, Ca	W, Fa	Tree–basket-shaped nest	M

It is worth noting that because a species has increased in range and/or abundance, or its status has remained unchanged since clearing began, does not mean that it may not become a cause for conservation concern. As conditions change, some aspect of their life history may result in their decline. The emu and regent parrot are examples of this phenomenon. Both were considered pests during earlier stages of development of the wheatbelt and both underwent major declines subsequently.

Development of the wheatbelt has also provided nest site opportunities for species which exploit sand and gravel pits, and buildings. These sites have assisted in the increase in abundance of the white-backed and welcome swallows. The yellow-throated miner and singing honeyeater are the only honeyeaters which have apparently increased in the wheatbelt. They occur in remnant vegetation, even badly degraded native vegetation along road verges, on farms and in towns. They are able to fly over open areas and can move between isolated remnants. For example, the singing honeyeater will cross more than 1 200 m of open agricultural land to move between remnants, something the similar-sized white-eared honeyeater may not do if distances between remnants are greater than 200 m (Cale 1994).

Four species (cattle egret, large egret, white ibis and stubble quail) have become established in the wheatbelt as part of general or widespread range expansions. In the case of the cattle egret, this is part of a universal phenomenon. Creation of suitable habitat in the wheatbelt has aided these invasions.

Most of the changes to the wheatbelt avifauna are the direct result of human-induced alterations to the landscape through clearing of native vegetation, provision of water, new sources of food, introductions of species, or a combination of factors connected with development, rather than the result of fluctuations in climate as suggested by Serventy (1977). He discussed the expansion of a number of species into the southwest and attributed these expansions to a decrease in rainfall and to the fact that the arid zone was expanding. Examination of changes in the distributions of three of the species cited by Serventy (crested pigeon, galah and pied butcher bird) show that human-induced changes are more likely to be responsible than climatically-induced changes. All three species were present in the Murchison District, which is adjacent to the wheatbelt, prior to European settlement. However, they were uncommon and did not occur in the dry areas between the water courses (Saunders and Curry 1990). With the provision of stock troughs as part of pastoral development in these areas, the birds expanded their ranges from the water courses across the hinterland and are now common throughout the Murchison. As mentioned earlier, they have also invaded the wheatbelt which resembles an extensive riverine plain.

Overall, more species of passerines have declined than non-passerines (Table 2). Eighty-seven species of passerine (excluding vagrants) have been recorded in the wheatbelt since settlement by Europeans. Fifty-eight (67%) have declined in range and/or abundance, 13 (15%) have increased, and there has been no apparent change in 16 (18%). One hundred and eight species of non-passerine (excluding vagrants) have been recorded from the wheatbelt; 37 (34%) have declined, 21 (19%) have increased and 50 (46%) have apparently not changed.

Resident species have suffered the greatest decline in number of species in the wheatbelt (Table 2). Overall, residents comprise nearly 65 per cent of the species recorded from the wheatbelt (excluding vagrants), yet they constitute 82 per cent of the species which have declined. More species of passerines have declined than non-passerines; 69 species (79%) of wheatbelt passerines are residents and 72 per cent of these have declined compared with 57 species (53%)

Table 2. Change in range and/or abundance of wheatbelt residents, nomads and migrants (— = declined; 0/? = no change or insufficient data to assess change; and + = increased). Percentages are given in brackets.

	—	0/?	+	Total
Non-passerines				
Resident	28 (49)	12 (21)	17 (30)	57
Nomad	5 (14)	26 (74)	4 (12)	35
Migrant	4 (25)	12 (75)	0	16
Total	37 (34)	50 (46)	21 (20)	108
Passerines				
Resident	50 (72)	7 (10)	12 (17)	69
Nomad	5 (42)	6 (50)	1 (8)	12
Migrant	3 (50)	3 (50)	0	6
Total	58 (67)	16 (18)	13 (15)	87

of resident non-passerines, of which half have declined.

Those families of birds mostly dependent upon aquatic habitats (ducks and wading birds) have been least affected by changes imposed on the landscape. Eight out of 12 species of the Anatidae (ducks and swans) and 14 out of 17 species of Charadriidae, Recurvirostridae and Scolopacidae (shore or wading birds) have apparently not changed in range and/or abundance.

The majority of species that declined in range and/or abundance are residents dependent on native vegetation. Extensive loss of habitat is a major factor in their decline (Table 3). Fifty-eight per cent of the 57 species of non-passerines resident in the wheatbelt were found in native vegetation

Table 3. Changes in distribution and/or abundance of wheatbelt birds in relation to habitat. Vagrants have been excluded from this analysis. — = declined in range and/or abundance; 0/? = no change or insufficient data to assess change; and + = increased in range and/or abundance. The figures in each column (— 0/? +) are percentage of birds in each habitat which have decreased, not changed or increased. The percentages have been calculated from data in Table 1. For example, the habitat preferences of all species of non-passerine classified as resident in the wheatbelt were scored against the changes in status and the results expressed as a percentage. To illustrate how the raw data were allocated, the following example is given: each species considered in the analyis was given 1 habitat point, which was allocated according to the number of habitats it frequents, e.g., the crested pigeon occurs in woodland, mallee, shrublands, heath and farmland and so 0.2 was placed against each of those habitats in the + column for resident non-passerines. Similarly, the Port Lincoln parrot is a resident which occurs in woodland, mallee, shrubland and farmland and each habitat scored 0.25 in the + column. After allocating scores the totals in each cell were expressed as a percentage with the total number of birds in each habitat given in bold type. Because some species occur in a number of categories, these are not necessarily whole numbers.

(A) Residents		**Non-passerines**				**Passerines**		
Habitat	—	0/?	+	Total	—	0/?	+	Total
Woodland	53.2	17.4	29.4	**17.2**	75.2	13.7	11.1	**25.31**
Mallee	63.1	21.1	15.8	**4.75**	83.8	9.8	6.4	**15.98**
Shrubland	68.6	11.7	19.7	**6.85**	84.7	7.4	7.9	**13.04**
Heath	65.4	22.2	12.4	**4.05**	75.9	10.5	13.6	**6.39**
Aquatic	14.2	37.2	18.6	**10.75**	0	0	0	**0**
Saltlakes/riparian	0	0	0	**0**	100	0	0	**1**
Farmland	29.5	18.1	52.4	**12.7**	9.5	3.2	87.3	**5.28**
Town	0	0	100	**0.7**	8.5	8.5	83.0	**2**
Total	Resident non-passerines			**57**	Resident passerines			**69**

(B) Nomads		**Non-passerines**				**Passerines**		
Habitat	—	0/?	+	Total	—	0/?	+	Total
Woodland	12.5	87.5	0	**2.4**	63.6	36.4	0	**2.2**
Mallee	42.9	57.1	0	**0.7**	0	0	0	**0**
Shrubland	50	50	0	**0.8**	50	50	0	**5.6**
Heath	0	100	0	**0.3**	36.4	40.9	22.7	**2.2**
Aquatic	16.7	70.8	12.5	**24.0**	0	100	0	**1**
Saltlakes/riparian	0	100	0	**4**	0	0	0	**0**
Farmland	0	64.3	35.7	**2.8**	0	50	50	**1**
Town	0	0	0	**0**	0	0	0	**0**
Total	Nomad non-passerines			**35**	Nomad passerines			**12**

(C) Migrants		**Non-passerines**				**Passerines**		
Habitat	—	0/?	+	Total	—	0/?	+	Total
Woodland	63.1	36.9	0	**3.25**	71.4	28.6	0	**1.75**
Mallee	40.6	59.4	0	**1.6**	100	0	0	**0.25**
Shrubland	51.7	48.3	0	**1.45**	100	0	0	**0.25**
Heath	35.7	64.3	0	**0.7**	100	0	0	**0.25**
Aquatic	0	100	0	**8**	0	100	0	**1**
Saltlakes/riparian	0	0	0	**0**	0	0	0	**0**
Farmland	30	70	0	**1**	40	60	0	**2.5**
Town	0	0	0	**0**	0	0	0	**0**
Total	Migrant non-passerines			**16**	Migrant passerines			**6**

and a further 19 per cent on aquatic habitats. Only 14 per cent of aquatic species declined, however, 53 per cent of species occurring in woodland, 63 per cent of those in mallee, 69 per cent in shrubland and 65 per cent in heath have declined. The declines were proportionally greater amongst the passerines. Eighty-eight per cent of the 69 resident species occurred in native vegetation and 75 per cent of those found in woodland, 84 per cent in mallee, 85 per cent in shrubland and 76 per cent in heath have declined (Table 3A). Amongst nomadic and migratory species, more of those dependent upon native vegetation have declined compared with other habitat categories (Tables 3B and C). Resident species inhabiting towns and farmland have increased.

These declines are illustrated by three resident species that are dependent on woodland; purple-crowned lorikeet, rufous treecreeper and yellow-plumed honeyeater. When land was released for agriculture, it was classed on the basis of its vegetation associations, with woodlands being indicative of first-class agricultural soils. As a result, woodland areas were taken up and were the earliest vegetation associations to be modified and cleared. Shrubland and heath were indicative of poorer soils and were cleared much later, when the addition of trace elements to fertilizers made those soils more productive. The purple-crowned lorikeet was abundant and widespread in woodland and mallee before 1937. It feeds on pollen, nectar, fruit and insects in the canopy. It is now uncommon or rare having been extirpated from large areas of the central and southern wheatbelt along with its habitat. The rufous treecreeper often forages on the ground for insects among litter and fallen timber. It is this layer which is destroyed when stock are allowed to graze native vegetation. The yellow-plumed honeyeater was once the most abundant honeyeater in the wheatbelt, feeding on nectar and insects in woodland and mallee. It has undergone a catastrophic decline in many districts over the last 50 years along with its habitat. That it is habitat fragmentation that is mainly responsible for the decline in these species, is shown by the fact that all three are common in the uncleared native woodlands to the east of the wheatbelt (Lynch and Saunders 1991).

Other species may have declined in abundance because of loss of habitat, but may still occur throughout the range they occupied during the early part of this century. Species like the grey-shrike thrush and rufous whistler may be locally common in remnant vegetation and able to move between patches, provided the patches are not too far apart, and possibly within sight of the bird. For example, the rufous whistler was seen flying over 450 m of cleared land to move between patches (Cale 1994). Unfortunately, we do not know the critical distance beyond which a species will not fly to move to another patch, although we do know that it differs for different species. Different species view fragmented agricultural landscapes in different ways. For example, the singing honeyeater, one of the most common honeyeaters in the wheatbelt, can move more than a kilometre between patches and the wheatbelt provides a continuous range of accessible patches. On the other hand, the white-eared honeyeater, which is a similar size, appears not to cross the same distances as readily, and to it the landscape may be fragmented with patches which are not accessible. Accordingly, this species is uncommon in many extensively cleared districts. It may require links of native vegetation to move between patches. The western yellow robin is an uncommon resident on some of the larger patches of native vegetation. It may be isolated on those patches if neighbouring patches are more than tens of metres apart. It may require vegetated links to avoid the isolation of populations with the attendant risks of extirpation.

An analysis of diet in relation to change in status of wheatbelt birds (Table 4) revealed that among resident non-passerines, the species that feed on invertebrates and vertebrates were most likely to have undergone a decline in range and/or abundance. For example, 27 per cent of resident non-passerines were listed as feeding on invertebrates for part or all of their diet, and 47 per cent of these declined while 29 per cent increased. Thirty per cent of resident non-passerines were listed as feeding on vertebrates, with reptiles, birds and mammals being listed most commonly in the diet. Between 50 and 75 per cent of those resident non-passerines feeding on different groups of vertebrates declined compared with 9 to 15 per cent which increased (Table 4A). Among resident passerines, 53 per cent were listed as feeding on invertebrates with 76 per cent of those declining compared with only 14 per cent increasing. Only 8 per cent were listed as feeding on vertebrates and sample sizes are too small to draw any inferences. Passerines listed as feeding on fruit, nectar, seed and other plant material (39% of resident passerines) also declined significantly (Table 4A). Amongst nomads and migrants, only the migrants feeding on invertebrates showed a clear trend. Sixty-five per cent of the 16 migrant non-passerines were listed as feeding on invertebrates and 25 per cent of these declined while none increased. Similarly, half of the six passerine migrants were listed as feeding on invertebrates; most declined and none increased.

Table 4. Changes in distribution and/or abundance of wheatbelt birds in relation to diet. Vagrants have been excluded from this analysis. — = declined in range and/or abundance; 0/? = no change or insufficient data to assess change; and + = increased in range and/or abundance. The figures in each column (— 0/? +) are percentage of birds in each diet category which have decreased, not changed or increased. The percentages have been calculated from data in Table 1 as explained for habitat data in Table 3.

(A) Residents		**Non-passerines**				**Passerines**		
Diet	—	0/?	+	Total	—	0/?	+	Total
Plant	21.7	45.5	32.8	**5.17**	75.5	24.5	0	**2.04**
Fruit	61.9	0	38.1	**5.25**	84.0	7.9	8.1	**3.82**
Nectar	100	0	0	**0.25**	85.4	3.7	10.9	**6.85**
Seed	42.0	6.9	51.1	**13.89**	70.7	8.3	21.0	**13.94**
Pollen	100	0	0	**0.25**	53.3	16.7	30.0	**1.50**
Invertebrate	47.4	24.0	28.6	**15.20**	76.4	9.7	13.9	**36.78**
Fish	73.2	14.9	11.9	**1.68**	0	0	100	**0.12**
Frog	74.5	10.2	15.3	**1.96**	26.4	0	73.6	**1.06**
Reptile	62.3	25.6	12.1	**3.71**	48.9	0	51.1	**0.92**
Bird	53.8	35.3	10.9	**4.11**	49.5	0	50.5	**0.93**
Mammal	50.0	40.6	9.4	**4.80**	48.4	0	51.6	**0.93**
Carrion	65.8	34.2	0	**0.73**	0	0	100	**0.11**
Total		Resident non-passerines		**57**		Resident passerines		**69**
(B) Nomads		**Non-passerines**				**Passerines**		
Diet	—	0/?	+	Total	—	0/?	+	Total
Plant	17.6	75.6	5.9	**5.1**	0	0	0	**0**
Fruit	0	100	0	**0.5**	61.5	38.5	0	**1.3**
Nectar	0	0	0	**0**	42.1	57.9	0	**1.9**
Seed	12.0	88.0	0	**7.52**	31.1	46.7	22.2	**2.25**
Pollen	0	0	0	**0**	0	100	0	**0.55**
Invertebrate	18.2	67.3	14.5	**12.07**	40.0	50.9	9.1	**5.5**
Fish	13.0	64.8	22.2	**3.83**	0	0	0	**0**
Frog	15.6	61.1	23.3	**1.93**	0	0	0	**0**
Reptile	0	78.3	21.7	**0.92**	0	0	0	**0**
Bird	10.5	76.4	13.1	**1.91**	100	0	0	**0.5**
Mammal	0	83.6	16.4	**1.22**	0	0	0	**0**
Carrion	0	0	0	**0**	0	0	0	**0**
Total		Nomad non-passerines		**35**		Nomad passerines		**12**
(C) Migrants		**Non-passerines**				**Passerines**		
Diet	—	0/?	+	Total	—	0/?	+	Total
Plant	0	100	0	**0.4**	0	0	0	**0**
Fruit	37.5	62.5	0	**0.8**	0	100	0	**0.14**
Nectar	0	0	0	**0**	100	0	0	**0.14**
Seed	14.3	85.7	0	**2.1**	41.2	58.8	0	**1.94**
Pollen	0	0	0	**0**	0	0	0	**0**
Invertebrate	24.9	75.1	0	**10.45**	61.2	38.8	0	**2.94**
Fish	40.0	60.0	0	**0.5**	0	0	0	**0**
Frog	23.5	76.5	0	**0.85**	0	100	0	**0.14**
Reptile	44.4	55.6	0	**0.45**	0	100	0	**0.14**
Bird	0	0	0	**0**	0	100	0	**0.15**
Mammal	44.4	55.6	0	**0.45**	0	0	0	**0**
Carrion	0	0	0	**0**	0	100	0	**0.15**
Total		Migrant non-passerines		**16**		Migrant passerines		**6**

A similar analysis of category of nest site in relation to change in status in the wheatbelt showed that amongst both resident non-passerines and passerines, declines were greatest among those species that nested on the ground, in trees or shrubs (Table 5A). Forty-three per cent of nomadic non-passerines nested on the ground and while most did not change in status, 13 per cent declined compared with only 3 per cent that increased. Twenty-five per cent of nomadic non-passerines nest in trees and most did not change in status, but 30 per cent increased while none declined. Sixty-seven per cent of nomadic passerines nested in shrubs and 25 per cent of these declined, while 13 per cent increased. Nest parasites account for half of the non-passerine migrants (four species) and three of these declined.

Habitat loss and fragmentation have not been the only causes of species decline and loss; other factors such as species introductions and changes

Table 5. Changes in distribution and/or abundance of wheatbelt birds in relation to nest site. Vagrants have been excluded from this analysis. — = declined in range and/or abundance; 0/? = no change or insufficient data to assess change; and + = increased in range and/or abundance. The figures in each column (— 0/? +) are percentage of birds in each nest site category which have decreased, not changed or increased. The percentages have been calculated from data in Table 1 as explained for habitat data in Table 3.

(A) Residents		**Non-passerines**				**Passerines**		
Nest site	—	0/?	+	Total	—	0/?	+	Total
Raft on water	0	0	0	**0**	0	0	0	**0**
Ground	46.9	34.4	18.7	**16.0**	75.0	0	25.0	**4.0**
Shrub	100	0	0	**1.0**	82.2	6.5	11.3	**29.2**
Tree	65.6	12.5	21.9	**16.0**	66.2	13.9	19.9	**27.2**
Hollow in tree	38.3	17.0	44.7	**23.5**	93.0	7.0	0	**4.3**
Burrow	0	0	0	**0**	33.3	33.3	33.4	**3.0**
Nest parasite	0	0	0	**0**	0	0	0	**0**
Reeds/rushes	0	0	0	**0**	0	0	0	**0**
Cliff	0	100	0	**0.5**	0	0	0	**0**
Buildings	0	0	0	**0**	0	0	100	**1.3**
Total		Resident non-passerines		**57**		Resident passerines		**69**

(B) Nomads		**Non-passerines**				**Passerines**		
Nest site	—	0/?	+	Total	—	0/?	+	Total
Raft on water	33.3	66.7	0	**3.0**	0	0	0	**0**
Ground	13.3	84.0	2.7	**15.0**	100	0	0	**1.0**
Shrub	0	81.8	18.2	**3.3**	25.0	62.5	12.5	**8.0**
Tree	0	70.5	29.5	**8.8**	80.0	20.0	0	**2.5**
Hollow in tree	0	100	0	**2.5**	0	0	0	**0**
Burrow	0	0	0	**0**	0	0	0	**0**
Nest parasite	100	0	0	**1.0**	0	0	0	**0**
Reeds/rushes	71.4	0	28.6	**1.4**	0	100	0	**0.5**
Cliff	0	0	0	**0**	0	0	0	**0**
Buildings	0	0	0	**0**	0	0	0	**0**
Total		Nomad non-passerines		**35**		Nomad passerines		**12**

(C) Migrants		**Non-passerines**				**Passerines**		
Nest site	—	0/?	+	Total	—	0/?	+	Total
Raft on water	0	0	0	**0**	0	0	0	**0**
Ground	0	100	0	**1.0**	50.0	50.0	0	**2.0**
Shrub	0	0	0	**0**	50.0	50.0	0	**1.0**
Tree	0	0	0	**0**	33.3	66.7	0	**1.5**
Hollow in tree	100	0	0	**0.5**	0	0	0	**0**
Burrow	20.0	80.0	0	**2.5**	0	0	0	**0**
Nest parasite	75.0	25.0	0	**4.0**	0	0	0	**0**
Reeds/rushes	0	0	0	**0**	0	0	0	**0**
Cliff	0	0	0	**0**	0	100	0	**0.5**
Buildings	0	0	0	**0**	100	0	0	**1.0**
Total		Migrant non-passerines		**8**		Migrant passerines		**6**

in ranges of other species, or a combination of factors have also had adverse effects. For example, the bustard and bush stone curlew are species which have declined in abundance. Both frequent open country and might be expected to have benefited from clearing for agriculture because of the provision of more open habitat. In fact, these species were common during the earlier phases of development (Hobbs *et al.* 1993). They became increasingly rare from the 1950s. Both species are ground dwellers and have almost certainly been adversely affected by foxes. Their declines coincided with the period when rabbits and foxes were common and myxomatosis was introduced into the rabbit population. Rabbit numbers fell rapidly and it is possible that foxes preyed more on native animals than when rabbits were common. Species like the bustard and bush stone curlew may have been affected by this predation. Other species, like the mallee fowl, which had been reduced in numbers as a result of loss of habitat and confined to remnants may have then been further reduced as predation eliminated isolated populations. As these populations disappeared the range of the species contracted.

Competition between species that have invaded the wheatbelt and those already present is another factor contributing to species decline. An example of this phemomenon is the interaction between the galah and Carnaby's cockatoo over nest sites. There is some overlap in the sizes of hollows used by these two species (Saunders *et al.* 1982) and the galah has been recorded entering hollows being used by black cockatoos and smashing their eggs. Because of food shortages induced by habitat reduction and fragmentation, the black cockatoos often do not have a chance to breed successfully that season, if their first breeding attempt fails (Saunders 1982). The galah is capable of taking over the hollow and breeding successfully. The regent parrot may be another species that has been affected by competition for nest sites with the galah.

The fact that remnant vegetation is not in one large patch, but is fragmented into thousands of small patches with varying degrees of connectivity has almost certainly been responsible for the decline of many species. For example, lack of connectivity between patches is strongly implicated in the failure of Carnaby's cockatoo to survive in heavily cleared and fragmented rural landscapes (Saunders 1990). While it is difficult to prove that a species needs corridors of vegetation to foster its movements between remnants in the wheat-belt, corridors may increase the chances of some species finding remnants during their dispersal phase. So it is better to take a conservative view of corridors and retain all existing links in the wheatbelt and prevent their further degradation.

As a result of habitat reduction and fragmentation many species, dependent on native vegetation, survive as a collection of small populations confined to remnant patches. These populations are too small to be viable and may be too isolated to allow the remnant to be recolonized if the population is lost. As these isolated populations disappear we will see a continuing loss of species. This will leave us biotically poorer, however, the long term implications of this impoverishment are not known. Some species may be functionally important and their loss may affect ecosystem function. For example, some species of honey-eaters are important pollinators of plants and their loss may affect the ability of native plants to regenerate or move around the landscape.

chapter six

Conservation actions

THE problem for those involved in nature conservation is to identify those species that are adversely affected by human-induced changes in the landscape and particularly those that are in danger of being eliminated. Having identified them, it is essential to work out what can be done to protect them, and where necessary, enhance their long term chances of survival. In the wheatbelt, the vulnerable species are mainly residents which are dependent on native vegetation, and which have declined in range and/or abundance as a result of development for extensive wheat-sheep farming. Although we can now identify them (species accounts in Chapter 4 and Table 1), we know little about their life history characteristics, population dynamics, resource use, movements or longevity. In order to manage them in rural landscapes, we need to know why they changed in range and/or abundance, how they are distributed at the landscape level, their population sizes, how they move within and between remnants, and the sort of linkages required to foster their movements between populations. This will involve more research at the species, community and landscape levels. However, we cannot simply postpone management while we wait for the results of the required research. We must start managing our agricultural landscapes if we are to prevent the continuing loss of our natural heritage.

We do know that major soil and nature conservation problems have developed in the wheatbelt as a result of the changes we have imposed on the landscape. We also know that there are obvious problems we can deal with immediately. The losses of bird populations isolated on remnants may be due to foxes, cats or the degradation of native vegetation by grazing of sheep and/or rabbits. We should begin widespread control of these exotic species. We have the technology and the resources, provided local communities accept responsibility for what is a community problem. Local action, aided by advice from government agencies like the Western Australian Department of Conservation and Land Management and the Agriculture Protection Board, can be taken to reduce the numbers of foxes, cats and rabbits. Attempts should be made to control all three rather than dealing with only one species, as control of one may see one or both of the other two species change in abundance with further adverse affects on the avifauna. For example, control of foxes may see rabbits increase as predation is reduced and this increase may lead to further degradation of remnant vegetation with concommittant loss of wildlife.

Domestic livestock should be prevented from grazing remnant vegetation. We should not allow further degradation of what amounts to the skeleton on which we will rebuild our degrading landscapes. It is necessary for community groups throughout the wheatbelt to take on the task of management of remnant vegetation and its associated fauna. If this task is left to conservation or other government agencies, most of the work will not be done, as they lack the resources to deal with the land directly under their control, let alone manage the remainder of the landscape. It is worth pointing out that plants and animals use the entire landscape and take no account of human concepts of ownership.

In southwestern Australia, the landcare movement has developed rapidly and landcare groups have been formed all over the wheatbelt (Goss and Chatfield 1993). The aim of these groups is to deal with problems of land degradation at an ecological scale, rather than at the scale of the individual farm. The basis of planning for these groups is a river catchment and revegetation plans are being developed to deal with soil conservation problems within those catchments. It is widely accepted that revegetation with perennial vegetation is a key strategy in tackling some aspects of land degradation. One example is that perennial vegetation uses more water than annual vegetation and has an important role in lowering water tables; an essential approach to tackling the increasing problem of soil salinity. Revegetation can also be used to address problems associated with wind and water erosion and will have a key role to play in ecologically sustainable agricultural practices. It is apparent that in the light of increasing land degradation, revegetation will continue. Given that we need to address the continuing loss of native species, we should examine revegetation strategies to see where they can achieve the dual goals of soil conservation and nature conservation. Considerations relating to soil conservation may dictate where revegetation is to take place and the extent of revegetation. However, nature conservation should then be considered with a view to using local native species for revegetation to create habitat and develop linkages between patches of remnant vegetation, in addition to their role in soil conservation. While it is not possible in a book like this to provide revegetation prescriptions for the range of different catchments covered by the wheatbelt, a series of guidelines is given to help planning for nature conservation.

It is important to realize that no patch of remnant vegetation in the wheatbelt is of sufficient size to function only on internal processes. Because of ecological changes resulting from clearing of native vegetation and from agricultural practices, all remnant vegetation patches are ecologically driven mainly by processes arising in the surrounding agricultural land. Most, if not all, remnant vegetation patches are degrading and will continue to degrade unless the processes leading to that degradation are managed. The management required will differ, depending on the major degrading processes.

No remnant is of sufficient size to support viable populations of the majority of the avifauna. Those species dependent on native vegetation exist in a landscape which involves a collection of remnants and a range of linkages between them. Revegetation of only tens of hectares or a link between only two remnants may appear pointless when trying to conserve a species, but the combination of these actions over the entire landscape, rather than one farm paddock, may achieve the desired result.

A series of steps which should be taken into account when planning nature conservation and habitat reconstruction are set out below.

- **Be informed.** A short course of reading is essential to find out what information is available and where to go for advice. There are a number of published works which may be of great assistance in planning to manage remnant vegetation and incorporating nature conservation into productive agricultural landscapes. It is worth reading some, or all of these works which are introduced briefly below in chronological order of publication dates.

  Wildlife in the Home Paddock. Nature Conservation for Australian Farmers by Roland Breckwoldt (1983). This excellent practical guide was written by a farmer in New South Wales who argues that there are economic benefits from learning to live with the land as well as on it. The book provides a number of steps that can be taken to enhance the conservation potential of farms, without losing agricultural production.

  Grow Your Own Wildlife. How to Improve Your Local Environment by Peter Johnston and Alan Don (1990). This little book provides a series of case studies of Australian and New Zealand farmers who are trying to incorporate nature conservation into farm management. The conservation principles and methods demonstrated provide a framework for others to follow to help maintain our natural heritage.

  Guidelines for Bush Corridors by Penny Hussey, Richard Hobbs and Denis Saunders (1991). This booklet sets out a series of steps to guide the management and creation of linear strips of vegetation that may serve to link remnant vegetation patches and help animals move between patches.

  A Revegetation Guide to the Central Wheatbelt by Ted Lefroy, Richard Hobbs and Lyn Atkins (1991). This book was written in conjunction with a number of people, including farmers, and was sent free to about 3 000 farmers in the central wheatbelt in late 1991 and early 1992. It provides descriptions of soil types, the native plants that grow on those soils and the birds associated with those plants. It provides a description of over 80 native species that may be used in revegetation and gives a range of practical advice on planning revegetation and how to go about it. It also provides a list of useful references regarding revegetation. It is essential reading for anyone farming in the wheatbelt. It was produced to help people interested in revegetating the central wheatbelt and while the details on native vegetation may not be accurate in the northern and southern wheatbelts, the principles set out in this guide have wide applicability.

  Reintegrating Fragmented Landscapes: Towards Sustainable Production and Nature Conservation edited by Richard Hobbs and Denis Saunders (1993). This book should be read by every wheatbelt farmer as it provides a description of the central wheatbelt and the ecological consequences of our development and current farming methods. It has chapters on the social history, soils, hydrologic balance, and flora and fauna, and also on how agricultural production and nature conservation may be integrated with benefits for both. It provides a review of current problems and solutions, and includes a comprehensive list of references.

Managing Your Bushland by Penny Hussey and Ken Wallace (1993). This excellent guide was written for farmers and other landholders in the southwest. It provides good, practical information on assessing the conservation value of remnant vegetation and on how to manage it.

Nature Conservation 3: the Reconstruction of Fragmented Ecosystems edited by Denis Saunders, Richard Hobbs and Paul Ehrlich (1993a). This is a scientific publication with chapters contributed by ecologists, farmers and agricultural scientists who were asked to apply their work to the practical problems associated with landscape reconstruction. This book was based on a workshop held at the Tammin Environment Centre in conjunction with the Tammin Land Conservation District Committee. Ecological problems facing the wheatbelt are the main subjects covered. It includes practical material, such as a chapter on the need for restoring patchiness in the landscape to cater for mobile species like honeyeaters which may require a range of species which flower at different times of the year (Lambeck and Saunders 1993). The book also includes a little booklet *Repairing a Damaged World* (Saunders *et al.* 1993b) which provides a set of guidelines to aid planning landscape reconstruction.

Growing Understorey Seed by Joanna Seabrook (1994). This practical book focuses on understorey plants and provides commonsense advice on establishing and managing seed orchards which will be necessary for the large areas of revegetation needed.

- **Define the nature conservation problem.** What are the environmental problems in the area to be revegetated and are there agricultural production reasons directing that the area be revegetated? Can habitat reconstruction be incorporated within revegetation plans for addressing degradation problems within the agricultural matrix? If so, is it aimed at a single species, or at a community? At what scale will the revegetation be conducted?
- **Establish what skeleton is available for landscape reconstruction.** What is the distribution of remnant vegetation over the landscape and what vegetation types are involved? What bird communities are associated with these remnants? This needs to be established at the smallest unit involved in the reconstruction and this might be at the level of a remnant of native vegetation or an individual farm. However, for the reconstruction to stand any chance of success it must be planned with neighbouring units over a considerable area, be it a river catchment, shire, or some other large unit. There are now many agencies who can assist in this planning stage by providing co-ordinators trained in facilitating these plans, together with supplying computer-generated maps of soils, remnant vegetation, roads, fences and drainage lines as well as supplying aerial photographs and satellite imagery. Such planning needs to be done in association with **all** of the land managers in the planning unit, e.g., farmers, conservation agencies, main roads agencies, shire officials, and other Crown land managers with interests in the area.
- **Retain and protect all remnant vegetation.** Habitat reconstruction will rely on an increasing supply of seed from plants of local provenance, together with information on where in the landscape the plants grew and what bird communities were associated with them. Remnant vegetation is essential for the provision of seed and information. Accordingly, there should be no further clearing of remnant vegetation. In addition, remnant vegetation will need to be managed to prevent further degradation. Grazing by domestic livestock is a major degrading influence and stock and feral animals should be kept away from remnant vegetation. Removal of weeds may also be necessary. Dead or decaying vegetation (both standing and fallen) should be retained as it provides habitat for a range of species.
- **Establish what the landscape looked like before development and degradation.** It is necessary to know what to aim at for habitat reconstruction. That can be established by examining remnant vegetation, particularly that found along linear features such as road reserves, railway reserves and fence lines. This vegetation will sample the pre-clearing vegetation associations more accurately than patches. Use these remnants to establish which vegetation grew on which soil types, and in which position in the landscape. Local knowledge among older residents and research workers can add to this information. The resulting information can then be used to plan revegetation. Existing

remnants may, however, be degraded to the extent that they no longer represent a useful baseline for reconstruction. It may, in fact, be impossible to return to exactly what was there before, and it may be a waste of resources to try.

- **Develop goals for habitat reconstruction.** Goals for reconstruction need to be developed at the smallest unit of action, be it a remnant of native vegetation or farm. These goals should be linked to goals at the catchment, landscape or regional level. To have any chance of success they must take cognisance of the predominant land use in the area and work within that land-use. For example, it is pointless attempting reconstruction in a remnant of native vegetation which is being affected by soil salination, unless the reconstruction addresses the rising water tables in the surrounding landscape. This will obviously involve revegetation in the area surrounding the remnant. Such revegetation will not be attempted unless it is in the interests of the surrounding landholders. The priorities will differ depending on the landholders involved. However, by planning at the landscape scale, revegetation for addressing some problem in the productive landscape may be useful for habitat reconstruction and bolster the area of available habitat.
- **Design a habitat reconstruction plan.** Working with all land managers in the area, design a habitat reconstruction plan that takes into account the various objectives of the diverse interests in the landscape. Revegetation can be used to increase the area of available habitat, to provide buffer strips around existing remnants, or to provide corridors linking otherwise isolated remnants. This can be done in ways compatible with other restoration goals. For example, wind breaks can be planted with a range of local species, where possible proportionally following the distribution, abundance and soil types such species occurred in before clearing. Such revegetation can provide linkages and in addition use water, reduce erosion and provide shelter. Our current state of knowledge does not allow an assessment of how useful such corridors might be for fostering movement, but if their development allows major land conservation problems to be tackled, nothing is lost in the attempt. It is better to try things that might work than do nothing because we do not know enough: we wil never know enough.
- **Aim at structural and patch diversity.** Unfortunately many revegetation schemes specify trees and ignore the need for structural diversity. Habitat reconstruction should aim at developing all the former structural layers, together with the former diversity within those layers. It should also provide a range of patches emulating the former uncleared landscape. For example, some birds feeding on nectar need a range of patches which flower at different times of year so that they have food available all year somewhere in the landscape. Some birds may require native vegetation to foster their movements around the landscape if they will not fly over cleared agricultural fields. These species will need vegetated linkages to allow them access to the full array of patches. As mentioned earlier, these sorts of linkages may also serve as wind breaks on farms as well as lowering the water table in areas subject to rising soil salinity.
- **Monitor and record results.** It is essential to record what is being planned, what has been carried out, as well as monitoring the results of revegetation and the use of revegetated areas by native birds. Monitoring is essential to allow the revegetation programme to be adjusted where necessary. Communicate these results so that others may learn from your experience.

Many areas of the wheatbelt have lost most of their native vegetation and hence most of their habitat for birds. What remains is not sufficient to retain functioning ecosystems which will maintain viable faunal and plant populations in the long term, even with considerable management input. Increasing the extent of areas useful for conservation is the only hope of reversing the current situation, and this involves large-scale habitat reconstruction. Extension of current revegetation programmes can form the basis for this, but this requires a radical rethink about how agricultural landscapes are planned and managed. It also requires a huge increase in revegetation/restoration activity, which can only come about via greater incentives. Given that the current level of activity is less than optimal, we need to make sure that it will have the maximum benefit.

It takes a salmon gum about 140 years to grow large enough to develop a hollow of sufficient size for a black cockatoo. Our success in landscape reconstruction can only be judged in decades, if not centuries. However, we need to start large-scale revegetation for nature conservation as well as soil conservation now, while we still have the skeleton of native vegetation on which to base our reconstruction (photo by D. A. Saunders).

References

Barker, R. D. and Vestjens, W. J. M., 1989. The Food of Australian Birds. 1. Non-passerines. CSIRO: Australia.

Barker, R. D. and Vestjens, W. J. M., 1990. The Food of Australian Birds. 2. Passerines. CSIRO: Australia.

Beard, J. S and Sprenger, B. S., 1984. Geographical data from the vegetation survey of Western Australia. Vegetation Survey of Western Australia, Occasional Paper No. 2. Vegmap Publications: Perth.

Beruldsen, G., 1980. A Field Guide to Nests and Eggs of Australian Birds. Rigby: Sydney.

Blakers, M., Davies, S. J. J. F. and Reilly, P. N., 1984. The Atlas of Australian Birds. Melbourne University Press: Melbourne.

Bradshaw, F. R., 1928. Fauna of the Stirling Range (WA). *Emu* **27:** 176–78.

Breckwoldt, R., 1983. Wildlife in the Home Paddock. Angus and Robertson: Sydney.

Brooker, M. G. and Brooker, L. C., 1989. Cuckoo hosts in Australia. *Aust. Zool. Rev.* **2:** 1–67.

Brooker, M. G. and Ridpath, M. G., 1980. The diet of the Wedge-tailed Eagle, *Aquila audax,* in Western Australia. *Aust. Wildl. Res.* **7:** 433–52.

Burbidge, A. A. and McKenzie, N. L., 1989. Patterns in the modern decline of Western Australia's vertebrate fauna: causes and conservation implications. *Biol. Conserv.* **50:** 143–98.

Burbidge, A. H. and Rose, A., 1993. Birds. Pp. 99–108 *in* Mountains of Mystery. A Natural History of the Stirling Range ed by C. Thomson, G. Hall and G. Friend. Department of Conservation and Land Management: Perth.

Cale, P. G., 1990. The value of road reserves to the avifauna of the central wheatbelt of Western Australia. *Proc. Ecol. Soc. Aust.* **16:** 359–67.

Cale, P. G., 1994. The effects of landscape fragmentation on the bird community of the Kellerberrin district of the Western Australian wheatbelt. M.Sc Thesis, University of Western Australia.

Carnaby, J. C., 1933. The birds of the Lake Grace District, WA. *Emu* **33:** 103–09.

Carter, T., 1923. Birds of the Broome Hill District. *Emu* **23:** 125–42, 223–35, 307–18.

Coates, A. M., 1987. Management of Native Vegetation on Farmland in the Wheatbelt of Western Australia. Report from the Voluntary Native Vegetation Retention Project.

Couper, D., 1993. What information do primary producers need from ecologists? Pp. 215–23 *in* Nature Conservation 3: Reconstruction of Fragmented Ecosystems ed by D. A. Saunders, R. J. Hobbs and P. R. Ehrlich. Surrey Beatty & Sons: Chipping Norton.

Crossman, A. F., 1909. Birds seen at Cumminin Station, Western Australia. *Emu* **9:** 84–90.

de Rebeira, C. P. S. and de Rebeira, A. M., 1977. Birds. Pp. 77–96 *in* The Natural History of the Wongan Hills ed by K. F. Kenneally. Western Australian Naturalists' Club: Perth.

Dell, J., 1978. Birds of Kodj Kodjin and Durokoppin Reserves. *Rec. WA Mus. Suppl.* **7:** 55–68.

Dell, J., 1979. Birds of Buntine and Nugadong Reserves. *Rec. WA Mus. Suppl.* **9:** 95–115.

Dell, J., 1980. Birds of Yorkrakine Rock, East Yorkrakine and North Bungulla Nature Reserves. *Rec. WA Mus. Suppl.* **12:** 55–67.

Ford, J. R. and Stone, P. S., 1957. Birds of the Kellerberrin/Kwolyin District, Western Australia. *Emu* **57:** 9–21.

Garnett, S., 1992. Threatened and extinct birds of Australia. RAOU Report 82. Royal Australasian Ornithologists Union and Australian National Parks and Wildlife Service: Victoria.

George, R. J., McFarlane, D. J. and Speed, R. J., 1995. The consequences of a changing hydrologic environment for native vegetation in southwestern Australia. In press *in* Nature Conservation 4: the Role of Networks ed by D. A. Saunders, J. L. Craig and E. M. Mattiske. Surrey Beatty & Sons: Chipping Norton.

Goss, K. F. and Chatfield, J., 1993. Landcare groups in Western Australia: the role of self help groups in restoring degraded farmland. Pp. 281–93 *in* Nature Conservation 3: Reconstruction of Fragmented Ecosystems ed by D. A. Saunders, R. J. Hobbs and P. R. Ehrlich. Surrey Beatty & Sons: Chipping Norton.

Halse, S. A., 1978. Feeding habits of six species of honeyeater in south-western Australia. *Emu* **78:** 145–48.

Hobbs, R. J. and Saunders, D. A. (eds), 1993. Reintegrating Fragmented Landscapes: Towards Sustainable Production and Nature Conservation. Springer-Verlag: New York.

Hobbs, R. J., Saunders, D. A., Lobry de Bruyn, L. A. and Main, A. R., 1993. Changes in biota. Pp. 65–106 *in* Reintegrating Fragmented Landscapes: Towards Sustainable Production and Nature Conservation ed by R. J. Hobbs and D. A. Saunders. Springer-Verlag: New York.

Hopper, S. D., 1992. Patterns of plant diversity at the population and species levels in south-west Australian mediterranean ecosystems. Pp. 27–46 *in* Biodiversity in Mediterranean Ecosystems in Australia ed by R. J. Hobbs. Surrey Beatty & Sons: Chipping Norton.

Hopper, S. D., van Leeuwen, S., Brown, A. P. and Patrick, S., 1990. Western Australia's Endangered Flora and Other Plants Under Consideration for Declaration. Department of Conservation and Land Management: Perth.

Hussey, B. J. M., Hobbs, R. J. and Saunders, D. A., 1991. Guidelines for Bush Corridors. CSIRO, Department of Conservation and Land Management, Main Roads Department and Roadside Conservation Committee: Perth.

Hussey, B. J. M. and Wallace, K. J., 1993. Managing Your Bushland. Western Australian Department of Conservation and Land Management: Perth.

Jaensch, R. P., Vervest, R. M. and Hewish, M. J., 1988. Waterbirds in nature reserves of Southwestern Australia 1981–1985: Reserve Accounts. Royal Australasian Ornithologists Union Report 30.

Jenkins, C. F. H., 1931. Birds of the Northam District, WA. *Emu* **31:** 30–35.

Johnston, P. and Don, A., 1990. Grow Your Own Wildlife. How to Improve Your Local Environment. Greening Australia: Canberra.

Kitchener, D. J., Chapman, A., Dell, J., Muir, B. G. and Palmer, M., 1980. Conservation value for mammals of reserves in the Western Australia wheatbelt — some implications for conservation. *Biol. Conserv.* **18:** 179–207.

Lambeck, R. J. and Saunders, D. A., 1993. The role of patchiness in reconstructed wheatbelt landscapes. Pp. 153–61 *in* Nature Conservation 3: Reconstruction of Fragmented Ecosystems ed by D. A. Saunders, R. J. Hobbs and P. R. Ehrlich. Surrey Beatty & Sons: Chipping Norton.

Lambeck, R. J. and Wallace, J. F., 1993. Assessment of the conservation value of remnant vegetation in the central wheatbelt of Western Australia using Landsat TM imagery. Report for Australian National Parks and Wildlife Service "Save the Bush" programme.

Lawson, F., 1905. A glance at the birds of the Moore River (WA). *Emu* **4:** 132–37.

Leake, B. W., 1962. Eastern Wheatbelt Wildlife. B. W. Leake: Perth.

Lefroy, E. C., Hobbs, R. J. and Atkins, L. A., 1991. A Revegetation Guide to the Central Wheatbelt. Western Australian Department of Agriculture, Greening Western Australia and CSIRO: Perth.

Leigh, J., Boden, R. and Briggs, J., 1984. Extinct and Endangered Plants of Australia. MacMillan: Melbourne.

Loaring, W. H. and Serventy, W. L., 1952. The birds of the Moore River Gorge country. *WA Nat.* **3** 107–16.

Long, J. L., 1981. Introduced Birds of the World. Reed: Sydney.

Long, J. L., 1984. The diets of three species of parrots in the south of Western Australia. *Aust. Wildl. Res.* **11:** 357–71.

Lynch, J. F. and Saunders, D. A., 1991. Responses of bird species to habitat fragmentation in the wheatbelt of Western Australia: interiors, edges and corridors. Pp. 143–58 *in* Nature Conservation 2: the Role of Corridors ed by D. A. Saunders and R. J. Hobbs. Surrey Beatty & Sons: Chipping Norton.

McArthur, W. M., 1993. History of landscape development. Pp. 10–22 *in* Reintegrating Fragmented Landscapes: Towards Sustainable Production and Nature Conservation ed by R. J. Hobbs and D. A. Saunders. Springer-Verlag: New York.

McFarlane, D. J., George, R. J. and Farrington, P., 1993. Changes in hydrological balance. Pp. 146–86 *in* Reintegrating Fragmented Landscapes: Towards Sustainable Production and Nature Conservation ed by R. J. Hobbs and D. A. Saunders. Springer-Verlag: New York.

Main, B. Y., 1993. Social history and impact on the landscape. Pp. 23–58 *in* Reintegrating Fragmented Landscapes: Towards Sustainable Production and Nature Conservation ed by R. J. Hobbs and D. A. Saunders. Springer-Verlag: New York.

Marchant, S. and Higgins, P. J. (eds), 1990. The Handbook of Australian, New Zealand and Antarctic Birds. Oxford University Press: Melbourne.

Masters, J. R. and Milhinch, A. L., 1974. Birds of the Shire of Northam, about 100 km east of Perth, WA. *Emu* **74:** 228–44.

Milligan, A. W., 1903. Notes on a trip to the Stirling Range. *Emu* **3:** 9–19.

Milligan, A. W., 1904. Notes on a trip to the Wongan Hills, Western Australia, with a description of a new *Ptilotis*. *Emu* **3:** 217–26; *Emu* **4:** 2–11.

Milligan, A. W., 1905. Notes on a trip to the Yandanooka District, Western Australia. *Emu* **4:** 151–57.

Nulsen, R. A., 1993. Changes in soil properties. Pp. 107–45 *in* Reintegrating Fragmented Landscapes: Towards Sustainable Production and Nature Conservation ed by R. J. Hobbs and D. A. Saunders. Springer-Verlag: New York.

Orton, C. L. E. and Sandland, P. T., 1913. Birds of Moora (WA) and district. *Emu* **13:** 75–80.

Pittock, A. B., 1988. Actual and anticipated changes in Australia's climate. Pp. 35–51 *in* Greenhouse. Planning for Climate Change ed by G. I. Pearman. CSIRO: Australia.

Readers' Digest, 1976. Complete Book of Australian Birds. Readers' Digest Services: Surry Hills.

Reid, D., 1951. A Bunketch bird list. *WA Nat.* **3:** 68–71.

Rowley, I., 1990. Behavioural Ecology of Galahs. Surrey Beatty & Sons: Chipping Norton.

Rowley, I. and Chapman, G., 1991. The breeding biology, food, social organization, demography and conservation of the Major Mitchell or pink cockatoo, *Cacatua leadbeateri,* on the margin of the Western Australian wheatbelt. *Aust. J. Zool.* **39:** 211–61.

Saunders, D. A., 1977. Red-tailed black cockatoo breeding twice a year in the south-west of Western Australia. *Emu* **77:** 107–10.

Saunders, D. A., 1979. Distribution and taxonomy of the White-tailed and Yellow-tailed Black Cockatoos, *Calyptorhynchus* spp. *Emu* **79:** 215–27.

Saunders, D. A., 1980. Food and movements of the short-billed form of the white-tailed black cockatoo. *Aust. Wildl. Res.* **7:** 257–69.

Saunders, D. A., 1982. The breeding behaviour and biology of the short-billed form of the white-tailed black cockatoo *Calyptorhynchus funereus. Ibis* **124:** 422–55.

Saunders, D. A., 1986. Breeding season, nesting success and nestling growth in Carnaby's cockatoo, *Calyptorhynchus funereus latirostris,* over 16 years at Coomallo Creek, and a method for assessing the viability of populations in other areas. *Aust. Wildl. Res.* **13:** 261–73.

Saunders, D. A., 1989. Changes in the avifauna of a region, district and remnant as a result of fragmentation of native vegetation: the wheatbelt of Western Australia. A case study. *Biol. Conserv.* **50:** 99–135.

Saunders, D. A., 1990. Problems of survival in an extensively cultivated landscape: the case of Carnaby's cockatoo *Calyptorhynchus funereus latirostris. Biol. Conserv.* **54:** 277–90.

Saunders, D. A. and Curry, P. J., 1990. The impact of agricultural and pastoral industries on birds in the southern half of Western Australia: past, present and future. *Proc. Ecol. Soc. Aust.* **16:** 303–21.

Saunders, D. A. and de Rebeira, C. P., 1986. Seasonal occurrence of members of the suborder Charadrii (waders or shorebirds) on Rottnest Island, Western Australia. *Aust. Wildl. Res.* **13:** 225–44.

Saunders, D. A. and de Rebeira, C. P., 1991. Values of corridors to avian populations in a fragmented landscape. Pp. 221–40 *in* Nature Conservation 2: the Role of Corridors ed by D. A. Saunders and R. J. Hobbs. Surrey Beatty & Sons: Chipping Norton.

Saunders, D. A. and de Rebeira, C. P., 1993. Birds of Rottnest Island. DAS and CPdeR: Guildford.

Saunders, D. A., Hobbs, R. J. and Ehrlich, P. R. (eds), 1993a. Nature Conservation 3: Reconstruction of Fragmented Ecosystems. Surrey Beatty & Sons: Chipping Norton.

Saunders, D. A., Hobbs, R. J. and Ehrlich, P. R., 1993b. Repairing a Damaged World. Surrey Beatty & Sons: Chipping Norton.

Saunders, D. A., Rowley, I. and Smith, G. T., 1985. The effects of clearing for agriculture on the distribution of cockatoos in the south-west of Western Australia. Pp. 309–21 *in* Birds of Eucalypt Forests and Woodland: Ecology, Conservation, Management ed by A. Keast, H. F. Recher, H. Ford and D. Saunders. Surrey Beatty & Sons: Chipping Norton.

Saunders, D. A., Smith, G. T. and Rowley, I., 1982. The availability and dimensions of tree hollows that provide nest sites for cockatoos (Psittaciformes) in Western Australia. *Aust. Wildl. Res.* **9:** 541–56.

Scougall, S. A., Majer, J. D. and Hobbs, R. J., 1993. Edge effects in grazed and ungrazed Western Australian wheatbelt remnants in relation to ecosystem reconstruction. Pp. 163–78 *in* Nature Conservation 3: the Reconstruction of Fragmented Ecosystems ed by D. A. Saunders, R. J. Hobbs and P. R. Ehrlich. Surrey Beatty & Sons: Chipping Norton.

Seabrook, J., 1994. Growing Understorey Seed. Greening Western Australia: Perth.

Sedgwick, E. H., 1937. Notes from Nangeenan, Western Australia. *Emu* **37:** 172–75.

Sedgwick, E. H., 1974. Birds of the Gibb Rock area. *WA Nat.* **13:** 7–14.

Sedgwick, E. H., 1986. Further notes on the avifauna of the Gibb Rock area, Western Australia. *Aust. Bird Watch.* **11:** 171–80.

Sedgwick, L., 1952. A Gabbin bird list. *WA Nat.* **3:** 152–55.

Sedgwick, L. E., 1964. Birds of the Stirling Ranges, Western Australia. *Emu* **64:** 6–19.

Serventy, D. L., 1977. The use of data on the distribution of birds to monitor climatic changes. *Emu* **77:** 162–66.

Serventy, D. L. and Whittell, H. M., 1962. Birds of Western Australia. 3rd ed. Paterson Brokensha: Perth.

Serventy, D. L. and Whittell, H. M., 1976. Birds of Western Australia. 5th ed. University of Western Australia Press: Nedlands.

Smith, G. T., 1991. Breeding ecology of the Western long-billed corella, *Cacatua pastinator pastinator. Aust. Wildl. Res.* **18:** 91–110.

Smith, R. C. G., Huang, X., Lyons, T. J., Hacker, J. H. and Hick, P. T., 1992. Change in land surface albedo and temperature in southwestern Australia following the replacement of native perennial vegetation: satellite observations. Paper No. IAF-92-0117. 43rd Congress of the International Astronautical Federation, Washington, DC (28 August–5 September 1992). International Astronautical Federation, France.

Smith, G. T. and Moore, L. A., 1992. Patterns of movement in the Western long-billed corella, *Cacatua pastinator* in the south-west of Western Australia. *Emu* **92:** 19–27.

Storr, G. M., 1991. Birds of the South-west Division of Western Australia. *Rec. WA Mus. Suppl.* **35.**

Whitlock, F. L., 1911. In the Stirling Ranges, Western Australia. *Emu* **10:** 305–17.

Index

D

E

F

G

T

V

W

Y

Z